Ultra High Pressure Treatments of Foods

FOOD ENGINEERING SERIES

Food Engineering Series

Jose M. Aguilera and David W. Stanley, *Microstructural Principles of Food Processing and Engineering,* Second Edition (1999)
Stella M. Alzamora, María S. Tapia, and Aurelio López-Malo, *Minimally Processed Fruits and Vegetables: Fundamental Aspects and Applications* (2000)
Gustavo Barbosa-Cánovas and Humberto Vega-Mercado, *Dehydration of Foods* (1996)
Pedro Fito, Enrique Ortega-Rodríguez, and Gustavo Barbosa-Cánovas, *Food Engineering 2000* (1997)
P.J. Fryer, D.L. Pyle, and C.D. Rielly, *Chemical Engineering for the Food Industry* (1997)
Richard W. Hartel, *Crystallization in Foods* (2001)
Marc E.G. Hendrickx and Dietrich Knorr, *Ultra High Pressure Treatments of Food* (2002)
S.D. Holdsworth, *Thermal Processing of Packaged Foods* (1997)
Michael J. Lewis and Neil J. Heppell, *Continuous Thermal Processing of Foods: Pasteurization and UHT Sterilization* (2000)
Rosana G. Moreira, M. Elena Castell-Perez, and Maria A. Barrufet, *Deep-Fat Frying: Fundamentals and Applications* (1999)
Rosana G. Moreira, *Automatic Control for Food Processing Systems* (2001)
M. Anandha Rao, *Rheology of Fluid and Semisolid Foods: Principles and Applications* (1999)

Ultra High Pressure Treatments of Foods

Edited by

Marc E.G. Hendrickx, Dr. ir.
Professor
Department of Food and Microbial Technology
Katholieke Universiteit Leuven
Leuven, Belgium

Dietrich Knorr, Dr. Dipl.-Ing.
Professor
Department of Food Biotechnology and Food Process Engineering
Berlin University of Technology
Berlin, Germany

with

Linda Ludikhuyze, Dr. ir.
Research Assistant
Department of Food and Microbial Technology
Katholieke Universiteit Leuven
Leuven, Belgium

Ann Van Loey, Dr. ir.
Research Assistant
Department of Food and Microbial Technology
Katholieke Universiteit Leuven
Leuven, Belgium

Volker Heinz, Dr.-Ing.
Assistant Professor
Department of Food Biotechnology and Food Process Engineering
Berlin University of Technology
Berlin, Germany

Springer Science+Business Media, LLC

Library of Congress Cataloging-in-Publication Data

Ultra High Pressure Treatments of Foods / edited by Marc E.G. Hendrickx, Dietrich Knorr with
Linda Ludikhuyze, Ann Van Loey, Volker Heinz.
p. cm.—(Aspen food engineering series)
Includes index.
ISBN 978-0-306-47278-7 ISBN 978-1-4615-0723-9 (eBook)
DOI 10.1007/978-1-4615-0723-9

1. Food industry and trade. 2. High pressure (Technology) I. Hendrickx, Marc E.G.
II. Knorr, Dietrich W. III. Food engineering series (Springer Science+Business Media, LLC)

TP371.75.U77 2001
664—dc21 2001022757

Library of Congress Catalog Card Number: 2001022757
ISBN 978-0-306-47278-7

1 2 3 4 5

Dedicated to Rutger, Anna, Fanny, and Hannah

Table of Contents

Contributors

Karin Autio, Dr.
Head of Food Structure Group
VTT Biotechnology
Espoo, Finland

Siegfried Denys, Dr. ir.
Research Assistant
Department of Food and Microbial Technology
Katholieke Universiteit Leuven
Leuven, Belgium

Grahame W. Gould, PhD
Visiting Professor of Microbiology
Department of Food Science
University of Leeds
United Kingdom

Volker Heinz, Dr.-Ing.
Assistant Professor
Department of Food Biotechnology and Food Process Engineering
Berlin University of Technology
Berlin, Germany

Johan C. Hellemons, Ing.
Unilever Research Vlaardingen
Vlaardingen, The Netherlands

Marc Hendrickx, Dr. ir.
Professor
Department of Food and Microbial Technology
Katholieke Universiteit Leuven
Leuven, Belgium

Karel Heremans, Dr.
Professor
Department of Chemistry
Katholieke Universiteit Leuven
Leuven, Belgium

H. Hoogland
Engineer
Unilever Research Vlaardingen
Vlaardingen, Netherlands

Indrawati, Dr. ir.
Research Assistant
Department of Food and Microbial Technology
Katholieke Universiteit Leuven
Leuven, Belguim

Dietrich Knorr, Dr. Dipl.-Ing
Professor
Department of Food Biotechnology and Food Process Engineering
Berlin University of Technology
Berlin, Germany

H.L.M. Lelieveld, Ing.
Senior Technologist
Unilever Research Vlaardingen
Vlaardingen, Netherlands

Linda Ludikhuyze, Dr. ir.
Research Assistant
Department of Food and Microbial Technology
Katholieke Universiteit Leuven
Leuven, Belgium

Martin Michel, PhD
Head of Food Structure Group
Department of Food Science and Process Research
Nestlé Research Center
Nestec Ltd.
Lausanne, Switzerland

Eric Needs
Editor
Journal of Dairy Research
Reading, UK
Formerly, Research Scientist
Institute of Food Research
Norwich, United Kingdom

Margaret Patterson, PhD
Reader
Food Science Division, Queens University, Belfast, and Department of Agriculture and Rural Development
Belfast, Northern Ireland, United Kingdom

Pierpaolo Rovere, Dr.
Food Technologist
Flow Italia
Parma, Italy

Oliver Schlüter, Dipl.-Ing.
Department of Food Biotechnology and Food Process Engineering
Berlin University of Technology
Berlin, Germany

Jan P. Smelt, Dr.
Food Microbiologist
Molecular Cell Biology
University of Utrecht
Utrecht, The Netherlands

Robert W. van den Berg, Dr. ir.
Head of Department of Preservation Technology and Food Safety
Agrotechnological Research Institute ATO
Wageningen, The Netherlands

Ann Van Loey, Dr. ir.
Research Assistant
Department of Food and Microbiol Technology
Katholieke Universiteit Leuven
Leuven, Belgium

Ludo Van Schepdael, ir.
Director
Solico
Oosterhout, The Netherlands

Preface

During the past decade, consumer demand for convenient, fresh-like, safe, high-quality food products has grown. The food industry has responded by applying a number of new technologies including high hydrostatic pressure for food processing and preservation. In addition, food scientists have demonstrated the feasibility of industrial-scale high pressure processing. This technology is of specific interest to the food industry because it provides an attractive alternative to conventional methods of thermal processing, which often produce undesirable changes in foods and hamper the balance between high quality (color, flavor, and functionality) and safety. In addition, it offers opportunities for creating new ingredients and products because of the specific actions of high pressure on biological materials and food constituents. It allows food scientists to redesign existing processes and to create entirely new ones using high pressure technology alone or in combination with conventional processes (e.g., pressure-temperature combinations).

Researchers have investigated high pressure processing for the past century. Scientists such as Hite and Bridgman did pioneering work at the turn of the 20th century. Then during the 1980s and 1990s, there was a large effort to investigate the effects of high pressure on biological materials, particularly foods. The initial research activities in the late 1980s and early 1990s focused on exploratory activities in the food area. However, researchers recognized the need for a more scientific approach to understand the benefits of high pressure and the interaction of high pressure with biological materials. To do this, they initiated a true food engineering approach that allowed design, validation, and optimization of these processes. This approach has led to research activities that focus on the mechanistic and kinetic effects of high pressure on biological materials and the synergistic and antagonistic effects when high pressure is applied in combination with other extrinsic and intrinsic variables. Research areas have included aspects of food safety and food quality, food structure engineering, phase transitions in foods, and predictive modeling. This research has attracted extensive funding from both private and public sources and has resulted in approximately 750 scientific publications and industrial applications in Japan, France, The United

States, and Spain. Commercially available products that are processed using high pressure include fruit- and vegetable-based products, fruit juices, meat products, seafood products, and rice-based products. Today, high pressure processing is available for use on an industrial scale by the food industry.

The exciting developments sketched above have provided us with the incentive to write this book. We have brought together key players involved in research on high pressure treatment of foods to explain state-of-the-art research and technologies, recent issues, and future challenges in a manner that is accessible to everyone in the field. The book has been organized into three parts: the fundamental aspects of treating foods with high pressure, the effects of high pressure on different food attributes, and the current status of industrial-scale high pressure applications.

From this book it should become clear that tremendous advances have been achieved over the past years. This compilation of knowledge now even allows food scientists to share knowledge and transfer technologies from the area of food science to other fields in applied biological sciences. At the same time, researchers still face a number of challenges ahead. For example, researchers hope to develop more insight into the effect of high pressure processing on nutrients and nutritional aspects of foods (such as toxins and allergens), to identify indicator organisms that can be used in a standard way to assess the impact of high pressure processing, and to create unique or improved functionality by using scientific insight to combine raw materials and processing technologies within the constraints of food safety, environmental impact, and other social and economic factors. Expected benefits for the consumer include the availability of gently processed, fresh-like, safe products with high sensory and nutritional quality and optimum physicochemical properties. We sincerely hope that this book contributes to further research on the development of safe and healthy foods that comply with consumer demands, industrial needs, and the concerns of authorities.

M. Hendrickx and D. Knorr

Acknowledgments

We would like to use this opportunity to thank a number of people who have made high-quality research on high pressure treatment of foods possible in the last decade. We hereby acknowledge the European Commission (EC) for its support of these research efforts since 1992. All contributions to this book are based on and made by key scientists who were involved in EC-funded projects—particularly Project FAIR-CT96-1113 ("High Pressure Treatment of Liquid Foods and Derived Products") and Project FAIR-CT96-1175 ("Combined High Pressure Thermal Treatments of Foods: A Kinetic Approach to Safety and Quality Evaluation"). The cooperation of many colleagues in the field, especially scientists from different laboratories and EC–funded projects who contributed to individual chapters of this book, and the cooperation of EC Scientific Officers, L. Breslin and X. Goenaga, have been an exciting and rewarding experience. The young scientists who are involved in these projects and conducting doctoral work on issues related to high pressure treatment of food as well as the scientists involved in European mobility programs should receive specific attention. We also wish to acknowledge the cooperation of Mary Anne Langdon, Senior Developmental Editor of Aspen Publishers, Inc.; our scientific colleagues for their contributions; and finally our young, highly professional coworkers involved in high pressure research who create the attraction and reward for working in our academic profession.

PART I

Fundamental Aspects of Treating Foods with High Pressure

CHAPTER 1

The Evolution of High Pressure Processing of Foods

Grahame W. Gould

INTRODUCTION

With few exceptions, foods lose quality at some rate or other following harvest, slaughter, or manufacture. The nature of the quality loss is dependent on the type of food; its composition, formulation, and processing; and conditions of storage. The most important quality-loss reactions, and thus the most important targets for preservation, include microbiological, enzymatic, chemical, and physical reactions (see Table 1–1) (Gould, 1989).

When preservation fails, the consequences range broadly—from extreme hazard (e.g., when toxicogenic microorganisms are present and able to multiply), to economically important spoilage (which results in loss of valuable products), to relatively trivial loss of quality (e.g., changes in color, flavor, or texture). The most serious forms of quality loss are those that result from the survival and/or growth of infectious pathogenic bacteria (*Salmonella* species, *Campylobacter jejuni* and *coli*, enteropathogenic and verocytotoxigenic [verotoxin producing, toxic to cells] strains of *Escherichia coli*, *Listeria monocytogenes*, *Yersinia enterocolitica*, *Vibrio*

Table 1–1 Quality-Loss Reactions of Foods

Microbiological	*Enzymatic*	*Chemical*	*Physical*
• Growth of toxinogenic microorganisms	• Hydrolytic reactions catalyzed by lipases, proteases, etc.	• Oxidative rancidity	• Mass transfer, movement of low-molecular-weight compounds
• Growth or presence of infective microorganisms	• Rancidity catalyzed by lipoxygenases	• Oxidative and reductive discoloration	• Loss of crisp texture
• Growth of spoilage microorganisms	• Enzymatic browning	• Nonenzymatic browning	• Loss of flavors
	• Destruction of nutrients	• Destruction of nutrients	• Freeze damage

parahaemolyticus, *Aeromonas hydrophila*, *Plesiomonas shigelloides*, enterotoxin-producing strains of *Bacillus cereus*, *Clostridium perfringens*) or the multiplication in foods of toxinogenic bacteria (*Staphylococcus aureus*, emetic toxin-producing strains of *Bacillus cereus*, *Clostridium botulinum*) (Lund et al., 2000).

A limited range of techniques are available to combat pathogenic and spoilage microorganisms and to delay other types of quality loss. Most of the antimicrobial techniques (e.g., chill storage, frozen storage, drying, curing, conserving, vacuum packing, modified atmosphere packing, acidifying, fermenting, adding preservatives) act by inhibiting, slowing down, or preventing the growth of microorganisms. In contrast, few available techniques act primarily by inactivating microorganisms in foods (e.g., heat pasteurization; heat sterilization; and, to a much smaller extent, irradiation). Inactivation techniques are fundamentally more desirable than inhibitory ones, particularly if the aim is to reduce the risks of food poisoning. Elimination of pathogens from foods is more likely to be an effective intervention than inhibition of growth, which assumes some contamination in the first place. Thus, it is encouraging that most of the techniques that are coming into use or are under development act by inactivation (e.g., high hydrostatic pressure, electroporation, manothermosonication, high-intensity laser and noncoherent light pulses, high-strength magnetic field pulses, addition of bacteriolytic enzymes, addition of bacteriocins, and surface decontamination techniques for carcasses).

Against this background, consumers in many countries are demanding foods that are more convenient (i.e., easier to store and prepare); fresher (e.g., chill-stored); more natural; and therefore not as heavily processed (e.g., processed with less heat), less heavily preserved (e.g., preserved with less acid, salt, and sugar), and less reliant on additive preservatives (e.g., sulfite, nitrite, sorbate, propionate, benzoate, and parabens) than foods that have previously been available. A potential consequence of these trends is a reduction in the intrinsic preservation of foods and in their stability and safety. As the food industry reacts to these trends (see Exhibit 1–1), it is important that new technologies retain or improve upon the effectiveness of preservation and ensure safety that might otherwise be lost.

The application of physical techniques, including high pressure, for food preservation has the potential to produce foods that meet many of these consumer demands. Physical techniques are based on the inactivation of microorganisms rather than on their inhibition, the techniques are essentially additive-free, they are mostly nonthermal or involve reduced heat treatments, they have minimal deleterious effects on product quality, and they are usually perceived as "natural."

PRESERVATION TECHNOLOGIES

Traditional Technologies

Of the limited range of preservation technologies currently employed to preserve foods, some (such as drying) have a long history of use, while others (such as modified atmosphere packaging) are recent (see Exhibit 1–2).

Exhibit 1–1 Trends in Consumers' Requirements and Food Industry Reactions

Trends in Consumers' Requirements

- Improved convenience (in preparation, storage, and shelf life)
- Higher quality (in flavor, texture, and appearance)
- Fresher
- More natural (with fewer additives)
- Nutritionally healthier
- Minimally packaged
- Safer

Food Industry Reactions

- Milder processing (minimal overheating, less intense heating, evaluation of nonthermal alternatives to heat)
- Fewer additives (fewer "chemical" additives, more "organic" foods)
- Use of hurdle technologies (combination of preservation systems)
- Evaluation of naturally occurring antimicrobial systems as preservatives
- Less use of salt, saturated fats, and sugar (more low-calorie foods)
- Reduced, environmentally friendly packaging
- Elimination of food poisoning microorganisms

A major trend has been to apply these techniques in new or improved combinations in order to minimize the extreme use of any single technology and, consequently, to improve the quality of the products. This strategy has formed the basis of the successful "hurdle technology" of Leistner and Gorris (1995), which has steadily fostered the development of new routes for food preservation around

Exhibit 1–2 Principal Existing Technologies for Food Preservation

Techniques That Inhibit the Growth of Microorganisms

- Reduction in temperature (chilled and frozen storage)
- Reduction in water activity (drying, curing with added salts, conserving with added sugar)
- Reduction in pH (acidification [e.g., use of acetic acid, citric acid, etc.], fermentation)
- Removal of oxygen (vacuum or modified atmosphere packaging)
- Modified atmosphere packaging (replacement of air with CO_2, O_2, N_2, noble gas mixtures)
- Addition of preservatives (inorganic [e.g., sulfite, nitrite], organic [e.g., propionate, sorbate, benzoate, parabens], bacteriocin [e.g., nisin], antimycotic [e.g., natamycin]
- Control of microstructure (water-in-oil emulsions)

Techniques That Inactivate Microorganisms

- Heat (thermization, pasteurization, sterilization)

Techniques That Restrict Access of Microorganisms to Products

- Packaging
- Aseptic processing and packaging

the world. While traditional hurdle technologies were developed empirically, new technologies are being developed supported by the use of mathematical models of microbial growth and survival (Baranyi & Roberts, 1995). These models, which are derived from large multifactorial experiments and validation in foods, allow the confident prediction of the effects of parameters such as pH, water activity (a_w), temperature, gas phase, and preservatives on the growth, survival, and thermal death of specific microorganisms in foods (Baranyi & Roberts, 2000). To date, little modeling has been done to predict the effects of high pressure or other new and emerging technologies on microorganisms. However, modeling will be valuable as these technologies come into greater use.

Lowered Temperature

As the temperature of a chilled food is reduced, the ability of microorganisms to multiply and grow is also reduced. Two particularly important temperatures are approximately 12°C, which is the lower limit for the growth of the strict anaerobes, *Clostridium perfringens* and *Clostridium botulinum* (type A and some types of B), and about 3°C, which is the lower limit for the growth of nonproteolytic strains of *Clostridium botulinum* (types E and F and some types of B). *Listeria monocytogenes* and *Yersinia enterocolitica* are able to multiply at temperatures below 1°C, so indicated shelf lives and sell-by dates have an important role in ensuring safety—particularly when temperature control cannot confidently be ensured (e.g., in the home) (Herbert & Sutherland, 2000). Many types of spoilage microorganisms may continue to multiply at temperatures down to about −7°C. At the temperature of correctly stored frozen foods, −18°C in many countries, the growth of microorganisms is completely prevented. However, slow loss of quality may occur through the activity of enzymes, chemical reactions, and physical changes (see Table 1–1).

Reduced Water Activity

Water activity (a_w) values are widely used to predict the stability of foods with respect to the growth of microorganisms and the enzymatic, chemical, and physical changes that lead to quality deterioration (Christian, 2000). Values range from 1 (pure water) to 0 (no water), equivalent to equilibrium relative humidities (ERH) on a scale from 100% to 0%. The a_w of foods is reduced by drying, by adding solutes such as salt (as in cured products), sugars (as in conserves), or by a combination of these treatments. Small reductions (to 0.98–0.97 or so) are sufficient to prevent the multiplication of some important spoilage microorganisms. For example, various species of *Pseudomonas* grow at high a_w and rapidly spoil fresh foods such as red meats stored in the presence of air. Because cured meats generally have a_w sufficiently reduced to ensure that they are *Pseudomonas*-free, they have longer shelf lives. Instead, they undergo slow souring, caused by more a_w-tolerant lactic acid bacteria. If the a_w is below about 0.95, as in some salamis and dry-cured meat products, even the lactic acid bacteria are inhibited. These products undergo slower spoilage by micrococci with lower a_w tolerance. These and similar relationships are widely used to explain and predict the storage stability and microbiological safety of foods. Of the food-poisoning bacteria, *Staphylococcus aureus* is the most tolerant to a_w; a_w must be reduced to about 0.86 to limit its multiplication in the presence of oxygen and 0.91 to limit its multiplication anaerobically. Thus, *Staphylococcus aureus* is able to grow and produce enterotoxin in relatively low-a_w foods if other conditions of time and temperature are conducive.

Microorganisms that grow at low a_w values are able to do so because they divert energy to the synthesis or accumulation of high levels of intracellular solutes such as glycine, betaine, proline, and polyols. In this way, they "osmoregulate" so as to avoid water loss by osmosis and maintain turgor (and therefore functionality) in their cytoplasmic membranes (Booth, 1998).

At a_w values below 0.86, few bacteria, and none of public health concern, can multiply. Food spoilage in products with a_w values below this level is caused by yeasts and molds, some of which can grow slowly at a_w values as low as 0.6. Shelf-stable, dried foods are generally formulated with a_w values around 0.3, where lipid oxidation and other enzymatic and chemical changes are minimal.

Although a low a_w efficiently preserves foods, it may also preserve bacteria in them, so that bacteria survive longer in low-a_w foods than in higher-a_w ones. Furthermore, the resistance of vegetative (Corry, 1974, 1976) and spore forms (Murrell & Scott, 1966) of microorganisms to heat may be greatly enhanced at low a_w values, probably as a result of osmotic dehydration of their cytoplasmic contents. The same phenomenon may be true for other antimicrobial physical treatments, including pressure.

Vacuum Packaging and Modified Atmosphere Packaging

Vacuum packaging and modified atmosphere packaging (MAP) are effective because they remove oxygen, and lack of oxygen inhibits the growth of strictly oxidative microorganisms. Fermentative microorganisms continue to multiply when oxygen is absent. However, they generally do so more slowly and, for some types of foods, they have less unpleasant consequences for food quality. It is important to take precautions against the growth of strictly anaerobic food-poisoning microorganisms, such as *Clostridium botulinum*. For this reason, it is recommended that foods such as "sous vide" products, which are vacuum packed and heat-pasteurized rather than sterilized, be subjected to minimal heat treatments and strict temperature controls in distribution (Notermans et al., 1990). Carbon dioxide is widely used in MAP foods because it has a specific antimicrobial activity, acting as a preservative that uniquely dissipates when the food package is opened (Molin, 2000). For example, many supermarket meats are packed in gas mixtures containing about 70% oxygen and 30% carbon dioxide. The oxygen maintains pigments in the meat in the bright-red oxymyoglobin state, which is the color that consumers prefer, and the carbon dioxide slows down the growth of Gram-negative spoilage bacteria so as to approximately double the product's shelf life.

Acidification

Few bacteria of significance in foods are able to multiply at pH values below 3.5 or so. Those that do so are adapted to acidic environments (e.g., the lactic acid bacteria). In addition to being spoilage microorganisms, some of these bacteria are employed in numerous acid-generating food and feed fermentations. A particularly important pH for food safety is 4.5, because this is the pH below which *Clostridium botulinum* is unable to multiply. Consequently, in thermal processing or other microorganism-inactivating procedures including pressure, it is not necessary to treat foods that are more acidic than this to the same extent as higher-pH, "low-acid" foods. When the pH is below about 4.2, other food-poisoning and spoilage bacteria are mostly controlled. However, recently, the spore-forming bacterium *Alicyclobacillus acidoterrestris*, capable of growth at pH values as low as 2, has caused spoilage problems in some acidic foods. In addition, many yeasts and molds are able to multiply at very low pH values (pH values well below 2), and these organisms predominate in the flora of spoiling acidic foods.

Preservation

Most of the preservatives that are used in foods are acids (see Table 1–2), such as the weak lipophilic organic acids (sorbate, benzoate, propionate) and the inorganic ones (sulfite, nitrite). All of these acids are more effective at low pH levels. Indeed, with the possible exception of the alkyl esters of p-hydroxybenzoate (parabens), there are no wide-spectrum antimicrobial food preservatives that are effective at near-neutral pH (Russell & Gould, 2000). There is a well-established rationale for the efficacy of the weak acids and for their synergy with hydrogen ions (i.e., in a low-pH environment). In their unionized forms, which are favored by low pH, they are able to readily equilibrate across the microbial cell membrane and enter the cell cytoplasm. The pH values of the common weak acid preservatives range from 4.2 (benzoic) to 4.87 (propionic). At pH values much above these levels, activity is greatly reduced. At the pH of most foods, microorganisms maintain an internal pH higher than that of their surroundings. Consequently, on entering the cytoplasm, the undissociated acids tend to dissociate, delivering hydrogen ions along with the particular anion. The additional hydrogen ions may be exported by the microorganism, but this is energy-demanding, so cell growth is restricted. If the energy supply is insufficient, then the pH of the cytoplasm eventually falls to a level that is so low that growth cannot continue. In addition, the accumulated anions may have specific antimicrobial effects (Eklund, 1983).

Combination Preservation and "Hurdle" Technologies

Combining lowered pH with a weak acid preservative is a good example of a widely used and effective combination preservation technique involving two "hurdles" (hydrogen ions and the particular acid). The technique also provides an example of how the use of additional hurdle technologies can improve preserva-

Table 1–2 Principal Food Preservatives

Preservatives	*Examples of Foods in Which They Are Used*
Weak lipophilic acids and esters	
Sorbate	Cheeses, syrups, cakes, dressings
Benzoate	Pickles, soft drinks, dressings
p-Hydroxybenzoate esters	Marinated fish products
Propionate	Bread, cakes, cheese, grains
Organic acid acidulants	
Acetic, lactic, citric, malic	Acidulants for low-pH sauces, mayonnaise, dressings, salads, drinks, fruit juices and concentrates
Mineral acid acidulants	
Phosphoric, hydrochloric	Acidified foods, same as for organic acid acidulants
Inorganic anions	
Sulfite (SO_2, metasulfite)	Fruit pieces, dried fruits, wine, meat (sausages)
Nitrite	Cured meats
Antibiotics	
Nisin	Cheese, canned foods
Natamycin	Soft fruit, dry-cured meat
Smoke	Meats and fish

tion (Gould, 1995b). For example, from the point of view of practical food preservation, it is sensible to include a weak acid whenever possible, then to acidify a food product as much as is organoleptically acceptable in order to capitalize on the weak acid-low pH synergy, then to vacuum pack to restrict the amount of energy that is available for the extrusion of hydrogen ions. It is then sensible to reduce the a_w value as much as possible, because this will place additional energy requirements on the cell for the synthesis of compatible solute and osmoregulation, and so on. In this way, many preservation technologies are being used, and new, logically based ones sought (Leistner, 1995; Tapia de Daza et al., 1996). To date, little research has been done on the inclusion of the new physical preservation technologies such as high hydrostatic pressure in such combinations. Use of physical preservation technologies will most probably lead to new and useful hurdle technologies.

Heat

Some of the world's largest food industries rely on pasteurization at times and heat to inactivate vegetative microorganisms, and sterilization at times and heat to inactivate bacterial spores (Pflug & Gould, 2000). With the slow acceptance of irradiation for food preservation in most countries, heat remains the only substantial means for inactivating microorganisms in foods. It is encouraging that most of the new and emerging technologies that have been investigated and promoted in recent years act by inactivation, but without the need for substantial heating.

New and Emerging Preservation Technologies

Most of the new and emerging technologies for food preservation act by inactivating microorganisms rather than inhibiting them. These new technologies include some that are chemically based, some that are enzymatic, and some that are physical (see Exhibit 1–3) (Gould, 1995a).

Although heat remains by far the most used of the inactivation techniques, new technologies are being introduced. Most of these represent nonthermal or mild heat alternatives to conventional heat processing. Some of these techniques are widely used already (Davidson & Brannen, 1993; Dillon & Board, 1994). For example, the bacteriolytic enzyme, hen egg white lysozyme, is employed at levels in excess of 100 tons per year to prevent gas formation ("blowing") in cheeses. The enzyme works by lysing vegetative cells of *Clostridium tyrobutyricum* that are outgrowing from contaminant spores. Activation of the milk-derived lactoperoxidase system is useful for the extension of the shelf life of bulk milk in countries in which pasteurization soon after milking is not possible and refrigerated transport systems are poorly developed (Ekstrand, 1994). The bacteriocin, nisin, is increasingly used to prevent spoilage in some cheeses and to prevent food spoilage by thermophilic, spore-forming bacteria, particularly in countries that have high ambient temperatures (Abee & Delves-Broughton, 2000). More than 40 other bacteriocins have been identified, and some are being evaluated for food use (Abee et al., 1995). Hundreds of compounds extracted from herbs and spices have been discovered and shown to have antimicrobial proper-

Exhibit 1–3 New and Emerging Technologies

Natural Additives

- Animal-derived antimicrobials (lysozyme, lactoperoxidase system, lactoferrin, lactoferricin)
- Plant-derived antimicrobials (herb and spice extracts)
- Microbial products (nisin, pediocin, other bacteriocins and culture products)

Physical Processes

- Gamma and electron beam irradiation
- High-voltage electric gradient pulses ("electroporation")
- High hydrostatic pressure
- Combined ultrasonics, heat, and pressure (manothermosonication)
- High-intensity laser and noncoherent light pulses
- High magnetic field pulses

ties in laboratory studies (Hoover, 2000). Although some have been shown to be effective in foods, the efficacy of these compounds is often reduced in food environments, as a result of binding of the compounds to food proteins, partition into lipids, etc.

Although food scientists will continue to find new uses for chemical- and enzyme-based techniques, particularly those derived from natural sources, it is probable that the new and emerging physical processes will offer the most exciting alternatives to heat for food preservation (Barbosa-Canovas et al., 1995). These processes include the now well-established application of high hydrostatic pressure that is the focus of this book, as well as other techniques that are in various stages of laboratory-scale research, pilot study, and commercial evaluation.

Ionizing Radiation

The use of ionizing radiation for the preservation of foods was first proposed in the early 1900s and, at "pasteurization" doses up to 10 kGy, is legal in about 40 countries now. The World Health Organization (WHO) has recently recommended that, because so much research has shown that there is no problem regarding safety, there is no longer any reason to impose an upper-dose limit; WHO recommends that sterilization treatments using ionizing radiation (e.g., 30 to 50 kGy, depending on the type of food) should be permitted (WHO, 1999; Patterson & Loaharanu, 2000). Food irradiation is technologically relatively simple to apply, with straightforward inactivation kinetics and geometry that make dose control and processing requirements much easier than for many thermal processes. The potential of food irradiation to prevent food poisoning by eliminating infectious pathogens such as *Escherichia coli* O157:H7, *Salmonella,* and *Campylobacter* from some foods of animal origin and some seafoods is substantial. However, slow acceptance by the public continues to hamper its introduction in most countries.

Electroporation

High-voltage electric shock treatments (electroporation) are most effective for the inactivation of vegetative cells of bacteria, yeasts, and molds, while bacterial spores are more tolerant. Electroporation has a lethal effect because it breaches the cell membrane, one of the most important structures controlling many of the vegetative cell's homeostatic mechanisms (Qin et al., 1996). High-voltage pulse machines have been shown to effectively pasteurize and to inactivate vegetative food-poisoning microorganisms in liquid foods such as fruit juices and milk (Zhang et al., 1995).

Manothermosonication

Ultrasonication at sufficiently high intensities has long been known to inactivate vegetative microorganisms and to reduce the heat resistance of spores, and the effect is amplified by increasing the temperature. However, as the temperature is increased, the relative magnitude of the amplification becomes reduced. It is thought that this occurs because, as the vapor pressure rises, it reduces the effect of cavitation, which is the cause of the killing effect. However, application of a slight overpressure, optimally about 500 kPa (Raso et al., 1998), has been reported to overcome this reduced effectiveness, so that the amplification is retained at the higher temperatures. Researchers claim that the combination procedure (manothermosonication) has the potential to reduce pasteurization and sterilization temperatures for pumpable liquid and semisolid foods (Sala et al., 1995).

High-Intensity Light

High-intensity laser and noncoherent light-pulse generators have been developed for the decontamination of surfaces of foods, and possibly of transparent foods (Dunn et al., 1988), as well as for applications in dentistry (Rooney et al., 1994). For some applications, the killing effect results from the ultraviolet content and for others it results from intense but local heating and removal of microorganisms by ablation. Additional nonultraviolet and nonthermal effects have been claimed by some researchers.

High-Strength Magnetic Fields

Exposure to high-strength, oscillating magnetic fields has been reported to have a variety of effects on biological systems, ranging from selective inactivation of malignant cells (Costa & Hoffman, 1987) to the inactivation of bacteria on packaging materials and in foods (Hoffman, 1985). Treatment times are very short (typically less than 25 milliseconds), and field strengths are very high (typically from 2 to 100 tessla at frequencies of 5 to 500 kHz). Treatment efficacy did not exceed about 10^2-fold reduction in the number of vegetative microorganisms inoculated into milk (*Streptococcus thermophilus*), orange juice (*Saccharomyces*), bread (mold spores), and no inactivation of bacterial spores has been reported. Thus, the practical potential for the technique as it has been developed so far seems to be limited (Barbosa-Canovas et al., 1995).

EVOLUTION OF HIGH PRESSURE PROCESSING

High pressure processing has advanced further than the other alternative physical technologies except for irradiation. To some extent, this is because of the efficacy of high pressure processing. It inactivates all vegetative and spore forms of microorganisms if the applied pressure is high enough. In addition, the engineering aspects of high pressure processing have advanced to such an extent that commercially economic processes have become viable within the last decade or so, at least for high-value niche market products. As with so many new technologies, the original observations on which high pressure processing was built were made a long time ago. The first reports that microorganisms could be inactivated by high pressure were made more than 100 years ago.

Historical Survey

The early development of high pressure technology was driven, not by food technologists, but by the needs of the military to improve the cannon. Crossland (1995) has summarized the early development of cannon designs that aimed to contain higher and higher pressures in order to allow gunpowder charges to project heavier and heavier missiles further and further. This work led to vessel designs that much later became available for laboratory-scale experiments, including experiments with foods. Perkins (1820) reported the first measurements of the compressibility of water and of other fluids (Perkins, 1826). The first persuasive experiments with microorganisms were reported at the end of the 19th century by Hite (1899), and effects of pressure on the physical properties of foods were reported soon after. For example, Bridgman (1914) studied the coagulation of egg albumin by high pressure and pointed out that the properties of the product differed from those resulting from coagulation by heat. Likewise, Basset et al. (1933) reported that pressure caused the gelation of horse serum proteins. Much later, Macfarlane (1973) reported the practical potential of high pressure technology with the demonstration of pressure-tenderization of meat. Although Hite (1899) had reported that some of the physical properties of milk changed on pressurization (e.g., it became more translucent), detailed investigations of the effects of pressure on the physicochemical properties of food biopolymers were not made until much later. For example, Payens and Heremans (1969) studied the effects of pressure on β-casein from milk, and Zipp and Kauzmann (1973) studied the effects of ionic environment on the pressure-denaturation of metmyoglobin.

Today, the effects of pressure on the physical properties of foods are increasingly seen as important and potentially exploitable. However, the earlier studies concentrated on the inactivation of microorganisms—particularly vegetative cells of bacteria and yeasts, with spores targeted later. Preempting some of the most successful modern developments, Cruess (1924), more than 75 years ago, predicted that pressure would be employed to preserve fruit juices (in which the low pH ensures that spore formers are not a problem). Although Giddings et al. (1929) reported the pressure inactivation of tobacco mosaic virus, there has been little interest in the pressure tolerance of the viruses that are of concern in foods.

Vegetative Microorganisms

Hite (1899), who reported the treatment and prevention of souring of milk, was the first investigator to show clearly that microorganisms can be inactivated by pressure. Larson et al. (1918) cited unsuccessful experiments by Certes as early as 1889 (see review by Cheftel, 1995) and successful experiments by Buchner that were reported in 1897.

In 1899, Hite pointed out some of the potential advantages of pressure treatment that we still recognize today: "antiseptics are always objectionable;" "heating milk at the temperature of boiling water . . . for about half an hour each day for three or four successive days" (Tyndallization); . . . or heating it to a somewhat higher temperature for a longer time . . . should keep indefinitely . . . but it has a 'cooked' odor and taste;" "pasteurization reduces the germ content—but as it does not kill all the germs the milk must be kept at a low temperature or it will sour in a very short time." He mentioned that, since late 1897, the chemist in his institute (West Virginia Experiment Station) had been studying the effects of pressure on the preservation of milk, meats, and fruit juices (which seems very relevant to modern interests, more than 100 years later). The first experiments, which involved driving a heavy screw into a cylinder to deliver a theoretical pressure that was "something tremendous" and holding it for periods of days, failed to postpone the souring of milk. Higher pressures were obtained when the researchers acquired a piston-cylinder device from the father of a student in Hite's agricultural chemistry class, who was connected with the National Transit Company of Oil City, Pennsylvania. Following many trials and errors and machine development, pressure-treated milk samples stayed sweet "twenty-four hours longer than untreated controls, some of them forty-eight to sixty hours longer. Two samples that had received ninety tons for one hour, remained sweet for a week. Thus encouraged, an attempt was made to submit other samples to ninety tons, when with a loud report the tool steel bush split from one end to the other" (Hite, 1899). Despite such difficulties, this seminal publication clearly showed that pressure inactivated microorganisms in milk and improved its keeping qualities. Pressures greater than 30 tons per square inch (approximately 600 MPa) applied for one hour generally postponed the souring of milk at ambient temperature for at least 24 hours. Occasionally, a sample did not sour even after two weeks, but there was much experiment-to-experiment and sample-to-sample variation.

It is interesting that microbiological safety issues were appreciated so early. Indeed, Hite (1899) reported that in experiments that today would require a high-category containment laboratory, "pure cultures of typhoid germs were obtained, but in the first attempt to subject an inoculated sample of milk to a pressure of ninety tons, the steel cylinder, the lead tube, the tin tube, and one of the pistons went to pieces, scattering the germs all over the room. Before another cylinder could be made, a member of the station staff contracted a violent case of typhoid and it was decided to discontinue this feature of the work." However, later on, with new presses, experiments were made with anthrax, typhoid, tuberculosis, *Proteus vulgaris*, and bubonic plague.

Although this early study reported by Hite did not reliably achieve complete preservation of milk, it is remarkable that so many original observations were made. The latter part of Hite's 1899 report even anticipated many modern investigations of the effect of pressure in combination with a slight elevation of temperature. More detailed and comprehensive studies followed, and these were reported in a 67-page report some 15 years later (Hite et al., 1914).

The earlier studies of the effects of high pressure on milk were abandoned to a large extent because of reasons that are still of concern today. Hite (1899) reported that "certain slow changes in the milk" continued, "due to enzymes that the pressure could not destroy"—even "by a pressure of 100,000 pounds per square inch at room temperature for seven days, nor by the same pressure applied for an hour each day for more than a week." Interestingly, these findings preempted modern attempts to exploit pressure cycling. "Cultures of bacteria have not been obtainable from any sample of milk so treated."

Unlike Hite's 1899 paper, the 1914 paper included microbial counts. Numbers of microorganisms in the region of 10^7 mL^{-1} were shown to be reduced to a few hundred or less by pressures in the region of 700 MPa applied at room temperature for 10 minutes. Experiments were performed with a wider range of foods including apple juice, grape juice, peaches, pears, plums, blackberries, raspberries, tomatoes, peas, beans, beets, and other vegetables. As with the earlier milk experiments, the investigators found a pattern of success mixed with failure. Some pressure-treated samples were effectively preserved, but others spoiled. (Today we would probably attribute many of the failures to surviving spores.) As a result of the variability, Hite and colleagues decided to undertake a systematic study using pure cultures of microorganisms; unfortunately, however, actual kinetic studies were not performed. Most data were derived from "end point-like" experiments. That is, results were recorded as "no growth," "several colonies," "good growth," etc., on agar plates. Most data were obtained with *Bacillus prodigiosus* (not a spore-forming bacillus, now named *Serratia marcescens*). Other, less extensive studies were done with lactic acid bacteria, *Bacillus fluorescens* (now *Pseudomonas fluorescens*), vegetative cells of *Bacillus subtilis*, yeasts (*Saccharomyces cerevisiae* and *Saccharomyces* [now *Candida*] *albicans*), as well as the pathogens *Bacillus typhosus* (*Salmonella typhimurium*) and *Bacillus diphtheriae* (*Corynebacterium diphtheriae*).

For Hite and colleagues (1914), the growth–no growth boundaries for the vegetative cells studied were in the region of 400 to 600 MPa, applied for 10 minutes, but considerable variability occurred. The authors considered whether decompression rate might influence results. They did experiments to investigate this, and concluded that there was no difference attributable to sudden or slow release of pressure. Hite et al. (1914) mention a paper by Chlopin and Tammann (1903) that claimed that pressures of about 300 MPa did not inactivate microorganisms, but forced them into a lethargic condition in which they might lose the power to reproduce (today we call this a "viable but nonculturable state"). However, Hite et al. performed long incubation experiments and concluded that pressurized microorganisms (*Serratia marcescens*) that did not grow on the first day of incubation failed to revive even after 89 days. The effect of medium composition was appreciated, but not studied in depth until Timson and Short (1965) showed the important protective effect of high concentrations of solutes (including NaCl, K_2HPO_4, K citrate, and glucose) on microorganisms pressurized in milk (a finding that has often been confirmed since) and the synergistic effects of extreme pH values.

When Hite and colleagues (1914) studied *Bacillus subtilus,* they employed young (12- to 18-hour) cultures so as to minimize the chance of spores being present. However, they pointed out that this could not be certain. Hite and colleagues did not deliberately study spores.

This comprehensive study by Hite and colleagues set the scene for later work. Most of the phenomena described have stood the test of time, and some have been "rediscovered" many times since. Soon after the comprehensive Hite et al. (1914) paper was published, Larson et al. (1918) reported investigations of the effect of pressure on microorganisms with the aim of improving vaccines. They pressurized *Bacillus typhosus* (*Salmonella typhimurium*), *Bacillus tuberculosis* (*Mycobacterium tuberculosis*), *Bacillus proteus* (*Proteus*), *Bacillus subtilis*, *staphylococci*, *streptococci*, and *pneumococci* at pressures from about 300 to 1200 MPa. Later, Basset and Macheboeuf (1933) similarly reported the pressure-inactivation and antigenicity of *Clostridium diphtheriae* and *tetani*. Larson et al. used long pressurization times (up to 14 hours). Although the initial number of microorganisms is unclear, they found that 300 MPa was insufficient to inactivate any of the bacteria studied, whereas 600 MPa inactivated them all. However, Larson et al. (1918) were the first to clearly show that spores (of *Bacillus subtilis*), in contrast to the vegetative cells, were not "regularly killed" at the highest pressure used (1200 MPa). Further experiments reported in this paper concentrated on the use of high gas pressures (carbon dioxide, hydrogen, and nitrogen) and the effects of decompression during sudden release of pressure. In particular, they reported that carbon dioxide under pressures as low as 5 MPa inactivated vegetative cells in 1.5 to 2.5 hours. Gram-negative microorganisms tended to disintegrate, while Gram-positive ones kept their shape. Hydrogen and nitrogen were relatively ineffective.

Interestingly, the filtrate from pressure-treated *Salmonella typhi* proved to generate high antibody titers in inoculated rabbits, much higher than when inoculated with living cells. Likewise, the cells inactivated with carbon dioxide pressure were highly antigenic.

Spores

The pressure tolerance of bacterial endospores was inferred in the report of Hite et al. (1914) and confirmed in more detail by Larson et al. (1918). Later, Basset and Macheboeuf (1932) investigated spore resistance in greater depth, and they recorded survival of spores of *Bacillus subtilis* following pressurization at up to 1750 MPa for 45 minutes at ambient temperature. This was reaffirmed by Timson and Short (1965). These authors repeated some of the early experiments on milk and found that failure to preserve was often the result of small numbers of surviving microorganisms (often at levels of only 0.05% or so of the initial population), which turned out to be spore forms. Again they studied combinations of pressure with temperatures up to 100°C. A larger kill resulted, but usually with low levels of survivors (1 to 10 or so, from starting numbers of about 10^5), even following treatment at 700 MPa and 100°C for up to 30 minutes.

The possibility of controlling spores in foods despite their extreme pressure tolerance received renewed interest following the surprising observation that, under certain conditions, inactivation of spores occurred more rapidly and completely at moderate pressures than at higher pressures (Clouston & Wills, 1969; Sale et al., 1970; Gould & Sale, 1970). Researchers learned that inactivation of spores by these moderate pressures, unlike the inactivation of vegetative microorganisms, occurs in two stages. First, moderate pressure causes the spores to germinate. Then the pressure, or heat if the temperature is high enough, inactivates the sensitive, germinated forms. Clouston and Wills (1970) and Wills (1974) showed that ionizing radiation, in place of heat, could inactivate the pressure-germinated spores in a similar manner. They reported that pressure germination of spores has a high temperature coefficient, so that pressure and temperature are highly synergistic. Pressure-induced germination is most effective under conditions of pH, a_w, etc. that are near optimal for normal germination, and is greatly amplified by low levels of conventional, biochemical germinants (Gould & Sale, 1970). However, as the pressure is raised, the relative effect of optimal germination conditions and germinants is reduced, and direct killing of the spores replaces the initiation of germination as the main route of inactivation.

Although pressure-heat synergy still offers a possible route to the use of pressure to sterilize high-pH/high-a_w foods, two problems were recognized that have still not been overcome. First, the pressure-temperature tolerance of spores of different strains and species varies greatly. Although some spores can be triggered to germinate by exposure to pressures as low as 50 MPa, others are far more tolerant and resist thousands of MPa. Second, the spores of important pathogens, including those of *Clostridium botulinum*, are highly tolerant to pressure. Effective control of the pathogenic spore formers (*Bacillus cereus, Clostridium perfringens*, and *Clostridium botulinum* in particular) is, of course, fundamental to any process for the production of ambient, stable, high-pH/high a_w foods and probably should be controlled to the same extent that it is controlled in thermal processing (i.e., equivalent to at least a D-12 process).

CONCLUSION

The use of high pressure for food preservation has the potential to address the requirements of modern consumers to prefer "minimally preserved" foods. The pace of development of commercially viable high pressure processes is accelerating, although many of the phenomena that claim our attention now were first noticed by researchers long ago. There is a wealth of information in the early reports of pressure effects on microorganisms that should be examined by anyone now working in or entering the field. However, three important aspects of the use of high pressure were absent from the early literature.

First, few of the early workers undertook careful kinetic studies of the effects of pressure on vegetative cells or spores. Such studies are essential for the intelligent application of pressure, and it is unfortunate that they have only recently been undertaken. Clouston and Wills (1970) showed convincing exponential inactivation kinetics of *Bacillus pumilus* spores. However, recent studies of some of the more important vegetative pathogens (Patterson et al., 1995a, 1995b) have shown kinetics that are commonly nonexponential, often with long "tails," and very much influenced by constituents of the food or medium (Patterson et al., 1995a, 1995b). Such information will become more and more important as new applications of pressure in foods and processing procedures are developed.

Second, the possibility of achieving spore control through the use of pressure-heat combinations has not been realized. A breakthrough in this area would lead to a quantum leap in the application of pressure technologies to foods. It is hoped that combination processes will introduce new options.

Third, a few of the early studies examined the effects of adjuncts on the pressure tolerance of microorganisms, but only superficially. Few preservation techniques are used alone, and the modern trend is to employ more and more adjuncts to improve food quality. Hence, the development and application of "combined preservation technologies," "minimal processing technologies," and "hurdle technologies" are popular and sensible targets in the area of food preservation. The application of high pressure for food preservation will most likely develop along similar lines, with the discovery of new combinations. For example, the use of high pressure with heat, gases, bacteriocins (e.g., nisin and pediocin AcH, already reported by Kalchayanand et al. [1994] and by Roberts and Hoover [1996]), enzymes such as lysozyme (already reported by Hauben et al. [1996]), and many other possible synergists are being studied and reported.

REFERENCES

Abee, T., & Delves-Broughton, J. (in press). Bacteriocins. In N.J. Russell, & G.W. Gould (Eds.), *Food Preservatives* (2nd ed.). Gaithersburg, MD: Aspen.

Abee, T., Krockel, L., & Hill, C. (1995). Bacteriocins: Modes of action and potential in food preservation and control of food poisoning. *Int. J. Food Microbiol., 28*, 169–185.

Baranyi, J., & Roberts, T.A. (1995). Mathematics of predictive microbiology. *Int. J. Food Microbiol., 26*, 199–218.

Baranyi, J., & Roberts, T.A. (2000). Principles and application of predictive modeling of the effects of preservative factors on microorganisms. In B.M. Lund, A.C. Baird-Parker, & G.W. Gould (Eds.), *The Microbiological Safety and Quality of Food* (pp. 342–358). Gaithersburg, MD: Aspen.

Barbosa-Canovas, G.V., Pothakamury, U.R., & Swanson, B.G. (1995). State of the art technologies for the sterilization of foods by nonthermal processes: Physical methods. In G.V. Barbosa-Canovas, & J. Welti-Chanes (Eds.), *Food Preservation by Moisture Control: Fundamentals and Applications* (pp. 493–532). Lancaster, PA: Technomic.

Basset, J., & Macheboeuf, M.A. (1932). Etude sur les effets biologiques des ultrapressions: Resistance de bacteries, de diastases et toxines aux pressions tres elevees. *Comptes Rendus, 196*, 1431–1442.

Basset, J., & Macheboeuf, M.A. (1933). Etudes sur les effets biologiques des ultrapressions. Etudes sur l'immunite: Influence des pressions tres elevees sur certain antigenes et anticorps. *Comptes Rendus, 197*, 67–68.

Basset, J., Macheboeuf, M., & Sandor, G. (1933). Etude sur les effets biologique des ultrapressions. Action des pressions tres elevees sur les proteides. *Comptes Rendus, 197*, 796–798.

Booth, I.R. (1998). Bacterial responses to osmotic stress: Diverse mechanisms to achieve a common goal. In D.S. Reid (Ed.), *The Properties of Water in Foods ISOPOW 6* (pp. 456–485). London: Blackie Academic & Professional.

Bridgman, P.W. (1914). The coagulation of egg albumin by pressure. *J. Biol. Chem., 19*, 511–512.

Cheftel, J.C. (1995). Review: High pressure, microbial inactivation and food preservation. *Food Sci. Technol. Int., 1*, 75–90.

Chlopin, G.W., & Tammann, G. (1903). Eber den einflus hoher drucke auf mikrooganismen. *Zeitschrift fur Hygiene und Infectionskrankheiten, 45*, 171.

Christian, J.H.B. (2000). Drying and reduction in water activity. In B.M. Lund, A.C. Baird-Parker, & G.W. Gould (Eds.), *The Microbiological Safety and Quality of Food* (pp. 146–174). Gaithersburg, MD: Aspen.

Clouston, J.G., & Wills, P.A. (1969). Initiation of germination and inactivation of *Bacillus pumilus* spores by hydrostatic pressure. *J. Bacteriol., 97*, 684–690.

Clouston, J.G., & Wills, P.A. (1970). Kinetics of germination and inactivation *of Bacillus pumilus* spores by hydrostatic pressure. *J. Bacteriol., 103*, 140–143.

Corry, J.E.L. (1974). The effect of sugars and polyols on the heat resistance of salmonellae. *J. Appl. Bacteriol., 37*, 31–43.

Corry, J.E.L. (1976). The effect of sugars and polyols on the heat resistance and morphology of osmophilic yeasts. *J. Appl. Bacteriol., 40*, 269–276.

Costa, J.L., & Hoffman, G.A. (1987). Malignancy treatment. U.S. Patent 4,665,898.

Crossland, B. (1995). The development of high pressure equipment. In D.A. Ledward, D.E. Johnston, R.G. Earnshaw, & A.P.M. Hasting (Eds.), *High Pressure Processing of Foods* (pp. 7–26). Nottingham: Nottingham University Press.

Cruess, W.V. (1924). *Commercial Fruit and Vegetable Products*. New York: McGraw-Hill.

Davidson, P.M., & Brannen, A.L. (Eds.). (1993). *Antimicrobials in Foods*. New York: Marcel Dekker.

Dillon, V.M., & Board, R.G. (Eds.). (1994). *Natural Antimicrobial Systems and Food Preservation*. Wallingford, Oxon: CAB International.

Dunn, J.E., Clark, R.W., Asmus, J.F., Pearlman, J.S., Boyer, K., & Parrichaud, F. (1988). Method and apparatus for preservation of foodstuffs. International Patent WO88/03369.

Eklund, T. (1983). The antimicrobial effect of dissociated and undissociated sorbic acid at different pH levels. *J. Appl. Bacteriol., 54*, 383–389.

Ekstrand, B. (1994). Lactoperoxidase and lactoferrin. In V.M. Dillon, & R.G. Board (Eds.), *Natural Antimicrobial Systems and Food Preservation* (pp. 15–63). Wallingford, Oxon: CAB International.

Giddings, N.J., Alland, A.H., & Hite, B.H. (1929). Inactivation of tobacco-mosaic virus by high pressures. *Phytopathology, 19*, 749–750.

Gould, G.W. (Ed.) (1989). *Mechanisms of Action of Food Preservation Procedures*. Barking, Essex: Elsevier Applied Science.

Gould, G.W. (Ed.). (1995a). *New Methods of Food Preservation*. Glasgow: Blackie Academic & Professional.

Gould, G.W. (1995b). Homeostatic mechanisms during food preservation by combined methods. In G.V. Barbosa-Canovas, & G. Welti-Chanes (Eds.), *Food Preservation by Moisture Control: Fundamentals and Applications* (pp. 397–410). Lancaster, PA: Technomic.

Gould, G.W., & Sale, A.J.H. (1970). Initiation of germination of bacterial spores by hydrostatic pressure. *J. Gen. Microbiol., 60*, 335–346.

Hauben, K.J.A., Wuytack, E.Y., Soontjens, C.C.F., & Michiels, C.W. (1996). High-pressure transient sensitization of *Escherichia coli* to nisin and lysozyme by disruption of outer-membrane permeability. *J. Food Prot., 59*, 350–355.

Herbert, R.A., & Sutherland, J.P. (2000). Chill storage. In B.M. Lund, A.C. Baird-Parker, & G. Gould (Eds.), *The Microbiological Safety and Quality of Food* (pp. 101–121). Gaithersburg, MD: Aspen.

Hite, B.H. (1899). The effect of pressure in the preservation of milk. *Bull. West Virg. Univ. Agric. Exp. Station, 58*, 15–35.

Hite, B.H., Giddings, N.J., & Weakley, C.W. (1914). The effect of pressure on certain microorganisms encountered in the preservation of fruits and vegetables. *Bull. of the West Virg. Univ. Agric. Exp. Station, 146*, 1–67.

Hoffman, G.A. (1985). Inactivation of microorganisms by an oscillating magnetic field. International Patent WO85/02094.

Hoover, D.G. (2000). Microorganisms and their products in the preservation of foods. In B.M. Lund, A.C. Baird-Parker, & G.W. Gould (Eds.), *The Microbiological Safety and Quality of Food* (pp. 251–276). Gaithersburg, MD: Aspen.

Kalchayanand, N., Sikes, T., Dunne, C.P., & Ray, B. (1994). Hydrostatic pressure and electroporation have increased bactericidal efficiency in combination with bacteriocins. *Appl. Environ. Microbiol., 60*, 4174–4177.

Larson, W.P., Hartzell, T.B., & Diehl, H.S. (1918). The effect of high pressure on bacteria. *J. Infect. Dis., 22*, 271–279.

Leistner, L. (1995). Use of hurdle technology in food: Recent advances. In G.V. Barbosa-Canovas, & G. Welti-Chanes (Eds.), *Food Preservation by Moisture Control: Fundamentals and Applications* (pp. 377–396). Lancaster, PA: Technomic.

Leistner, L., & Gorris, L.G.M. (1995). Food preservation by hurdle technology. *Trends Food Sci. Technol., 6*, 41–46.

Lund, B.M., Baird-Parker, A.C., & Gould, G.W. (Eds.). (2000). *The Microbiological Safety and Quality of Food.* Gaithersburg, MD: Aspen.

Macfarlane, J.J. (1973). Pre-rigor pressurization of muscle: Effects on pH, shear value and taste panel assessment. *J. Food Sci., 38*, 294–298.

Molin, G. (2000). Modified atmospheres. In B.M. Lund, A.C. Baird-Parker, & G.W. Gould (Eds.), *The Microbiological Safety and Quality of Food* (pp. 214–234). Gaithersburg, MD: Aspen.

Murrell, W.G., & Scott, W.J. (1966). Heat resistance of bacterial spores at various water activities. *J. Gen. Microbiol., 43*, 411–425.

Notermans, S., Dufrenne, J., & Lund, B.M. (1990). Botulism risk of refrigerated processed foods of extended durability. *J. Food Prot., 53*, 1020–1024.

Patterson, M., & Loaharanu, P. (2000). Food irradiation. In B.M. Lund, A.C. Baird-Parker, & G.W. Gould (Eds.), *The Microbiological Safety and Quality of Food* (pp. 65–100). Gaithersburg, MD: Aspen.

Patterson, M.F., Quinn, M., Simpson, R., & Gilmour, A. (1995a). Effects of high pressure on vegetative pathogens. In D.A. Ledward, D.E. Johnston, R.G. Earnshaw, and A.P.M. Hasting (Eds.), *High Pressure Processing of Foods* (pp. 47–63). Nottingham: University Press.

Patterson, M.F., Quinn, M., Simpson, R., & Gilmour, A. (1995b). Sensitivity of vegetative pathogens to high hydrostatic pressure treatment in phosphate-buffered saline and in foods. *J. Food Prot., 58*, 524–529.

Payens, T.A.J., & Heremans, K. (1969). Effect of pressure on the temperature-dependent association of β-casein. *Biopolymers, 8*, 335–345.

Perkins, J. (1820). On the compressability of water. *Philosphical Trans. R. Soc., 110*, 324–329.

Perkins, J. (1826). On the progressive compression of water by a high degree of force, with trials on the effect of other fluids. *Philosophical Trans. R. Soc., 116*, 541–547.

Pflug, I.J., & Gould, G.W. (2000). Heat treatment. In B.M. Lund, A.C. Baird-Parker, & G.W. Gould (Eds.), *The Microbiological Safety and Quality of Food* (pp. 36–64). Gaithersburg, MD: Aspen.

Qin, B.L., Pothakamury, U.R., Barbosa-Canovas, G.V., & Swanson, B.G. (1996). Nonthermal pasteurization of liquid foods using high-intensity pulsed electric fields. *Crit. Rev. Food Sci. Nutr., 36*, 603–627.

Raso, J., Palop, A., Pagan, R., & Condon, S. (1998). Inactivation of *Bacillus subtilis* spores by combining ultrasonic waves under pressure and mild heat treatment. *J. Appl. Microbiol., 85*, 849–854.

Roberts, C.M., & Hoover, D.G. (1996). Sensitivity of *Bacillus coagulans* spores to combinations of hydrostatic pressure, heat, acidity and nisin. *J. Appl. Bacteriol., 81*, 363–368.

Rooney, J., Midda, M., & Leeming, J. (1994). A laboratory investigation of the bactericidal effect of a Nd:Yag laser. *Br. Dent. J., 176*, 61–64.

Russell, N.J., & Gould, G.W. (Eds.). (2000). *Food Preservatives* (2nd ed.). Gaithersburg, MD: Aspen.

Sala, F.J., Burgos, J., Condon, S., Lopez, P., & Raso, J. (1995). Effect of heat treatment and ultrasound on microorganisms and enzymes. In G.W. Gould (Ed.), *New Methods of Food Preservation* (pp. 176–204). Glasgow: Blackie Academic & Professional.

Sale, A.J.H., Gould, G.W., & Hamilton, W.A. (1970). Inactivation of bacterial spores by hydrostatic pressure. *J. Gen. Microbiol., 60*, 323–334.

Tapia de Daza, M.S., Alzamora, S.M., & Welti-Chanes, G. (1996). Combination of preservative factors applied to minimal processing of foods. *Crit. Rev. Food Sci. Nutr., 36*, 629–659.

Timson, W.J., & Short, A.J. (1965). Resistance of microorganisms to hydrostatic pressure. *Biotechnol. Bioeng., 7*, 139–159.

WHO. (1999). High-dose irradiation: Wholesomeness of food irradiated with doses above 10 kGy. Report of a Joint FAO/IAEA/WHO Study Group. WHO Technical Report Series 890. Geneva: World Health Organization.

Wills, P.A. (1974). Effects of hydrostatic pressure and ionizing radiation on bacterial spores. *Atomic Energy Aust., 17*, 2–10.

Zhang, Q., Qin, B.L., Barbosa-Canovas, G.V., & Swanson, B.G. (1995). Inactivation of *E. coli* for food pasteurization by high strength pulsed electric fields. *J. Food Proc. Preserv., 19*, 103–118.

Zipp, A., & Kauzmann, W. (1973). Pressure denaturation of metmyoglobin. *Biochemistry, 12*, 4217–4228.

CHAPTER 2

The Effects of High Pressure on Biomaterials

Karel Heremans

INTRODUCTION

The study of the effects of pressure on food macromolecules and colloids has received a great deal of attention in recent years. Bridgman (1914) made considerable progress in the development of equipment not long after the initial attempts of Hite (1899), who referred to technical problems with high pressure equipment. Developments in France, primarily in the biomedical field between the two World Wars, are reviewed by Johnson et al. (1974). Modern studies started with the seminal papers by Suzuki (1960) in Japan. He conducted detailed studies of the effects of pressure and temperature on the kinetics of ovalbumin and hemoglobin denaturation.

While observing phase transitions on various compounds, Bridgman (1914) observed that the white of an egg coagulates after a pressure treatment for 30 minutes at 700 MPa. He observed that the appearance of the pressure-induced coagulum is quite different from that induced by temperature. In addition, he observed that the ease of the pressure-induced coagulation increases at low temperatures. In other words, he observed a negative activation energy for a chemical process. In addition he observed that, when the egg white is taken to 1.2×10^3 MPa into the ice VI phase (the phase of ice that has a density higher than normal ice), the coagulum does not seem to be affected by the freezing. Recently Hayashi et al. (1989) analyzed the effect of pressure on egg yolk. It is well known from the preparation of hard-boiled eggs that the yolk becomes solid at a slightly higher temperature than the white. The reverse is true for the effect of pressure: the yolk becomes solid at a lower pressure than the white. We now know that these observations are the consequence of the unique behavior of proteins.

It is a pleasure to acknowledge contributions, results of experiments as well as discussions, from a large number of collaborators and colleagues. The research in our laboratory was supported by the Research Fund of the Katholieke Universiteit Leuven; F.W.O. Flanders, Belgium; and the European Community (FAIR-CT96-1175).

If the conditions for equilibrium or isokineticity are plotted against temperature and pressure, a stability phase diagram is obtained with an elliptical shape (Suzuki, 1960; Hawley, 1971; Zipp & Kauzmann, 1973; Taniguchi & Suzuki, 1983; Heinisch et al., 1995; Weingand-Ziadé et al., 1997; Ludikhuyze et al., 1998). One of the practical consequences of this phenomenon is the stabilization against heat denaturation by low pressures. Not only has this been observed in several proteins and enzymes, but it also applies to the effect of pressure on the heat gelation of starch (Thevelein et al., 1981; Rubens et al., 1999). Of special interest is the observation that the inactivation kinetics of microorganisms results in diagrams similar to those of proteins (Sonoike et al., 1992; Ludwig et al., 1996). This observation also holds true for the cold inactivation of microorganisms (Hashizume et al., 1995). In addition to possible applications in food science, this discovery may be useful in the development of immunodesorption (Degraeve et al., 1996), vaccine development, and virus sterilization (Silva et al., 1996). A number of books and reviews can be consulted for more details (Balny et al., 1992; Ledward et al., 1995; Hayashi & Balny, 1996; Michels et al., 1996; Heremans, 1997; Isaacs, 1998; Ludwig, 1999).

This chapter concentrates on the basic effects of pressure and temperature on proteins and starch. It considers thermodynamic as well as kinetic concepts. Phase diagrams of water, lipids, and proteins are discussed in order to provide a coherent physical picture of the differences between temperature and pressure treatment. It emphasizes infrared spectroscopy, a technique that can give molecular details on the pressure behavior of proteins and other biopolymers for observations in situ on pressurized samples. Nuclear magnetic resonance imaging, fluorescence spectroscopy, and other experimental approaches are reviewed elsewhere (Heremans and Smeller, 1998).

PRESSURE VERSUS TEMPERATURE EFFECTS

This section explores the effect of temperature and pressure on the conformation, the dynamics, and the reactions of biomolecules. The unique properties of biomolecules are determined by the delicate balance between internal interactions, which compete with interactions with the solvent. The primary source of the dynamical behavior of biomolecules is the free volume of the system, and this may be expected to decrease with increasing pressure. As temperature effects act via an increased kinetic energy as well as increased free volume, it follows that the study of the combined effect of temperature and pressure is a prerequisite for a full understanding of the dynamic behavior of biomolecules. By intuition, pressure effects should be easier to interpret than temperature effects. This follows from the Le Châtelier-Braun principle: Pressure affects primarily the volume of a system, while temperature changes cause volume as well as energy changes.

If χ is a quantity characteristic of an equilibrium or a rate process, then the influence of temperature and pressure can be written as:

$$\left(\frac{\delta \ln \chi}{\delta T}\right)_p = \left(\frac{\text{A characteristic energy}}{RT^2}\right) \tag{1}$$

$$\left(\frac{\delta \ln \chi}{\delta P}\right)_T = \left(\frac{\text{A characteristic volume}}{RT}\right) \tag{2}$$

These formulas summarize the basic thermodynamic and kinetic concepts in our interpretation of the effects of pressure and temperature on equilibriums as well as rate processes.

Compression and Phase Transitions

The effect of compression of biomaterials can best be illustrated by first considering the effects on liquids. Water and hexane are of particular interest because they represent typical examples of polar and nonpolar materials, respectively. The compressibility of water is considerably less than that of most organic liquids. The volume of water is reduced by 10% at 300 MPa and by 15% at 600 MPa. The volume of hexane is reduced by 20% at 300 MPa and by 25% at 600 MPa.

If one performs the compression under adiabatic conditions, then the following equation gives the temperature increase for a given pressure increase:

$$\left(\frac{\delta T}{\delta P}\right)_S = \left(\frac{T}{C_p}\right)\left(\frac{\delta V}{\delta T}\right)_p = \frac{\alpha T}{\rho C_p} \tag{3}$$

In this equation, α is the thermal expansion, ρ the density, and C_p the heat capacity of the system. Using this equation for water, the temperature increases by 2 K/100 MPa at 25°C, and there is no temperature increase at 4°C. For hexane, the temperature increases by 40 K/100 MPa at 18°C. It is clear that a decrease in pressure results in a temperature decrease of the same order of magnitude. At higher temperatures, adiabatic compression gives larger temperature increases.

Pressure has a minor effect on the viscosity of pure water up to about 600 MPa. For the same pressure range, the viscosity of hexane increases tenfold.

The effect of pressure on the melting temperature (T_m) of compounds is given by the Clausius-Clapeyron equation,

$$\frac{dT_m}{dP} = \frac{T_m \Delta V}{\Delta H} \tag{4}$$

Because the change in volume (ΔV) and enthalpy (ΔH) on melting are generally positive, one can expect an increase in melting temperature with increasing pressure. For many organic compounds, dT_m/dP is approximately 15 K/100 MPa. As a first approximation, the effect of a pressure increase by 100 MPa would correspond to the physical effect of a decrease in temperature by 15 K. A notable exception to this general rule is water. At room temperature, a pressure of about 1000 MPa is needed to obtain the phase of ice (VI) that differs from normal ice

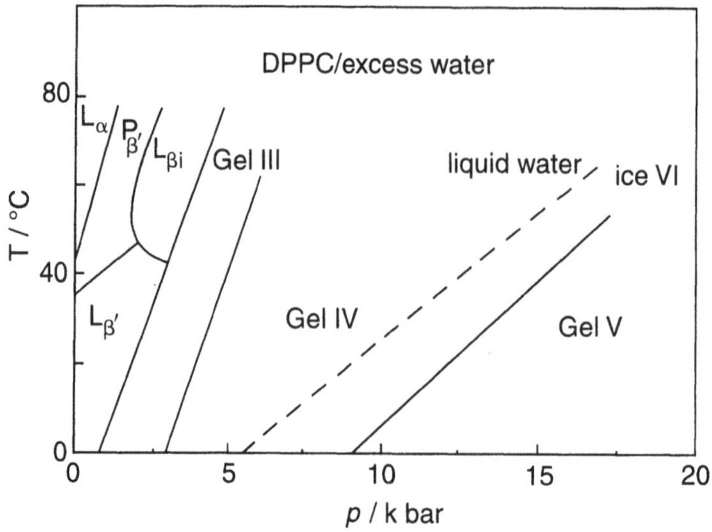

Figure 2–1 Pressure-temperature phase diagram for the transitions in dipalmitoylphosphatidylcholine (DPPC). The dotted line represents the transition from liquid water to ice VI.

by its higher density. This can be seen from the phase diagram of water (Klug et al., 1989).

The Clausius-Clapeyron equation can be illustrated by the effect of pressure on the temperature of the transitions in phospholipids, as shown in Figure 2–1 (Czeslik et al., 1998).

Pressure favors the crystalline state, as a result of the Le Châtelier-Braun principle. Whereas the transition temperature of the lipids depends on the length of the hydrocarbon chain, the rate at which the temperature changes with pressure is almost independent of the chain length (dT/dP = 20 K/100 MPa) (Heremans, 1982). A higher degree of unsaturation of the hydrocarbon chain lowers the dT/dp (dT/dP = 14 K/100 MPa) values of the lipids (Winter et al., 1996). The dT/dP values have very little relationship to the pressure, except for the formation of lipid interdigitated phases.

When proteins are incorporated in membranes, the situation is more complex. In biomembranes, the physical state of the lipids that surround the membrane proteins plays a crucial role in the activity of membrane-bound enzymes. In addition, the integrity of the membrane of living organisms such as bacteria is very sensitive to pressure. This explains the sterilization effect of pressure. It has been observed that the temperature and pressure dependence of the activity of membrane-bound enzymes such as the Na^+-K^+-ATPase and the Ca^{++}-ATPases are nonlinear (Heremans & Wuytack, 1980).

Chemical Reactions

For a chemical reaction $N = D$, where N represents an enzyme in its active (native) conformation and D in its inactive (denatured) conformation, the equilibrium constant, $K(D/N)$, is related to the Gibbs free energy by

$$\Delta G^0 = -RT\ln K \tag{5}$$

The temperature and pressure dependence is given by

$$\frac{d\Delta G^0}{dT} = -\Delta S^0 \qquad \frac{d\Delta G^0}{dP} = \Delta V^0 \tag{6}$$

Because the changes in Gibbs free energy for a substantial shift from N to D are of the order of a few kcal, it is clear that at ambient pressure the $P\Delta V$ will have a very small effect on chemical reactions. For a reaction volume, $\Delta V = -82$ mL, which is a large value for a chemical reaction, $P\Delta V = 2$ cal. Therefore pressures of the order of several hundred MPa are needed to affect chemical reactions in the liquid phase. The relation between ΔV and the change in equilibrium constant with pressure was first given by Planck:

$$-\Delta V^0 = \frac{RTd\ln K}{dP} \tag{7}$$

A positive ΔV^0 implies a shift toward the reactants at higher pressure. A higher temperature favors the reaction products for an endothermic reaction. The actual shift in the equilibrium depends not only on ΔV^0 or ΔH^0, but also on the position of equilibrium.

A similar expression may be derived to describe the effect of pressure on the reaction rate:

$$-V_a = \frac{RTd\ln k}{dP} \tag{8}$$

where V_a is the activation volume, a concept first proposed by Evans and Polanyi (1935).

An important point with relation to the effects of pressure on reaction velocities is that the reaction, depending on the mechanism, may be either accelerated or retarded by high pressure (i.e., V_a may be negative or positive). For example, the Maillard reaction is strongly inhibited by high pressure. On the other hand, oxidation of nonsaturated lipids is accelerated by high pressure (Tauscher, 1995). Negative activation energies are meaningless from a physical point of view. The fact that they are nevertheless observed indicates a coupling of the rate process with a strongly exothermic preequilibrium. This is observed in the temperature dependence of the pressure-induced denaturation of proteins.

Molecular Interpretation of Volume Changes

The observed reaction and activation volumes are often interpreted in terms of intrinsic and solvent effects:

$$\Delta V(observed) = \Delta V(intrinsic) + \Delta V(solvent) \tag{9}$$

Intrinsic effects arise from packing contributions (e.g., the increase in molecular ordering as a consequence of the volume decrease at high pressure and the changes in volume due to the formation or rupture of covalent bonds). The role of the solvent is taken into account in the concept of the partial molar quantities.

The partial molar volume, V_i, of a solute molecule or ion is defined as the volume change of the solution by the addition of a small amount of the solute over the number of moles of added solute keeping the other variables constant.

$$V_i = \left(\frac{\delta V}{\delta n_i}\right)_{n_j, p, T} \tag{10}$$

Here V is the volume of the solution and n_i is the number of moles of the solute added at constant temperature and pressure. It is not equal to the volume of the molecule or the ions, since it includes also the interaction with the solvent. This may be seen from the fact that the partial molar volume of the uncharged glycolamide is larger (56.2 mL/mol) than that of the amino acid glycine (43.5 mL/mol).

The thermal expansion is defined as the relative change of the partial molar volume, V_i, with temperature:

$$\alpha = \left(\frac{\delta \ln V_i}{\delta T}\right)_p \tag{11}$$

The isothermal compressibility is defined as the relative change of the volume with pressure:

$$\beta = -\left(\frac{\delta \ln V_i}{\delta P}\right)_T \tag{12}$$

For a biopolymer, V_i can be written as the sum of three terms:

$$V_i = V_{atom} + V_{cavities} + \Delta V_{hydration} \tag{13}$$

where V_{atom} and $V_{cavities}$ are the volumes of the atoms and the cavities, respectively, and $\Delta V_{hydration}$ is the volume change resulting from the interactions of the biopolymer with the solvent.

Because the volumes of the atoms may be considered to be temperature and pressure independent, it follows that both the thermal expansion and the compressibility are composed of two main terms, cavity volume and hydration volume. An estimate of the contribution of each factor relies on assumptions that are not easy to check experimentally. However, the compressibility of amino acids is negative because of changes in the solute-solvent interaction (i.e., the amino acid solution is less compressible than the pure solvent).

Table 2–1 gives the thermal expansion and the compressibility for a number of small molecules, proteins, and dextran. Note that the compressibility of amino acids is negative whereas that of proteins is positive. This suggests that the contribution of cavities compensates for the hydration effects, which causes the compressibility of proteins in solution to become positive. The situation is similar if one compares glucose with dextran. Although less negative, the sign of the compressibility suggests a hydration behavior like that of strong electrolytes.

The formation of ions in solution is accompanied by an electrostriction of the solvent (i.e., the ordering of the solvent around the electric charge of the ion).

Table 2–1 Thermal Expansion and Isothermal Compressibility of Water and Hexane at 25°C and 0.1 MPa

Substance	*Thermal Expansion (10^{-6}/K)*	*Compressibility (1/100 GPa)*
Water	210	45.8
Hexane	1380	166
Amino acids*	1000–1550	−0.28–−0.62
Proteins*	40–110	2–15
Glucose*	1200	−13
Dextran*	1100	−10–−5

*Dilute solution in water.

This electrostriction explains the changes in volume for the autoionization of water.

$$2\,H_2O = H_3O^+ + OH^- \qquad \Delta V = -\,22\ mL \tag{14}$$

Although water is a special case, this reaction shows that the formation of a monovalent ion is accompanied by a negative volume change of approximately −10 mL. Electrostriction also explains the effect of pressure on the pH of aqueous solutions. Table 2–2 illustrates that the shift in pH per 100 MPa (ΔpH/100 MPa) is considerable in the case of acetic and phosphoric acid. More extensive data can be found in the paper of Kitamura and Itoe (1987). Hayert et al. (1999) have recently suggested measuring the pressure dependence of the pH of solutions using fluorescent indicators.

For fruit juices, which are generally quite acidic, a treatment at 500 MPa would cause a pH shift of about one unit to the acid side. In practice, the effect may be less pronounced because the volume change of ionization becomes smaller at high pressure. In addition, buffers that are pressure dependent show a very small temperature dependence. Likewise, buffers that are temperature dependent show a small pressure dependence. From the reaction volume of the ionization of water, it follows that when the pH remains constant as a function of pressure, the pOH changes substantially.

Table 2–2 Volume, Enthalpy, and Pressure-Induced Shift in pH and pOH for Common Acid and Base Functions at 25°C and 0.1 MPa

	ΔV/mL	*ΔH/kJ*	*ΔpH/100 MPa*	*ΔpOH/100 MPa*
$R{-}NH_3^+ \longleftrightarrow R{-}NH_2 + H^+$	1	55	0	−0.3
$R{-}COOH \longleftrightarrow R{-}COO^- + H^+$	−12	0	−0.2	−0.2
$H_2PO_4^- \longleftrightarrow HPO_4^{-2} + H^+$	−25	0	−0.4	+0.1

Studies on various model systems have given information on the effect of pressure on noncovalent intermolecular interactions. One of the important results is that hydrogen bonds are stabilized by high pressures. Hydrophobic interactions that play a substantial role in the stabilization of the tertiary structure and in protein-protein interactions are destabilized under high pressure. On the other hand, stacking interactions between aromatic rings show negative volume changes and are stabilized by pressure (Mozhaev et al., 1996a).

STABILITY PHASE DIAGRAMS OF FOOD MACROMOLECULES

As mentioned in the introduction, the first systematic observation about the denaturation (coagulation) of proteins by high pressure was made by Bridgman in 1914. From what we know today, it is clear that he observed the essential features of the phase diagram for the stability of proteins. The next step forward was made by Suzuki (1960), who studied the kinetics of denaturation of ovalbumin and carbonylhemoglobin as a function of temperature and pressure. He plotted his results in a *P*, *T*, *k*-diagram. He connected the points with the same rate constants (*k*) at a given *P* and *T*. He noted that there was a change in sign for the activation energy (E_a) and volume (V_a) from positive at high temperature and low pressure to negative at high pressure and low temperature, as shown in Table 2–3. The rate of pressure-induced denaturation goes faster at lower temperatures. Suzuki explained the negative activation energy in domain I with the following mechanism:

$$\begin{aligned} &P + n\,H_2O \longleftrightarrow P\,(H_2O)n \longrightarrow Pd \\ &mPd \longrightarrow (Pd)m \end{aligned} \tag{15}$$

In the first step, water is pressed into the protein, which then unfolds. In the second step, the unfolded protein forms aggregates.

The observed negative activation energies are not unique for the high pressure conditions. Negative activation energies are also observed for the denaturation of proteins in high urea concentrations (Simpson & Kauzmann, 1953). To account for this observation, these authors proposed a mechanism including the exothermic binding of urea to the native protein. Thus, it seems that the binding of urea at ambient pressure and the binding of water at high pressure may account for the negative activation energies that are observed in these processes.

Table 2–3 Pressure and Temperature Dependence of the Sign of the Activation Energy (E_a) and Volume (V_a) for the Denaturation of Ovalbumin and Hemoglobin

	Temperature	*Pressure*	E_a	V_a
High pressure/low temperature	<30°C	>400 MPa	−	−
High pressure/high temperature	>40°C	>300 MPa	+	−
Low pressure/high temperature	>60°C	<300 MPa	+	+

A fairly complete list of all the experimental kinetic data available on the protein denaturation/unfolding and enzyme inactivation is given in Table 2–4. The same table includes complex systems that show similar behavior: the destruction of phages, viruses, bacteria, and cells; the cloud formation in synthetic polymers;

Table 2–4 Survey of Experimental Data That Show Reentrant Behavior in the *P,T* Plane

Process	*System*	*Reference*
Protein denaturation	Egg white	Bridgman (1914)
	Ovalbumin	Suzuki (1960)
	Hemoglobin	Suzuki (1960)
	Ribonuclease	Hawley (1971)
	Chymotrypsinogen	Hawley (1971)
	Myoglobin	Zipp and Kauzmann (1973)
	Ribonuclease	Zhang et al. (1995)
	Egg white	Kanaya et al. (1996)
Enzyme inactivation	Chymotrypsin	Taniguchi and Suzuki (1983)
	Lipoxygenase	Heinisch et al. (1995)
	Butylcholinesterase	Weingand-Ziadé et al. (1997)
	Lipoxygenase	Ludikhuyze et al. (1998)
	Polyphenoloxidase	Weemaes et al. (1998)
	Amylase	Ludikhuyze et al. (1997)
	Pectinmethylesterase	Van den Broeck (2000)
	Pectinmethylesterase	Crelier et al. (1999)
	Polygalacturonase	Crelier et al. (1999)
Protein unfolding	Amylase	Rubens (1999)
	Lipoxygenase	Rubens (1999)
Bacteria, viruses, and phages	*Escherichia coli*	Ludwig et al. (1996)
	Deep-sea bacteria	Yayanos (1998)
	Saccharomyces cerevisiae	Hashizume et al. (1995)
	Escherichia coli	Sonoike et al. (1992)
	Lactobacillus casei	Sonoike et al. (1992)
	Listeria monocytogenes	Hellemons and Smelt (1999)
	Bacteriophage T4	Ludwig et al. (1996)
Cell inactivation	Fruit fly eggs	Butz and Tauscher (1995)
Reaction kinetics	Myosin + ATP	Saldana and Balny (1992)
Cloud point	Synthetic polymers in H_2O	Cook et al. (1992), Sun and King (1996), Kunugi et al. (1997)
Phase separation	Polystyrene/cyclohexane	Van Opstal and Koningsveld (1992)
Starch gelation	Potato	Thevelein et al. (1981)
	Potato, wheat	Muhr et al. (1982)
	Wheat	Muhr et al. (1982)
	Rice	Douzals (1999), Rubens (1999)

and the gelation of starch. In all cases, kinetic data are presented, although in a few cases a thermodynamic analysis is presented. It is also clear from these publications that, under certain conditions of temperature and pressure, the kinetic behavior of the system is controlled by the thermodynamics of an equilibrium process.

A Physical Model for the Interpretation of the Stability Phase Diagram

A general thermodynamic model for the analysis of the effect of temperature and pressure on phase transition in pure compounds was first presented by Bridgman (1915). The first experimental data on the kinetics of the denaturation of proteins were presented by Suzuki (1960). The first thermodynamic analysis that is most often quoted in the biochemical literature was given by Hawley (1971).

Figure 2–2 shows three common ways to denature a protein: heat, pressure, and cold denaturation. In the thermodynamic description, the denaturation is a reversible process. However, in most cases the denatured protein tends to produce a gel stabilized by an intermolecular network, which prevents the refolding to the native state.

This stabilization gives rise to irreversible effects that can be described by kinetic theory. However, given the equilibrium assumption in the transition state theory, the thermodynamic theory can be extended and adapted to kinetic data. Instead of a phase diagram in the P, T plane, an isokinetic (isopleths) stability diagram is obtained for P, T, and k. This simple physical model provides the basis for a mathematical description. This model, which has been developed for proteins, can easily be extended to more complex situations such as the gelation of starch and the killing of bacteria.

The change in Gibbs free energy as a function of pressure and temperature is given by

$$d\,(\Delta G) = -(\Delta S)\,dT + (\Delta V)\,dP \tag{16}$$

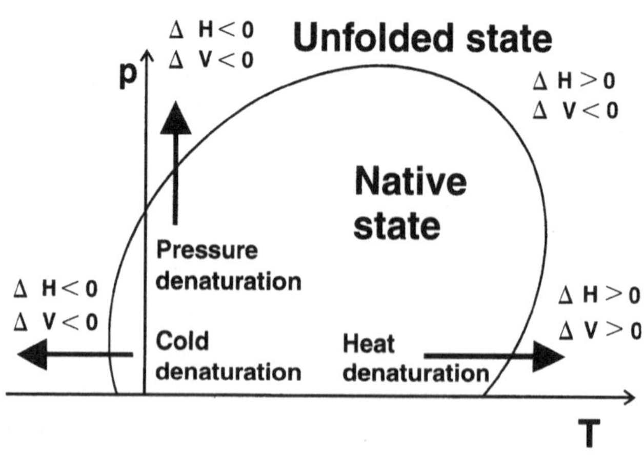

Figure 2–2 The relation between heat, cold, and pressure denaturation of proteins. Note the change in sign for ΔH and ΔV between the heat, pressure, and cold denaturation.

where ΔG represents the difference in partial molar Gibbs free energy between the denatured and the native state. An equivalent expression for the kinetic data would be:

$$d(Ga) = -(Sa)\,dT + (Va)\,dP \tag{17}$$

where Ga ($= -RT\ln k$) expresses the difference in the Gibbs free energy of activation between the transition state and the native state. In the following equation, we use only the equilibrium expressions. Integration of the equation with the assumption that both ΔS and ΔV are temperature as well as pressure dependent leads to the following:

$$\Delta G = a + bT + cT^2 + dP + eP^2 + fTP \tag{18}$$

Note that higher-order terms in T^3, P^3, PT^2, P^2T, etc., are neglected in this equation. Neglecting these higher-order terms is equivalent to the assumption that the parameters c, e, and f are independent of temperature and pressure. This equation has an advantage over purely empirical equations because the constants can be given a simple physical interpretation. Note that the thermodynamic quantities are all partial molar quantities.

$a = \Delta G^0$ at given reference conditions: $\Delta G = \Delta G\,(P_0, T_0)$
$b = \Delta S^0$ at reference conditions: $\Delta S\,(P_0, T_0)$
$c = T\,(\delta\Delta S^0/\delta T)$: change in the heat capacity (ΔC_p)
$d = \Delta V^0$ at reference conditions: $\Delta V\,(P_0, T_0)$
$e = \delta\Delta V^0/\delta P$: change in the compressibility factor ($\Delta\beta$)
$f = \delta\Delta V^0/\delta T\,(= -\,\delta\Delta S^0/\delta P)$: change in the thermal expansion factor ($\Delta\alpha$)

From the Maxwell equation $(\delta\Delta S/\delta P) = -(\delta\Delta V/\delta T)$, the constant f has two equivalent interpretations. The equation may be written in the following final form:

$$\Delta G = \Delta G^0 - \Delta S^0\,(T - T_0) - \frac{\Delta C_p}{2T_0}\,(T - T_0)^2 + \Delta V^0(P - P_0) + \frac{\Delta\beta}{2}\,(P - P_0)^2 + \Delta\alpha\,(P - P_0)(T - T_0) \tag{19}$$

with the definitions of ΔC_p, $\Delta\beta$, and $\Delta\alpha$ as given above. Except for the heat capacity, C_p, the compression and the thermal expansion do not have their conventional definition as given in the discussion of molecular interpretation of volume changes.

An equation similar to the Clausius-Clapeyron equation can be obtained for the slope of the phase boundary:

$$\frac{\delta T}{\delta P} = \frac{\Delta V^0 + \Delta\beta\,(P - P_0) + \Delta\alpha\,(T - T_0)}{\Delta S^0 - \Delta\alpha\,(P - P_0) + \Delta C_p\,(T - T_0)\,/\,T_0} \tag{20}$$

This expression reduces to the classical Clausius-Clapeyron equation when the differences in compressibility, thermal expansion, and heat capacity vanish as is observed for most phase transitions in lipids. The shape of the phase diagram for proteins is of considerable interest because it contains information on the volume and entropy fluctuations and on the coupling between volume and entropy fluctuations. The values for dT/dP and dP/dT become zero at two points in the diagram. A more detailed discussion of the consequences of the thermodynamic model is given by Smeller and Heremans (1997). The basic assumption of the model is that it is a two-state model. Other problems with the two-state model have been discussed by Franks (1995).

The free-energy equation reduces to simpler expressions at constant temperature or pressure. At constant pressure, we get the familiar expression that describes heat denaturation as well as cold denaturation:

$$\Delta G = \Delta G^0 - \Delta S^0 (T - T_0) - \frac{\Delta C_p}{2T_0} (T - T_0)^2 \tag{21}$$

It can be shown that this is equivalent with the Gibbs-Helmholtz equation, an equation often found in the biochemical literature.

$$\Delta G = \Delta G^0 - \Delta S^0 (T - T_0) + \Delta C_p \left[(T - T_0) - T \ln \frac{T}{T_0} \right] \tag{22}$$

At constant temperature, we obtain an equation that describes the pressure denaturation.

$$\Delta G = \Delta G^0 + \Delta V^0 (P - P_0) + \frac{\Delta \beta}{2} (P - P_0)^2 \tag{23}$$

It is important to emphasize the power but also the limits of equations 19 to 23. The power of the equation resides in the fact that it gives a consistent thermodynamic description of the heat, pressure, and cold denaturation. The cross term in the equation finds its origin in the pressure and temperature dependence of the volume changes:

$$\Delta V = \Delta V^0 + \Delta \alpha (T - T_0) - \Delta \beta (P - P_0) \tag{24}$$

The limit of the equation is that only the differences in the thermal expansion factor and compressibility factor and heat capacity between the denatured and the native state are obtained. For absolute quantities, one needs additional data from densitometry, calorimetry, and ultrasonic velocimetry. Because the equation gives a relation between partial molar quantities, the usual difficulties arise for a molecular interpretation of these quantities. It is the common practice to rely on low molecular mass model systems. However, in many model systems the cavity term is not considered.

A final point is the applicability of the equation for the analysis of kinetic data on more complex systems than proteins. From Table 2–4, it can be seen that reentrant P,T diagrams have been observed for microorganisms and starch. Reentrant behavior has not been observed for nucleic acids, and in the case of lipids it can be found under limited conditions where interdigitation of the hydrocarbon chains takes place (Winter et al., 1996). It may then be concluded that the observation of elliptic stability diagrams in the case of microorganisms can be attributed to the denaturation of a critical enzyme in the metabolism. In the case of starch, a preequilibrium as proposed for proteins can be assumed. No other polysaccharides show the reentrant behavior, which suggests that the packing of amylose and amylopectin is an important part of the mechanism.

Compressibility and Protein Flexibility

The compressibility is a thermodynamic quantity of interest from a dynamic as well as a static point of view. Its relevance to the biological function of a protein can be understood through the statistical mechanical relation between the isothermal compressibility β_T and volume fluctuations:

$$(V - [V])^2 = \beta_T k T_B V \tag{25}$$

where the expression between parentheses indicates an average for the isothermal-isobaric ensemble. Because of the small size of the protein, the volume fluctuations are relatively large and of the order of the volume changes for the denaturation. Similar expressions exist for the relation between the thermal expansion and volume-entropy fluctuations and between the heat capacity and entropy fluctuations (Heremans & Smeller, 1998).

A relation between the adiabatic compressibility of proteins and their surface activity has been suggested by Razumovsky and Damodaran (1999). In addition, researchers using the ultrasound technique to determine the adiabatic compressibility have found some interesting applications for the characterization of food colloids (Dickinson & McClements, 1996).

Role of Water and Cosolvents

Water is a prerequisite for the reentrant behavior. This is clear from the fact that dry proteins, dry starch, and dry bacterial spores may be compressed to very high pressures in the diamond anvil cell without showing any irreversible changes as judged from spectroscopic observations such as infrared spectroscopy. As soon as water is available, the systems behave in the normal way (i.e., pressure-induced changes take place that are in most cases irreversible). This suggests that a possible mechanism includes at least one reversible step followed by an irreversible step.

The role of the solvent composition (water activity) in the denaturation process of proteins can be understood if one keeps in mind that the structure of a protein in solution is the result of a delicate balance between the intramolecular interactions in the polypeptide chain, which compete with the solvent interactions. This also applies to other biopolymers including starch. The effect of the water activity for the biotechnology application of the high pressure technique is also documented (Mozhaev et al., 1996b). Timasheff (1998) has recently discussed the stabilizing effect of osmolytes (amino acids, sugars, and polyols) on the temperature denaturation of proteins. These compounds are preferentially excluded from the protein surface and thereby affect the binding of water to proteins. Their presence shifts the conformational equilibrium in macromolecules toward a state with the least amount of bound water. In a number of cases, the effects are quite nonspecific as expected for a pure osmotic effect.

Small organisms called tardigrades provide a fascinating example of the central role of water in the behavior of organisms under extreme conditions. These animals, composed of about 40,000 cells, become immobile and shrink into a special state when the humidity of the surroundings decreases. In such a state, they can survive temperatures from −253°C up to 151°C and pressures up to 600 MPa (Seki & Toyoshima, 1998). In the normal state, they are killed at 200 MPa. This behavior is similar to that of bacterial spores and dry proteins, in which pressures of more than 1000 MPa do not provoke any changes (Sojka & Ludwig, 1997; Goossens et al., 1996). The fact that the tardigrades can undergo a transformation to an extreme dry state may be much more exceptional than the fact that they are resistant to extremes of temperature and pressure in the dry state.

Effects of Pressure on Glass Transition Temperature

The glass transition temperature is an important concept in food science. For pure liquids, glass transition is a transition, not between equilibrium states, but between two metastable states—a supercooled liquid and a disordered solid state. Glass transition is a dynamic concept, not an equilibrium concept. Because the structure of a protein is the result of a delicate balance between the intramolecular interactions in the polypeptide chain and the solvent interactions, it is clear that the water activity plays an important role in glass transition temperature. Proteins in the dry state are extremely resistant to temperature and pressure unfolding. The pressure effect on dry proteins may easily be studied in the diamond anvil cell (Goossens et al., 1996). These experiments emphasize the role of water as a plasticizer in biomacromolecules via the reduction of the glass-rubber transition temperature. These experiments also suggest that proteins in the dry state are below their glass transition temperature. The fundamental similarities (and differences) between synthetic polymers and biomacromolecules have stimulated research in this field because of the importance of the glass transition on all aspects of the functional properties. In view of the nonequilibrium nature of the glass transition temperature, this transition is governed by activation rather than equilibrium parameters. However it may be assumed that the kinetics are driven by thermodynamic conditions.

Limited studies on the effect of pressure on the glass transition temperature in polymers suggest a change of approximately 22 K/100 MPa for nonhydrogen bonded systems. The effect of pressure on the hydrogen bonded system sorbitol shows a much weaker dependence of 4 K/100 MPa (Atake & Angell, 1979). This suggests that similar orders of magnitude may be expected for glass transitions in proteins.

STRUCTURE PROPERTY RELATIONSHIP IN FOOD BIOPOLYMERS

The structure property relationship of food can be attributed to one of the main macromolecular components. Pressure- and/or temperature-induced changes of these components give rise to changes in properties that often have functional significance. However, functionality is a higher-order concept that includes sensory aspects that are relevant to food quality. The concepts of functionality and quality, which do not have a physicochemical definition, are not considered here.

Concentration effects play a very important role in structure property relations. This is especially true for gel formation and phase separation phenomena that take place only in concentrated solutions of the order of 10% or higher in the case of proteins.

In the following sections, we discuss some selected effects of pressure and temperature on food hydrocolloids (high molecular-weight, water-soluble macromolecules such as proteins, starches, and mixtures of biopolymers).

Proteins

This section concentrates on the relation between unfolding of proteins by temperature or pressure and the possible effects on aggregation and gel formation.

Protein Unfolding and Denaturation

Proteins may be unfolded or denatured by heat, pressure, or cold treatment. Protein denaturation can be studied by spectroscopy, enzyme activity, thermal analysis, chromatography, enzymatic digestion, immunology, electrophoresis, microscopy, and solubility. However, in order to provide a better understanding of the structure property relation, it is desirable to have more details of the events that take place at the molecular level.

In recent years, several spectroscopic techniques have been used to obtain a deeper insight into the difference between temperature- and pressure-induced unfolding. These include nuclear magnetic resonance (NMR) (Zhang et al., 1995), ultraviolet (UV) spectroscopy (Mombelli et al., 1997), fluorescence, and infrared spectroscopy combined with small-angle X-ray scattering (Panick et al., 1999b). Whereas circular dichroism can be used to study cold and heat denaturation, for technical reasons, it cannot be used to study pressure denaturation. The high pressure diamond anvil cell may be used to follow in situ structural changes of biomacromolecules during the compression and decompression phase. With this technique, pressures of 1000 MPa and more can easily be obtained. The technique can be used with Fourier transform infrared spectroscopy and has distinct advantages for the observation of pressure-induced changes in biomolecules and in living tissues (Wong et al., 1993). Infrared spectroscopy is an absorption technique. Fluorescence impurities present in many samples of biological origin do not interfere, but the strong absorption of water in the region of interest for the protein studies necessitates its replacement by heavy water (D_2O). Proteins are dissolved in D_2O or any desired buffer solution and mounted in a stainless steel gasket of the diamond cell. Two gem-cut diamonds, with polished culets, are compressed against each other with a metal gasket (0.1 mm thick) leaving a hole (0.5 mm in diameter) between them. The sample volume is of the order of 50 nL, and the total amount of protein needed per run is about 1 mg. The mini-cell from Diacell Products (United Kingdom), which has a rated maximum of about 5000 MPa, is convenient.

Infrared spectroscopy probes the vibration of chemical bonds within the molecules. The bands in the spectrum are characteristic for functional groups, and the band position depends on local molecular interactions. For the carbon-oxygen (CO) bond of the peptide group of proteins, hydrogen bonding and coupling of CO vibrations (dipole-dipole coupling) are the predominant factors that determine the features of the amide I' band of proteins. The frequency of the CO vibration is dependent on the secondary structures of the proteins (Jackson & Mantsch, 1995).

Figure 2–3 shows the spectrum of lipoxygenase and the analysis of its components. A plot of the width (or the maximum) of the amide I band as a function of pressure and temperature usually shows a cooperative transition that can be associated with a sudden change in the conformation of the protein. This is illustrated in Figures 2–4 and 2–5.

The infrared spectroscopic experiments must be performed in D_2O because of the overlapping of the amide I band with the strong water vibration at 1640 cm^{-1}. The use of D_2O involves the possibility of hydrogen-deuterium exchange. While this effect is significant with conformational changes, where new amino acids are exposed to the solvent, it can also be used to probe the dynamics of proteins by estimating the internal accessibility of the polypeptide chain.

The phase stability diagram of proteins may also be studied by infrared spectroscopy. For the enzyme lipoxygenase, Rubens (1999) found an excellent correlation between the changes in secondary structure and the inactivation of the enzyme as observed by Ludikhuyze et al. (1998). The kinetics of hydrogen-deuterium exchange gives important information on the protein dynamics as shown by the dependence of the accessibility of the protein interior and its pressure and temperature dependence. A correlation is therefore found between the rate of hydrogen-deuterium exchange from the changes in the amide II′ region and the inactivation kinetics (Rubens, 1999).

Protein Unfolding and Aggregation

In many in vitro studies of protein folding, the formation of aggregates is considered to be an undesirable side effect that obscures the folding process. However, because of the important role of aggregation in a number of molecular diseases and in the formation of gels, the mechanism of the formation of aggregates and the possible role of folding intermediates deserve closer attention. Recent studies on lysozyme mutants have indicated that model systems may help us to obtain a better insight into the molecular mechanisms underlying the formation of aggregates (Booth et al., 1997). The general picture that is emerging

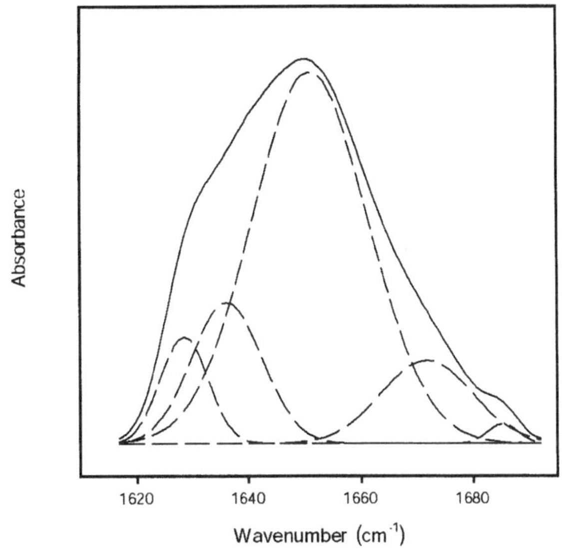

Figure 2–3 Analysis of the amide I band of the infrared spectrum of lipoxygenase in contributions from secondary structures. The predominance of the band at 1655 cm^{-1} indicates the presence of a large amount of helix.

from a number of studies is that, under partial denaturation conditions, proteins may acquire conformational intermediates that have a strong tendency to aggregate (Fink, 1998).

The phase diagram as we have discussed it in previous sections does not give any information about the mechanism of the unfolding or the aggregation of the unfolded state. Nor does it give information on the possible role of intermedi-

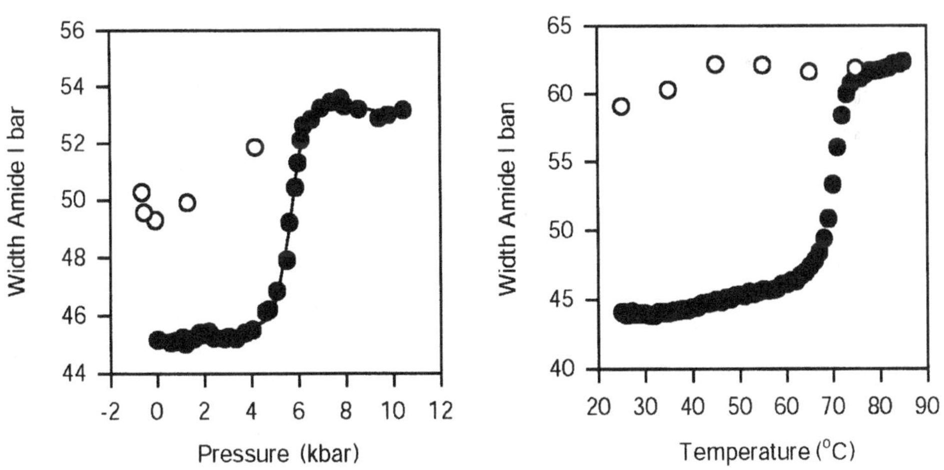

Figure 2–4 Cooperative transitions in the secondary structure of lipoxygenase. The width of the amide I band at half height is obtained from the data in Figure 2–3 as a function of pressure and temperature. Similar transitions can be observed from plots of the wavenumber maximum of the amide I band.

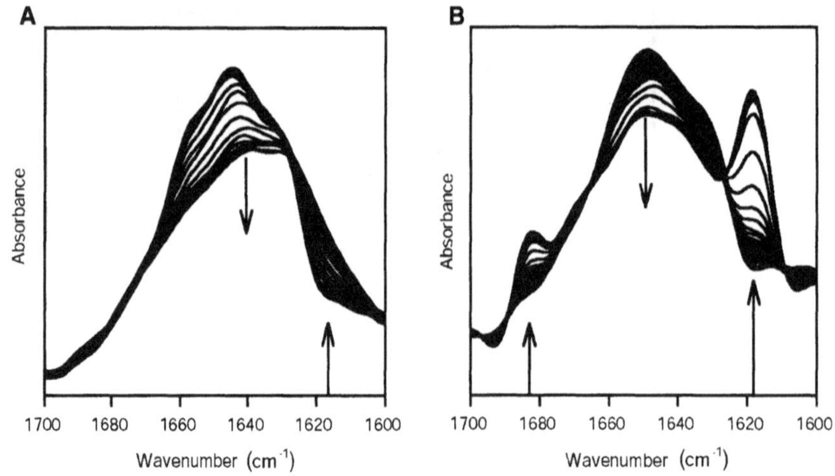

Figure 2–5 Spectral changes in the amide I band of lipoxygenase as a function of pressure (**A**) and temperature (**B**). Note the new bands arising from the formation of intermolecular aggregates in the temperature-induced spectral changes.

ates, which have been observed in a number of temperature and high pressure studies. Such a phase diagram, based on a two-state model, suggests that the heat-, pressure-, and cold-unfolded states of the protein do not differ from one another qualitatively. This speculation, however, is not supported by the recent NMR data for ribonuclease A, which indicate different structural changes for heat, cold, and pressure unfolding (Zhang et al., 1995).

The effect of pressure on protein-protein interactions can been studied with light scattering or turbidity methods (Gorovits & Horowitz, 1998). These methods give information on the degree of association and aggregation. Information on conformational changes or on the nature of the intermolecular interactions may be obtained from spectroscopic studies. Recent reports indicate differences between the structural changes induced by temperature and pressure from Fourier transform infrared spectroscopy (Smeller et al., 1999) and from small-angle X-ray scattering (Panick et al., 1999a).

In general, hydrogen bonds stabilize the intermolecular network of the heat-unfolded protein. The formation of the network after heat unfolding is an irreversible process, because the specific bands remain visible even after the cooling of the protein to room temperature. Fourier transform infrared studies of many proteins leads to the conclusion that pressure unfolding can be reversible, while temperature unfolding is almost always irreversible.

Recent results for myoglobin suggest that pressure induces partially unfolded states that may play a very important role in the aggregation of proteins. The high pressure unfolding of horse heart metmyoglobin results in an intermediate form that shows a strong tendency to aggregate after pressure release. These aggregates are similar to those that are usually observed upon temperature denaturation (Smeller et al., 1999). Similar effects are observed for lipoxygenase as shown in Figure 2–6. In part A of the figure, the infrared spectra in the amide I′ region indicate the formation of intermolecular antiparallel β-sheet stabilized by hydrogen bonding as the temperature is increased to 75°C. The formation of the aggregates is temperature dependent. Part B shows the effect of temperature after a 1000 MPa pressure treatment at room temperature. Intermolecular hydrogen bonding takes place well below 75°C, which suggests that pressure induces a conformation that is more prone to aggregate by temperature. It remains to be investigated whether the effects observed in myoglobin and lipoxygenase are a general property of proteins. Thus, a pressure pretreatment may have profound effects on the temperature-induced gel formation in food materials.

Protein Denaturation and Gel Formation

Pressure-induced denaturation with gel formation has been observed in a number of cases (Heremans et al., 1997). The question arises whether the gel formation is also accompanied by the appearance of bands indicative for intermolecular hydrogen bonds, as is the case for temperature denaturation. Clark and coworkers (1981) have observed aggregation bands in the temperature-induced gels of bovine serum albumin, insulin, lysozyme, and chymotrypsin, but not in ribonuclease. All these proteins form gels, and the correlation between the appearance of the high- and low-frequency bands in the infrared spectrum may therefore be used as a probe for the type of intermolecular interactions in the gel. The dissimilar behavior between the oxidized and reduced forms of horseradish peroxidases indicate that very subtle differences in protein conformation may be responsible for the presence of the aggregation bands (Holzbaur et al., 1996). The

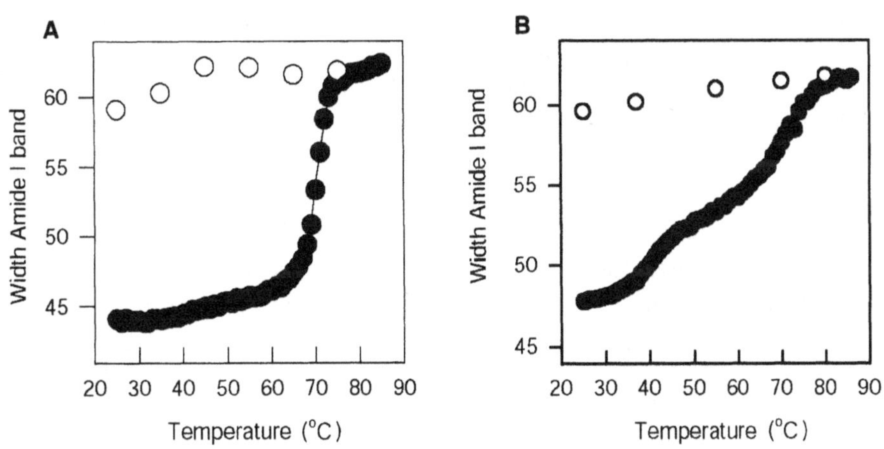

Figure 2–6 The effect of a pressure pretreatment on the development of intermolecular hydrogen bond network in lipoxygenase. **(A)** Without pressure pretreatment, the temperature-induced broadening of the amide I band develops at about 75°C. **(B)** With a pressure pretreatment at 1000 MPa, the broadening starts at 45°C.

Raman spectra of lysozyme (Heremans & Wong, 1985) and the infrared spectra of chymotrypsinogen (Wong & Heremans, 1988) do not show these bands when taken during or after compression. Similarly, the spectrum of the pressure-induced gels of whey proteins at 1000 MPa shows no extra bands, although a pressure-induced transition is observed at about 200 MPa. The spectrum of the pressure-denatured protein has more of the features of the spectrum of the native protein than does the spectrum of temperature-denatured protein. This suggests that the unfolding for the pressure denaturation is less extensive than unfolding for temperature denaturation. Just as in the case of the temperature-induced gel, the formation of the pressure-induced gel produces a change in turbidity that can be followed, in situ, in the diamond anvil cell on the same sample as that used for recording the infrared spectrum. Recent studies with small-angle X-ray scattering confirm this (Panick et al., 1999a).

In the majority of literature reports discussing the high pressure-induced gel formation of food proteins, rheological techniques are used to characterize the gel networks formed. In solid foods rheology, the relationship between the deformation of a material and the forces that act on this material is a function of time during which the force or deformation is applied. Forces are usually transformed into a stress, while the deformation is expressed in relative terms as a strain. The deformations are usually imposed by shear (i.e., parallel to the sample surface) or by uniaxial compression (i.e., perpendicular to the sample surface).

Rheological measurements performed at large sample deformations are often destructive and influence the textural properties of the material (e.g., mouthfeel). Nondestructive rheological measurements performed at small sample deformations are used to obtain basic information on the type of interactions stabilizing the sample material (Van Camp & Huyghebaert, 1995).

A comparison of the results obtained from infrared spectroscopy with those obtained from rheological techniques reveals some interesting correlations. The storage modulus and the loss modulus are higher for the temperature-induced gels than for the pressure-induced gels. Scanning electron microscopy shows that the temperature-induced gels have a higher number of links. Creep and stress relaxation, on the other hand, are higher for the pressure-induced gels. The porosity of these gels is higher. Taken together, these results suggest stronger interactions in the temperature-induced gels than in the pressure-induced gels. This finding correlates rather well with the presence of intermolecular hydrogen bonding in the temperature-induced gels as is observed in the infrared spectra (Heremans et al., 1997).

Starches

Starch granules are semicrystalline particles mainly composed of linear amylose and highly branched amylopectin, which are high molecular mass polymers. Recently, many studies have focused on the ultrastructure of the starch granule to get a deeper insight into the stability and the gelatinization mechanism. Small-

and wide-angle X-ray scattering and the recently developed X-ray microfocus diffraction allow a more detailed determination of the structure of the starch granule (Buléon et al., 1998; Waigh et al., 1999). Heat-induced changes in starches are well examined and characterized. More detailed analysis by small-angle X-ray scatterings and wide-angle X-ray scatterings reveals that the amorphous growth rings undergo hydration during gelation (Jenkins et al., 1994; Jacobs et al., 1998).

Thevelein et al. (1981) showed that the application of 200 MPa induces an upward shift of 5°C for the gelatinization temperature. Muhr et al. (1982) examined changes up to 400 MPa. The gelatinization temperature first increases by a few degrees, becomes constant at 150 to 200 MPa, and then decreases slightly at higher pressures. Hayashi and Hayashida (1989) reported that pressure-treated starches show an increased amylase digestibility at room temperature. Pressure-induced gelatinization has recently regained attention from in situ studies (Douzals et al., 1998; Rubens et al., 1999). A gelatinization mechanism similar to the temperature-induced mechanism can be proposed. High pressure induces hydration of the amorphous phase followed by a distortion of the crystalline region leading to destruction of the granular structure.

Even when high pressures and high temperatures induce gelatinization, these two techniques have differing effects on granule structure. Microscopic observations show a different degree of swelling for some pressure-treated starches than for temperature-treated ones (Stute et al., 1996). It is assumed that differences in gelatinization mechanism may cause these effects. Figure 2–7 shows the swelling of potato starch granules after a 5-minute treatment at 660 MPa.

Starch shows absorption bands in the infrared region between 1200 and 950 cm^{-1} that are sensitive to polymer conformation and hydration but are difficult to assign as they arise from highly coupled carbon-oxygen and carbon-carbon stretching modes. Fourier Transform-infrared spectroscopy and the high pressure

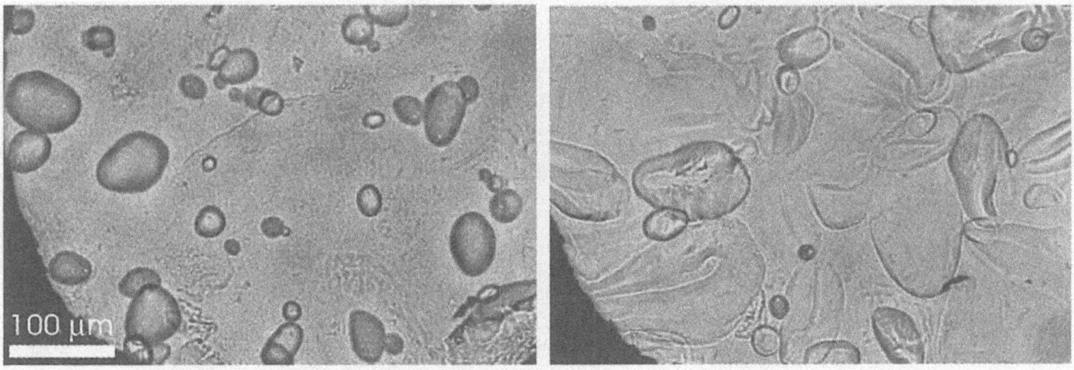

Figure 2–7 Optical micrographs of an aqueous suspension of potato starch granules in the diamond anvil cell at 660 MPa. (**A**) At time zero. (**B**) After 5 minutes.

diamond anvil cell can be used to follow in situ the gelatinization of starch granules as a function of temperature and pressure.

Temperature as well as pressure induces cooperative changes in the infrared spectra of starch granules that are related to the mechanism of gelatinization. By using light microscopy, researchers observed that the extensive swelling, characteristic of temperature-induced gelatinization, was not always present for pressure-treated samples. In some cases, the granule structure is maintained at high pressures.

The shape of the gelatinization diagram of starch (Rubens, 1999) is very similar to the stability diagram for the denaturation of proteins in Figure 2–2 (Heremans & Smeller, 1998), the cloud point of water-soluble synthetic polymers (Cook et al., 1992; Sun & King, 1996; Kunugi et al., 1997). This is remarkable, given the major differences in composition between these polymers. Given the applicability of the same thermodynamic model for starch gelatinization, protein denaturation, and the changes in cloud point of synthetic polymers, the molecular basis for this analogy needs to be considered.

For the water-soluble synthetic polymers, a pressure- and temperature-induced phase separation is observed resulting in a *P*, *T* diagram with a similar shape as that observed for starch. This is interpreted as arising from pressure and temperature effects on hydrogen bonding and hydrophobic interactions (Sun & King, 1996). This gives rise to a change from good solvent to poor solvent conditions at high pressure and/or high temperature. A change in conformation of the polymer from coil in solution to a more compact form is proposed in the phase separated form (Kunugi et al., 1997). The situation for starch and proteins is exactly the opposite. Here, a change from the helical to the coil form is proposed to be induced by high temperature or high pressure. In both cases, it is assumed that the solvation is the first step in the unfolding. For proteins, the role of hydrogen bonding and hydrophobic interactions has been assumed to be the primary source of the reentrant behavior. The role of cavities that are the consequence of the imperfect packing of the atoms in the interior of the macromolecule has been underestimated.

For starch, it seems that the imperfect packing of amylose and amylopectin chains in the starch granule plays an important role in the observed changes. When examined separately, neither component shows cooperative changes upon increase in heat or pressure. The combination of these components may give rise to specific interactions that determine the particular behavior of starch. In this respect, the phase diagram of starch is unusual in comparison with other polysaccharides (Gekko, 1992).

Biopolymer Mixtures

An important aspect of macromolecular behavior is phase separation and its relation to unfolding, aggregation, and gel formation. High pressure treatment of protein polysaccharide mixtures has led to interesting possibilities for the creation of new textures that cannot be obtained by heat treatment. This phenomenon is discussed in more detail in other chapters in this volume.

Aqueous solutions of biopolymers can undergo phase separation by two possible mechanisms: complex coacervation and thermodynamic incompatibility. In the case of coacervation, the system separates into a solvent-rich and a solvent-poor phase. The solvent-poor phase contains the precipitate of the polymer complex. In the case of thermodynamic incompatibility, the two phases are solvent rich; one phase is enriched in protein and the other is enriched in polysaccharide. The conditions for complete mixing, phase separation, and coacervation can be rationalized by the Flory Huggins equation (Dickinson & McClements, 1996), which was developed for synthetic polymers. The main difference between synthetic polymers and biopolymers is the monodispersity of the latter. In addition there is the highly specific structure of proteins in contrast with random coil polymers. Also, in real food systems we usually deal with at least a ternary system, and the complexity of the medium composition has an important effect on the behavior of the macromolecules.

The change in Gibbs free energy on mixing two polymers, 1 and 2, in solvent, 0, is given by:

$$\Delta Gmixing/RT = n_0\, ln v_0 + n_1 ln v_1 + n_2\, ln v_2 + X_{01}\, n_0 v_1 + X_{02} n_0\, v_2 + X_{12}\, n_1 v_2 \quad (26)$$

In this equation, v_0 is the volume fraction of solvent 0, v_1 is the volume fraction of polymer 1, and v_2 is the volume fraction of polymer 2. X_{ij} is the Flory-Huggins interaction parameter between the solvent and polymer 1 (X_{01}), the solvent and polymer 2 (X_{02}), and the polymer 1–polymer 2 interaction (X_{12}). X_{ij} depends on the concentration as well as on the polymer mass. In spite of the difficulties that are encountered with the application of this theory to aqueous polymer solutions, the equation has the advantage of being based on a simple physical concept so that it usually gives correct qualitative results for characteristic features of the systems. The difficulties associated with the theory are attributed to the small size and the complex behavior of water.

Schematically, one can interpret phase separation as repulsive interaction between the two polymers (e.g., between a protein and a polysaccharide), whereas complex coacervation is the consequence of strong polymer 1–polymer 2 interaction. It should be noted that this classification is rather arbitrary.

Thermodynamic incompatibility is also at the origin of the globular structure of proteins (Tolstoguzov, 1999). In many cases, the folded conformation of a protein is miscible with other native proteins whereas the unfolded conformation gives rise to insolubility and phase separation or complex coacervation. The separation threshold is a factor of approximately 10 times higher for globular proteins.

Temperature- and pressure-induced phase separation are observed when temperature or pressure induces an unfolding in one of the polymers, usually the protein. One such system, the pressure-induced phase separation in a mixture of bovine serum albumin and methylcellulose, is shown in Figure 2–8. The Flory-Huggins equation gives essentially an equilibrium description. In practice, kinetics plays a very important role in phase separation phenomena, particularly when conditions for gel formation prevail in one of the phases. Depending on the relative rate of the phase separation and the gel formation, different textures can be obtained.

An overview of these concepts and their application in the dairy industry has been given by Syrbe et al. (1998). Recent comparisons between temperature- and pressure-induced gels have been done with ovalbumin and sulfated polysaccharides (Galazka et al., 1999) and β-lactoglobulin and alginate and pectins (Dumay et al., 1999). Dumoulin and Hayashi (1998) have reviewed several strategies that demonstrate the usefulness of high pressure as a tool to texturize food and provide products with unique properties.

CONCLUSION

The use of pressure as an alternative to temperature treatment has brought about the need for fundamental studies on the pressure-temperature behavior of macromolecular food constituents. The mechanisms of protein gelation and the sol-gel behavior of polysaccharides is far from being understood. A further exploration of the correlation between the results of spectroscopic techniques and the results from other approaches might be a fruitful area of research.

Pressure-induced effects in food components can be studied in the laboratory with a number of experimental techniques. This chapter emphasized in situ, infrared spectroscopy and demonstrated that a combined approach helps us to interpret the phenomena observed. In the future, new experimental and theoretical approaches will add to our understanding of the physical and chemical behavior of food biomaterials.

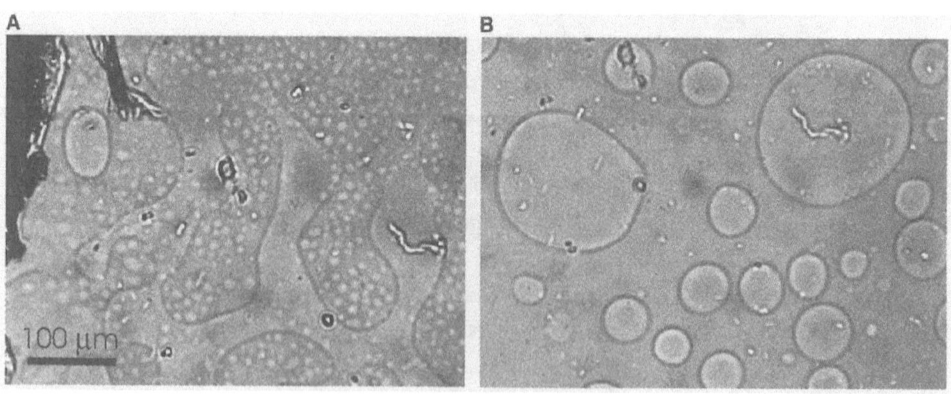

Figure 2–8 Pressure-induced phase separation in a mixture of bovine serum albumin (12.5%) and methylcellulose (2.5%) in water at 200 MPa. (**A**) Onset of the phase separation. (**B**) Complete separation after 20 minutes at the given pressure.

REFERENCES

Atake, T., & Angell, C.A. (1979). Pressure dependence of the glass transition temperature in molecular liquids and plastic crystals. *J. Phys. Chem., 83,* 3218–3223.

Balny, C., Hayashi, R., Heremans, K., & Masson, P. (Eds.). (1992). *High Pressure and Biotechnology,* Colloque INSERM (Vol. 224). Montrouge: John Libbey Eurotext.

Booth, D.R., Sunde, M., Bellotti, V., Robinson, C.V., Hutchinson, W.L., Fraser, P.E., Hawkins, P.N., Dobson, C.M., Radford, S.E., Blake, C.C.F., & Pepys, M.B. (1997). Instability, unfolding and aggregation of human lysozyme variants underlying amyloid fibrillogenesis. *Nature, 385,* 787–793.

Bridgman, P.W. (1914). The coagulation of albumen by pressure. *J. Biol. Chem., 19,* 511-512.

Bridgman, P.W. (1915). Changes of phase under pressure, II: New melting curves with a general thermodynamic discussion of melting. *Phys. Rev., 6,* 94–112.

Buléon, A., Colonna, P., Planchot, V., & Ball, S. (1998). Starch granules: Structure and biosynthesis. *Int. J. Biol. Macromol., 23,* 85–112.

Butz, P., & Tauscher, B. (1995). Inactivation of fruit fly eggs by high pressure treatment. *J. Food Proc. Preserv., 19,* 147–150.

Clark, A.H., Saunderson, D.H.P., & Suggett, A. (1981). Infrared and Laser-Raman spectroscopic studies of thermally-induced globular protein gels. *Int. J. Peptide Protein Res., 17,* 353-364.

Cook, R.L., King, H.E., Jr., & Peiffer, D.G. (1992) Pressure-induced crossover from good to poor solvent behavior for polyethylene oxide in water. *Phys. Rev. Lett., 69,* 3072–3075.

Crelier, S., Robert, M.-C., & Juillerat, M.-A. (In press). Tomato pectin methylesterase and polygalacturonase display opposite behaviours regarding heat- and pressure-induced inactivation.

Czeslik, C., Reis, O., Winter, R., & Rapp, G. (1998). Effect of high pressure on the structure of dipalmitoylphosphatidylcholine bilayer membranes: A synchrotron-X-ray diffraction and FT-IR spectroscopy study using the diamond anvil technique. *Chem. Phys. Lipids, 91,* 135-144.

Degraeve, P., Delorme, P., & Lemay, P. (1996). Pressure-induced inactivation of E. coli β-galactosidase: Influence of pH and temperature. *Biochim Biophys. Acta, 1292,* 61–68.

Dickinson, E., & McClements, D.J. (1996). *Advances in Food Colloids.* London: Chapman & Hall.

Douzals, J.P. (1999). Effets des hautes pressions isostatiques sur les modifications physico-chimiques de l'amidon. Ph.D. Thesis, Dijon.

Douzals, J.P., Perrier-Cornet, J.M., Gervais, P., & Coquille, J.C. (1998). High-pressure gelatinization of wheat starch and properties of pressure-induced gels. *J. Agric. Food Chem., 46,* 4824–4829.

Dumay, E., Laligant, A., Zasypkin, D., & Cheftel, J.C. (1999). Pressure- and heat-induced gelation of mixed β-lactoglobulin/polysaccharide solutions: Scanning electron microscopy of gels. *Food Hydrocolloids, 13,* 339–351.

Dumoulin, M., & Hayashi, R. (1998). High pressure, a unique tool for food texturization. *Food Sci. Technol. Int. Tokyo, 4,* 99–113.

Evans, M.G. & Polanyi, M. (1935). Some applications of the transition state method to the calculation of reaction velocities, especially in solution. *Trans. Faraday Soc., 31,* 875–894.

Fink, A.L. (1998). Protein aggregation: Folding aggregates, inclusion bodies and amyloid. *Folding and Design, 3,* R9–R23.

Franks, F. (1995). Protein destabilisation at low temperatures. *Adv. Prot. Chem. 46,* 105-139.

Galazka, V.B., Smith, D., Ledward, D.A., & Dickinson, E. (1999). Interactions of ovalbumin with sulphated polysaccharides: Effects of pH, ionic strength, heat and pressure treatment. *Food Hydrocolloids, 13,* 81-88.

Gekko, K. (1992). Effects of pressure on the sol-gel transition of food macromolecules. In C. Balny, R. Hayashi, K. Heremans, & P. Masson (Eds.), *High Pressure and Biotechnology,* Colloque INSERM (Vol. 224, pp. 105–113). Montrouge: John Libbey Eurotext.

Goossens, K., Smeller, L., Frank, J., & Heremans, K. (1996). Conformation of bovine pancreatic trypsin inhibitor studied by Fourier transform infrared spectroscopy. *Eur. J. Biochem., 236,* 254–262.

Gorovits, B.M., & Horowitz, P.M. (1998). High hydrostatic pressure can reverse aggregation of protein folding intermediates and facilitate acquisition of native structure. *Biochemistry, 37,* 6132–6135.

Hashizume, C., Kimura, K., & Hayashi, R. (1995). Kinetic analysis of yeast inactivation by high pressure treatment at low temperatures. *Biosci. Biotech. Comm., 59,* 1455–1458.

Hawley, S.A. (1971). Reversible pressure-temperature denaturation of chymotrypsinogen. *Biochemistry, 10,* 2436–2442.

Hayashi, R., & Balny, C. (Eds.). (1996). *High Pressure Bioscience and Biotechnology.* Amsterdam: Elsevier Science.

Hayashi, R., & Hayashida, A. (1989). Increased amylase digestibility of pressure-treated starch. *Agric. Biol. Chem., 53,* 2543–2544.

Hayashi, R., Kawamura, Y., Nakasa, T., & Okinaka, O. (1989). Application of high pressure to food processing: Pressurization of egg white and yolk, and properties of gels formed. *Agric. Biol. Chem., 53,* 2935-2939.

Hayert, M., Perrier-Cornet, J.M., & Gervais, P. (1999). A simple method for measuring the pH of acid solutions under high pressure. *J. Phys. Chem. A, 103,* 1785-1789.

Heinisch, O., Kowalski, E., Goossens, K., Frank, J., Heremans, K., Ludwig, H., & Tauscher, B. (1995). Pressure effects on the stability of lipoxygenase: Fourier transform infrared spectroscopy and enzyme activity studies. *Z. Lebensm. Unters. Forsch., 201,* 562–565.

Hellemons, J.C., & Smelt, J.P.P.M. (1999). Building fail-safe models describing the effect of temperature and pressure on the kinetics of inactivation of infectious pathogens in foods. In H. Ludwig (Ed.), *High Pressure Bioscience and Biotechnology* (pp. 109–112). Heidelberg: Springer Verlag.

Heremans, K. (1982). High pressure effects on proteins and other biomolecules. *Ann. Rev. Biophys. Bioeng, 11,* 1–21.

Heremans, K. (Ed.). (1997). *High Pressure Research in the Biosciences and Biotechnology.* Leuven: Leuven University Press.

Heremans, K., & Smeller, L. (1998). Protein structure and dynamics at high pressure. *Biochim. Biophys. Acta, 1386,* 353–370.

Heremans, K., Van Camp, J., & Huyghebaert, A. (1997). High pressure effects on proteins. In S. Damodaran & A. Paraf (Eds.), *Food Proteins and Their Applications* (pp. 473–502). New York: Marcel Dekker.

Heremans, K., & Wong, P.T.T. (1985). Pressure effects on the Raman spectra of proteins: Pressure-induced changes in the conformation of lysozyme. *Chem. Phys. Lett., 118,* 101-104.

Heremans, K., & Wuytack, F. (1980). Pressure effect on the Arrhenius discontinuity in Ca^{++}-ATPase from sarcoplasmic reticulum: Evidence for lipid involvement. *FEBS Lett., 117,* 161-163.

Hite, B.H. (1899). The effect of pressure in the preservation of milk. *Bull. West. Virg. Univ. Agric. Exp. Station, 58,* 16–35.

Holzbaur, I.E., English, A.M., & Ismail, A.A. (1996). FT-IR study of the thermal denaturation of horseradish and cytochrome c peroxidases in D_2O. *Biochemistry, 35,* 5488–5494.

Isaacs, N.S. (Ed.). (1998). *High Pressure Food Science, Bioscience and Chemistry.* London: Royal Society of Chemistry.

Jackson, M., & Mantsch, H.H. (1995). The use and misuse of FT-IR spectroscopy in the determination of protein structure. *Crit. Rev. Biochem. Mol. Biol., 30,* 95-120.

Jacobs, H., Mischenko, N., Koch, M.H.J., Eerlingen, R.C., Delcour, J.A., & Reynaers, H. (1998). Evaluation of the impact of annealing on gelatinization at intermediate water content of wheat and potato starches: A differential scanning calorimetry and small angle X-ray scattering study. *Carbohydr. Res., 306,* 1–10.

Jenkins, P.J., Cameron, R.E., Donald, A.M., Bras, W., Derbyshire, G.E., Mant, G.R., & Ryan, A.J. (1994). In situ simultaneous small and wide angle x-ray scattering: A new technique to study starch gelatinization. *J. Polym. Sci. Pol. Phys., 32,* 1579–1583.

Johnson, F.H., Eyring, H., & Stover, B.J. (1974). *The Theory of Rate Processes in Biology and Medicine.* New York: Wiley & Sons.

Kanaya, H., Hara, K., Nakamura, A., & Hiramatsu, N. (1996). Time-resolved turbidimetric measurements during gelation process of egg white under high pressure. In R. Hayashi & C. Balny (Eds.), *High Pressure Bioscience and Biotechnology* (pp. 343–346). Amsterdam: Elsevier Science.

Kitamura, Y., & Itoh, T. (1987). Reaction volume of protonic ionization for buffering agents: Prediction of pressure dependence of pH and pOH. *J. Sol. Chem., 16,* 715-725.

Klug, D.D., Handa, Y.P., Tse, J.S., & Whalley, E. (1989). Transformation of ice VIII to amorphous ice by "melting" at low temperature. *J. Chem. Phys., 90,* 2390–2392.

Kunugi, S., Takano, K., Tanaka, N., Suwa, K., & Akashi, M. (1997). Effects of pressure on the behavior of the thermoresponsive polymer poly(N-vinylisobutyramide). *Macromolecules, 30,* 4499–4501.

Ledward, D.A., Johnston, D.E., Earnshaw, R.G., & Hasting, A.P.M. (Eds.). (1995). *High Pressure Processing of Foods.* Nottingham: University Press.

Ludikhuyze, L.R., Van den Broeck, I., Weemaes, C.A., Herremans, C.H., Van Impe, J.F., Hendrickx, M.E., & Tobback, P.P. (1997). Kinetics for isobaric-isothermal inactivation of *Bacillus subtilis* α-amylase. *Biotechnol. Prog., 13,* 532–538.

Ludikhuyze, L., Indrawati, I., Van den Broeck, I., Weemaes, C., & Hendrickx, M. (1998). Effect of combined pressure and temperature on soybean lipoxygenase. 1. Influence of extrinsic and intrinsic factors on isobaric-isothermal inactivation kinetics. *J. Agr. Food Chem., 46,* 4074–4080.

Ludwig, H. (Ed.). (1999). *Advances in High Pressure Bioscience and Biotechnology.* Berlin: Springer.

Ludwig, H., Scigalla, W., & Sojka, B. (1996). Pressure- and temperature-induced inactivation of microorganisms. In J.L. Markley, C. Royer, & D. Northrup (Eds.), *High Pressure Effects in Molecular Biophysics and Enzymology* (pp. 346–363). Oxford: University Press.

Michels, P.C., Hei, D., & Clark, D.S. (1996). Pressure effects on enzyme activity and stability at high temperatures, *Adv. Prot. Chem., 48,* 341–376.

Mombelli, E., Afshar, M., Fusi, P., Mariani, M., Tortora, P., Connelly, J.P., & Lange, R. (1997). The role of phenylalanine 31 in maintaining the conformational stability of ribonuclease P2 from *Sulfolobus solfataricus* under extreme conditions of temperature and pressure. *Biochemistry, 36,* 8733–8742.

Mozhaev, V.V., Heremans, K., Frank, J., Masson, P., & Balny, C. (1996a). High pressure effect on protein structure and function. *Proteins: Struct. Funct. Genet., 24,* 81–91.

Mozhaev, V.V., Lange, R., Kudryashova, E.V., & Balny, C. (1996b). Application of high hydrostatic pressure for increasing activity and stability of enzymes. *Biotechnol. Bioeng., 52,* 320–331.

Muhr, A.H., Wetton, R.E., & Blanshard, J.M.V. (1982). Effect of hydrostatic pressure on starch gelatinisation, as determined by DTA. *Carbohydr. Polym., 2,* 91–102.

Panick, G., Malessa, R., & Winter, R. (1999a). Differences between the pressure- and temperature-induced denaturation and aggregation of β-lactoglobulin 1, B and AB monitored by FT-IR spectroscopy and small-angle X-ray scattering. *Biochemistry, 38,* 6512–6519.

Panick, G., Vidugiris, G, Malessa, R., Rapp, G., Winter, R., & Royer, C. (1999b). Exploring the temperature-pressure phase diagram of staphylococcal nuclease. *Biochemistry, 38*, 4157–4164.

Razumovsky, L., & Damodaran, S. (1999). Surface activity-compressibility relationship of proteins at the air-water interface. *Langmuir, 15*, 1392–1399.

Rubens, P. (1999). Exploring the pressure-temperature phase diagrams of biopolymers. Ph.D. Dissertation, Faculty of Sciences, Katholieke Universiteit Leuven, Leuven, Belgium.

Rubens, P., Snauwaert, J., Heremans, K., & Stute, R. (1999). In-situ observation of pressure-induced gelation of starches studied with FT-IR in the diamond anvil cell. *Carbohydr. Polym., 39*, 231–235.

Saldana, J.L., & Balny, C. (1992). Device for optical studies of fast reactions in solution as a function of pressure and temperature. In C. Balny, R. Hayashi, K. Heremans, & P. Masson (Eds.), *High Pressure and Biotechnology,* Colloque INSERM (Vol. 224, pp. 529–531). Montrouge: John Libbey Eurotext.

Seki, K., & Toyoshima, M. (1998). Preserving tardigrades under pressure. *Nature, 395,* 853–854.

Silva, J.L., Foguel, D., Da Poian, A.T., & Prevelige, P.E. (1996). The use of hydrostatic pressure as a tool to study viruses and other macromolecular assemblages. *Curr. Opin. Str. Biol., 6*, 166–175.

Simpson, R.B., & Kauzmann, W. (1953). The kinetics of protein denaturation. I. The behavior of the optical rotation of ovalbumin in urea solutions. *J. Am. Chem. Soc., 75*, 5139–5152.

Smeller, L., & Heremans, K. (1997). Some thermodynamic and kinetic consequences of the phase diagram of protein denaturation. In K. Heremans (Ed.), *High Pressure Research in the Biosciences and Biotechnology* (pp. 55–58). Leuven: Leuven University Press.

Smeller, L., Rubens, P., & Heremans, K. (1999). Pressure effect on the temperature-induced unfolding and tendency to aggregate of myoglobin. *Biochemistry, 38,* 3816–3820.

Sojka, B., & Ludwig, H. (1997). Effects of rapid pressure changes on the inactivation of *Bacillus subtilis* spores. *Pharm. Ind., 59*, 436–438.

Sonoike, K., Setoyama, T., Kuma, Y., & Kobayashi, S. (1992). Effect of pressure and temperature on the death rates of *L. casei* and *E. coli.* In C. Balny, R. Hayashi, K. Heremans, & P. Masson (Eds.), *High Pressure and Biotechnology,* Colloque INSERM (Vol. 224, pp. 297–300). Montrouge: John Libbey Eurotext.

Stute, R., Klinger, R.W., Boguslawski, S., Eshtiaghi, M.N., & Knorr, D. (1996). Effects of high pressure treatment on starches. *Starch, 48*, 399–408.

Sun, T., & King, H.E. (1996). Pressure-induced reentrant phase behavior in the poly (N-vinyl-2-pyrrolidine)-water system. *Phys. Rev. E., 54*, 2696–2703.

Suzuki, K. (1960). Studies on the kinetics of protein denaturation under high pressure. *Rev. Phys. Chem. Japan, 29*, 91–98.

Syrbe, A., Bauer, W.J., & Klostermeyer, H. (1998). Polymer science concepts in dairy systems—An overview of milk protein and food hydrocolloid interaction. Int. Dairy J., 8, 179–193.

Taniguchi, Y., & Suzuki, K. (1983). Pressure inactivation of α-chymotrypsin. *J. Phys. Chem., 87*, 5185–5193.

Tauscher, B. (1995). Pasteurization of food by hydrostatic pressure: Chemical aspects. *Z. Lebensm. Unters. Forsch., 200*, 3–13.

Thevelein, J., Van Assche, J. A., Heremans, K., & Gerlsma, S.Y. (1981). Gelatinisation temperature of starch as influenced by high pressure. *Carbohydrate Res., 93,* 304–307.

Timasheff, S.N. (1998). Control of protein stability and reactions by weakly interacting cosolvents: The simplicity of the complicated. *Adv. Prot. Chem.*, 51, 355–432.

Tolstoguzov, V. (1999). The origin of globular structure in proteins. *FEBS Lett., 444,* 145–148.

Van Camp, J., & Huyghebaert, A. (1995). A comparative rheological study between heat and high pressure-induced whey protein gels. *Food Chem., 54,* 357–364.

Van den Broeck, I. (2000). Kinetics of temperature and pressure inactivation of pectinesterase from oranges and tomatoes. Ph.D. Dissertation N. 415, Faculty of Agricultural and Applied Biological Sciences, Katholieke Universiteit Leuven, Leuven, Belgium.

Van Opstal, L., & Koningsveld, R. (1992). Mean-field lattice equations of state: 4. Influence of pressure on the phase behaviour of the system polystyrene/cyclohexane. *Polymer, 33*, 3433–3444.

Waigh, T.A., Donald, A.M., Heidelbach, F., Riekel, C., & Gidley, M.J. (1999). Analysis of the native structure of starch granules with small angle X-ray microfocus scattering. *Biopolymers, 49*, 91–105.

Weemaes, C.A., Ludikhuyze, L.R., Van den Broeck, I., & Hendrickx, M.E. (1998). Kinetics of combined pressure-temperature inactivation of avocado polyphenoloxidase. *Biotechnol. Bioeng., 60*, 292–300.

Weingand-Ziadé, A., Renault, F., & Masson, P. (1997). Combined pressure/heat-induced inactivation of butyrylcholinesterase. *Biochim. Biophys. Acta, 1340*, 245–252.

Winter, R., Landwehr, A., Brauns, T.H., Erbes, J., Czeslik, C., & Reis, O. (1996). High pressure effect on the structure and phase behavior of model membrane systems. In J.L. Markley, C. Royer, & D. Northrup (Eds.), *High Pressure Effects in Molecular Biophysics and Enzymology* (pp. 274–297). Oxford: University Press.

Wong, P.T.T., & Heremans, K. (1988). Pressure effect on protein secondary structure and deuterium exchange in chymotrypsinogen: A Fourier transform infrared spectroscopic study. *Biochim. Biophys. Acta, 956,* 1–9.

Wong, P.T.T., Lacelle, S., & Yadzi, H.M. (1993). Normal and malignant human colonic tissues investigated by pressure-tuning FT-IR spectroscopy. *Applied Spectroscopy, 47,* 1830–1836.

Yayanos, A. (1998). Empirical and theoretical aspects of life at high pressure in the deep sea. In K. Horikoshi, & W.D. Grant (Eds.), *Extremophiles* (pp. 47–92). New York: Wiley-Liss.

Zhang, J., Peng, X., Jonas, A., & Jonas, J. (1995). NMR study of the cold, heat and pressure unfolding of ribonuclease A. *Biochemistry, 34*, 8631–8641.

Zipp, A., & Kauzmann, W. (1973). Pressure denaturation of metmyoglobin, *Biochemistry, 12*, 4217–4228.

Part II

Effects of High Pressure on Food Attributes

CHAPTER 3

Effects of High Pressure on Vegetative Microorganisms

Jan P. Smelt, Johan C. Hellemons, and Margaret Patterson

INTRODUCTION

Increasing consumer demand for minimally processed, additive-free, shelf-stable products prompted food scientists to explore other physical preservation methods as alternatives to traditional heat treatments. Ultra high pressure is one of the most promising of these techniques. Ideally, the pressure treatment should result in microbiologically safe and stable foods over as wide a range of conditions as possible. Thus, there is a strong need for pressure treatments that inactivate not only vegetative cells but also bacterial spores. The application of high pressure as a sporicidal treatment is still in an exploratory state. However, pressure treatment has been known as a preservation technique for more than a century, since Hite (1899) demonstrated that microbial spoilage of milk could be delayed by the application of high pressure. Currently, a number of pressure-treated products are commercially available. Generally these are relatively acidic foods, such as fruit juices, which are pressure treated to reduce the number of vegetative bacteria, molds, and yeasts. Many other potential applications involving meat and dairy products are now being investigated.

High pressure is no different from other physical preservation methods in that its effectiveness is influenced by various intrinsic and extrinsic factors. These factors, which include the effect of pressure on water, adiabatic heat and heat dissipation, the composition of the food, and the physiological state of the microorganisms to be inactivated, must be taken into account when optimizing pressure treatments for the production of safe, high-quality foods. This chapter explores the interaction of these factors and discusses the possibilities and limitations of high pressure as a means of killing vegetative microorganisms. For more general information on food preservation by high pressure, readers are referred to the reviews by Hoover et al. (1989), Cheftel (1995), Earnshaw et al. (1995), Leadly

The authors would like to thank P.F. ter Steeg and S. Brul for their critical remarks as well as the European Community, which supported part of the research (FAIR-CT96-1175).

and Wiliams (1997), Thakur and Nelson (1998), Smelt (1998), and Palou et al. (1999).

MODE OF ACTION OF TEMPERATURE AND PRESSURE ON MICROORGANISMS

Because temperature almost always plays a role in pressure treatments, the mode of action of heat is briefly addressed here. The combined effect of heat and pressure have been studied by many investigators (Ludwig et al., 1992; Sonoike et al., 1992; Patterson & Kilpatrick, 1998). Heat processes and pressure processes are mainly based on empirical data on inactivation of microbes. Although the design of combined heat and pressure processes could be based on these data, the effects of specific food components on heat and pressure inactivation could be better predicted if the microbial physiology of the combined process was better understood. However, few comparative studies of the physiological effects of pressure and temperature on specific functions of the cell have been conducted. Elevated temperature combined with pressure does not affect one specific site, but hits a number of targets. Moreover, the physiological effects are dependent on the temperature and pressure level. For instance, at 64°C the membrane seems the main target for damage, whereas at higher temperatures enzyme inactivation is more pronounced (Earnshaw et al., 1995). It can be expected that the basal heat and pressure resistance may be due to the intrinsic stability of macromolecules (ribonucleic acid [RNA]; ribosomes; nucleic acids; enzymes; and proteins in the cell, the membrane, and in some cases the cell wall). It seems that many of the targets for temperature and pressure inactivation are the same, but that the changes at the molecular level vary.

Morphological Changes Brought about by High Pressure

In general, no morphological changes can be observed in the cells of prokaryotes or lower eukaryotes by observation under a light microscope. Perrier-Cornet et al. (1995) measured the diameter of yeast cells before, during, and after pressure treatment by image analysis. During pressure treatment, the average volume gradually decreased at 250 MPa. After pressure treatment, the cells did not totally recover their original volume. Shimada et al. (1993) observed a slight alteration in the outer shape of *Saccharomyces cerevisiae* with pressures below 400 MPa. However, the inner structure of the cell began to be affected as shown by transmission microscopy. Osumi et al. (1992) studied changes in the ultrastructure and the cytoskeleton of *Candida tropicalis* and found that the outer shape of the yeast was unaffected after treatment at 100 MPa. At 200 MPa the cell wall was damaged and the mitochondrial cristae were damaged. Kobori et al. (1995) observed changes in the nuclear apparatus of *Saccharomyces cerevisiae* by means of immuno electron microscopy. Hamada et al. (1992) noticed changes in colony form after pressure treatment of *Saccharomyces cerevisiae*.

Gene Expression, Protein Synthesis, and Stress Response

Microorganisms contain mechanisms to resist and repair the effects of heat and other unfavorable environmental conditions, such as starvation and low pH, which enable them to survive subsequent, potentially lethal temperatures. It is not known whether they also contain mechanisms to counteract pressure. Generally, there is a narrow range between optimum temperature for growth and maximum temperature where stress proteins are induced and the cell acquires an increased resistance to heat. It is unclear if a causal relationship exists between the synthesis of heat-shock proteins and the development of thermotolerance. There are reports of increased thermotolerance after sublethal heat shock in *Escherichia coli* (Tsuchido et al., 1984), *Salmonella* (Mackey & Derrick, 1986, 1987a, 1987b), *Legionella pneumophila* (Lema et al., 1988), and *Listeria monocytogenes* (Fedio & Jackson, 1989; Linton et al., 1990; Appleyard & Gaze, 1993).

The optimum pressure for growth of most food-spoilage and food-poisoning organisms is probably near atmospheric conditions. Many biochemical activities of bacteria such as protein synthesis, glucose utilization, and L-phenylalanine utilization are affected by pressures of about 50 MPa (Albright, 1975). Transcription and translation are highly sensitive to pressure, and culture pressure can influence the types of proteins that are synthesized (Somero, 1992). Protein synthesis is known to be one of the most pressure-sensitive cell activities (Landau, 1967). In *Escherichia coli* cell free systems, protein synthesis is completely inhibited at 68 MPa, but the inhibition is completely reversible after pressure release (Schwarz & Landau, 1972; Smith et al., 1975). Elevated hydrostatic pressures between 30 and 50 MPa can influence gene and protein expression. *Yarrowia lipolytica* showed enhanced proteolytic activity after pressure treatment of 80 MPa, presumably not caused by lysis but by activation of proteolytic enzymes by high pressure (Lanciotti et al., 1996). Sato et al. (1996) showed that gene expression was initiated from promotors malK-lam and mal-EFG in *Escherichia coli*. They found that the expression of the genes for osmoregulatory membrane proteins was markedly reduced by high pressure, most likely at the transcriptional level. In general, hydrostatic pressure can induce tetraploidy in *Saccharomyces cerevisiae* (Hamada et al., 1992), indicating that high pressure can interfere with replication of deoxyribronucleic acid (DNA).

Pressure-inducible proteins have been found in *Methanococcus thermolyticus, Rhodotorula rubra,* and *Escherichia coli,* three organisms representing archaeobacteria, yeast, and bacteria (Bartlett et al., 1995). A pressure of 53 MPa could induce proteins in *Escherichia coli* similar to those found at elevated temperature (Welch et al., 1993). Resistance to heat can be brought about by sublethally low pH and vice versa, and sublethal heat can enhance pressure resistance in yeasts (Iwahashi et al., 1991) and in bacteria (Smelt et al., 1998). Increased resistance to heat by sublethal pressure has not yet been observed. When yeasts (Fernandes et al., 1997) or *Lactobacillus plantarum* cells (Rijke &

Smelt, 1996) were subjected to a range of sublethal pressures, no increased heat resistance or pressure resistance was found. In view of the above-mentioned findings that starvation or sublethal pH increases resistance, it is not surprising that exponentially growing cells are more sensitive to pressure than stationary-phase cells (Mackey et al., 1995; Smelt et al., 1998). A prolonged stationary phase can even enhance resistance to pressure (Hellemons & Smelt, 1997). Hydrostatic pressure also inhibits the synthesis of some membrane proteins (Nakashima et al., 1995).

Ribosomes as a Target for Heat and Pressure Damage

Ribosomes are affected both by heat and pressure. It may be expected that the death of the individual cell occurs when the number of functional ribosomes has dropped below a critical threshold level, beyond which the cell cannot recover. In the microbial cell, ribosomal RNA (rRNA) ensures that ribosome proteins maintain their specific secondary and tertiary structure, enabling ribosomes to function during protein synthesis. RNA modification has been associated with the early events of heat damage in a cell (Coote et al., 1991). The same seems to hold for pressure damage. Gross et al. (1993) investigated the effects of high pressure and observed ribosome dissociation starting at 40 to 60 MPa. Hurst (1984) reported that heating produces membrane damage, resulting in Mg^{++} depletion in the cells. This, in turn, can lead to destabilization of 70S ribosomes. This phenomenon might be similar for high pressure; Hauben et al. (1997c) reported that divalent cations such as Ca^{++}, Mg^{++}, Mn^{++}, and Fe^{++} reduced inactivation by high pressure. It has been demonstrated that ribosomes dissociate in vitro under high pressure (Gross & Jaenicke, 1990; Gross et al., 1993). Mackey et al. (1991) demonstrated in in vivo studies with differential scanning calorimetry that the 50S subunit is more heat stable than the 30S subunit; however, both subunits are more heat labile than the complete 70S particle. By comparing the behavior of the 30S, 50S, and 70S subunits in isolation and in whole cells, Mackey and colleagues (1991) demonstrated that heating in the region between 50°C and 80°C initiated denaturation of the 30S subunit. A relation between the maximum temperature and melting points of the ribosomes would suggest a relationship between ribosome stability and thermal resistance. Niven et al. (1999) carried out a similar study of the change of conformation of ribosomes by pressure in vivo and found a correlation between cell death and ribosome damage.

Susceptibility of Proteins and Enzymes to Heat and Pressure

Both pressure and temperature have significant effects on proteins including enzymes. There is an optimum temperature at which proteins are most resistant to pressure (Hawley, 1971; *cfr.* Chapter 4). A similar pattern has been observed for the inactivation of microorganisms (Sonoike et al., 1992; Hashizume et al., 1995; Hayakawa et al., 1998). Unfortunately, most enzymes have been investigated in isolation and not in whole cells. Hei and Clark (1994) investigated enzymes of extreme thermophiles and found that these enzymes are not only more heat resistant but also more pressure resistant than those from mesophilic microorganisms. Hydrostatic pressure can presumably directly affect enzymes and carriers of transport systems. Lactic dehydrogenase from rabbit muscle and glyceraldehyde-3-phosphate dehydrogenase from baker's yeast were inactivated by pressures of 200 and 100 MPa, respectively. This provides some circumstantial evidence that enzyme inactivation plays an important role in pressure inactivation of microorganisms. However, kinetic data on the inactivation of the various sites such as glycolytic enzymes, membrane-bound enzymes, ribosomes, or membranes are lacking. Moreover, the study of single compounds such as enzymes and nucleic acids can more easily be carried out in isolation, but most biopolymers behave differently in the cell. Pressure brings about changes in the quaternary structure rather than the tertiary structure because the quaternary structure is mainly maintained by hydrophobic interactions, which are pressure sensitive (Balny & Masson, 1993). As a result, monomeric enzymes are usually more resistant to pressure than multimeric enzymes (Masson, 1992). Schmid et al. (1975) studied the effect of pressure on glycolytic enzymes with coenzymes in vitro and found that rabbit muscle lactic dehydrogenase was more stable against pressure than yeast glyceraldehyde-3-phosphate dehydrogenase. However, little attention has been paid to the extent to which the weakest link in a system is affected (e.g., in glycolysis). Degraeve et al. (1996) studied the effects of pH and temperature on pressure inactivation of *Escherichia coli* β-galactosidase. No inactivation took place below 250 MPa. Cioni and Strambini (1997) studied the dissociation of yeast glyceraldehyde-3-phosphate dehydrogenase. The main effect was due to dissociation of the tetramer. Simpson and Gilmour (1997b) studied the effect of high hydrostatic pressure on the activity of 13 metabolic enzymes of *Listeria monocytogenes*. They isolated the enzymes before and after pressure treatment of whole cells. Phosphoglucomutase and aconitase were particularly susceptible to pressure, and even pressures of around 200 MPa were sufficient to inactivate these enzymes. Because *Listeria monocytogenes* was hardly affected by this amount of pressure, these enzymes cannot be considered as critical sites.

Specific adaptations of thermophiles and of deep-sea bacteria might indicate which sites in the organism are extremely sensitive to pressure. As in heat inactivation, it can be expected that there is more than one specific target that can affect the whole organism depending on the pressure level.

Nucleic Acids

Nucleic acids, especially DNA, are relatively resistant to heat and very resistant to pressure. However, an extreme condensation of the nuclear material has been found for *Listeria monocytogenes, Salmonella typhimurium* (Mackey et al., 1994), and *Lactobacillus plantarum* (Wouters et al., 1998). The hypothesis is that at elevated pressures DNA comes into contact with endonucleases, which cleave DNA (Chilton et al., 1997). This condensation, which has been found in many other instances, is reversible and presumably an enzyme responsible for renaturation is also involved. If this enzyme is deactivated by high pressure, the cell is no longer able to multiply.

Membrane-Related Events

Pressure treatment almost always involves a perturbation of the bacterial membrane. The same is true for heat treatment, but usually to a lesser extent. Pressure resistance is probably inversely related to the rigidity of the membrane. As an increase in pressure and a decrease in temperature reduce membrane fluidity, it can be expected that cells respond to this stress by altering the composition of the membrane. There is considerable evidence correlating the production of increased proportions of membrane unsaturated acids with bacterial growth at low temperature or high pressure. Fatty acids of barophilic microorganisms become more polyunsaturated with increasing growth pressure (DeLong & Yayanos, 1985; Kamimura et al., 1993). Wirsin et al. (1987) reported that in a psychrophilic and barophilic bacterium of the genus *Alteromonas*, the ratio of unsaturated to saturated fatty acid increased with growth pressure between 1 and 45 MPa at 2°C. The ratio of unsaturated fatty acids was higher when cells were grown at 2°C than when grown at 10°C. According to Allen et al. (1999), monounsaturated rather than polyunsaturated fatty acids would be required for growth of the deep-sea bacterium *Photobacterium profundum*.

Exponentially growing cells of *Lactobacillus plantarum* were more resistant to pressure when the cells were grown at suboptimal temperatures (Smelt et al., 1994; Mackey et al., 1995). Under these conditions, the ratio of unsaturated fatty acids was higher in cells grown at optimum temperatures. Lanciotti et al. (1996) obtained similar results with *Listeria monocytogenes* and *Yarrowia lipolytica,* but sometimes observed different trends for *Escherichia coli*. Unfortunately, from their study it is not known whether they used exponentially growing cultures or stationary-phase cultures. This could explain the exception of enhanced pressure resistance in *Escherichia coli* when grown at 30°C. It has also been shown that, when cholesterol is included, the fluidity of cell membranes of eukaryotes decreases and the cells become more sensitive to pressure (MacDonald, 1992).

The pressure resistance of cells in the stationary phase is not higher after incubation at suboptimum temperatures. This discrepancy might be due to the complicating fact that cells grown in the stationary phase are subjected to a number of stress conditions such as starvation and low pH. We observed (Hellemons & Smelt, 1998) that longer incubation in the stationary phase results in increased resistance to pressure. This factor has to be taken into account when stationary phase cells grown at different temperatures are compared.

The protective effect of different carbohydrates on the membrane is listed from highest to lowest protective effect: trehalose, sucrose, fructose, glucose, glycerol.

The same order was found for the protective effect of these carbohydrates against pressure (Crowe et al., 1984; Smelt et al., 1998). Propidium iodide and ethidium bromide bind to nucleic acids, but they can only penetrate into the cell when the membrane is damaged. Unlike untreated cells, pressure-treated bacteria can be stained with propidium iodide or ethidium iodide, indicating that membrane damage has occurred (Smelt et al., 1994; Benito et al., 1999). Pressure inactivation is also accompanied by an increase in extracellular ATP (adenosine triphosphate), again showing leakage of the membrane (Smelt et al., 1994). Integral and peripheral proteins become more detached from the plasma membrane when the membrane bilayer is sufficiently perturbed by pressure.

Events Related to Maintenance of Intracellular pH

For microbial cells to maintain essential cell functions, their internal pH must remain constant. High pressure treatment can result in reduced intracellular pH, while membrane damage may impair acid efflux. A decrease in intracellular pH after pressure treatment has been found in *Lactobacillus plantarum* (Wouters et al., 1998). The change in intracellular pH is accompanied by impaired acid efflux, but this change might also be caused by impaired glycolysis. Observations of membrane damage, protein inactivation, and decreased intracellular pH, along with observations on yeasts, suggest that membrane-bound enzymes associated with efflux of protons might be a target in high pressure inactivation. Membrane-bound F_0F_1 ATPase may be such a target, as it can be inactivated or dislocated by pressure. As with heat, very severe pressure stress causes considerable damage to the cell, and pressure-treated cells become more sensitive to adverse environmental conditions. Abe and Horikoshi (1995) observed that a pressure of 40 to 60 MPa reduced the vacuolar pH in yeasts. They concluded that the yeast vacuole served as a sequestrant for protons to prevent the cytosol from acidification. Iwahashi et al. (1997) found that mutants of *Saccharomyces cerevisiae*, which lacked the ability to accumulate trehalose and/or heat-shock protein 104, showed both lower barotolerance and thermotolerance. However, they found that trehalose seems to be more important for barotolerance.

CLASSES OF HEAT RESISTANCE AND PRESSURE INACTIVATION

It is difficult to compare pressure and heat resistance, because test strains vary widely and test conditions are often not comparable. However, as shown in Table 3–1, heat-resistant vegetative cells tend to be relatively resistant to pressure. Pressures between 200 and 300 MPa are sufficient to inactivate vegetative forms of prokaryotes such as yeasts and molds. Yeasts are also relatively sensitive to heat. However, *Salmonella senftenberg*, which is notoriously resistant to heat, is not particularly resistant to pressure. Psychrotrophic bacteria are generally sensitive to heat and also to pressure (Yuste et al., 1998). Gram-positive bacteria are often more resistant to heat and pressure than are Gram-negative bacteria.

For example, Table 3–1 shows that the Gram-positive *Listeria monocytogenes* and *Staphylococcus aureus* are more resistant than the other organisms mentioned, which are Gram-negative. However, there are numerous exceptions. Benito et al. (1999) tested five strains of *Escherichia coli* O157:H7 and found that some were very pressure resistant. The most pressure-resistant strains were also the most heat-resistant strains. In another study, 80 strains of *Listeria monocytogenes* and 98 different *Salmonella* serotypes were tested; a weak correlation between heat resistance and pressure resistance was found for *Listeria monocytogenes,* but there was no correlation between heat resistance and pressure resistance in *Salmonella* (Hellemons & Smelt, 1997).

Marquis and Bender (1980) selected a variant of *Streptococcus faecalis* with increased barotolerance by gradually increasing the growth pressure. Hauben et al. (1997a) have shown that pressure-resistant mutants of *Escherichia coli* can be obtained by repeatedly exposing subcultures of pressure-treated cells to increasing pressure levels. Finally they obtained mutants that could withstand a treatment of 750 MPa for 15 minutes. They found that these mutants were not particularly heat resistant.

Vibrio are one of the most pressure-sensitive groups of bacteria reported to date. A population of around 10^6 colony-forming units (CFU)/mL of *Vibrio parahaemolyticus* was killed by exposure to 170 MPa within 10 minutes at 25°C in clam juice and 30 minutes in phosphate buffer (Styles et al., 1991). Berlin et al. (1999) also treated various *Vibrio* species (*V. parahaemolyticus, V. vulnificus, V. cholerae* O:1, *V. cholerae* non-O:1, *V. hollisae,* and *V. mimicus*) to pressures of 200 to 300 MPa for 5 to 15 minutes at 25°C in artificial sea water containing 23.4% NaCl. A treatment of 250 MPa for 10 minutes was sufficient to reduce all strains by at least 10^6 CFU/mL. The pressure treatment did not induce a viable

Table 3–1 Approximate Heat Resistance and Pressure Resistance of Some Pathogenic Bacteria

Species	*D Value at 60°C (Minutes)*	*Inactivation (Log Cycles) after 15 Minutes at Different Pressures (MPa)*			
		300	*400*	*500*	*600*
Aeromonas hydrophilia	0.1–0.2	>6			
Pseudomonas aeruginosa	0.1–0.2	>6			
Campylobacter	0.1–0.2	>6			
Salmonella	0.1–2.5	1–4.5			
Yersinia enterocolitica	2–3	>6			
Escherichia coli	4–6	1–2			
Escherichia coli O157:H7	2.5				2.5
Salmonella senftenberg	6–10	3			
Staphylococcus aureus	1–10		0.1	1.9	2.1
Listeria monocytogenes	3–8			1–3	>6

but nonculturable (VBNC) state; however, cells already existing in a VBNC state appeared to be more pressure resistant.

THE EFFECTS OF FOOD CONSTITUENTS ON PRESSURE RESISTANCE

The effects of food constituents on pressure resistance is complicated. Some effects are the result of pressure on the molecules and especially water (Bridgeman, 1958; Makita, 1992). For instance, high pressure causes water to become solid by formation of different types of ice. The ice formation can be altered by the presence of high concentrations of solutes. Little attention has been paid to the behavior of complex foods in this respect. In addition, the pH shift due to pressure is dependent on the type of buffer. The situation is further complicated by the fact that during pressure treatments temperature changes almost always occur. As mentioned above, we have a poor understanding of the biochemical and molecular basis of the effects of heat and high pressure on microorganisms. To date, mainly empirical data are available.

In real food situations there are always two effects that determine microbiological safety and stability: the influence of the food during pressure treatment and the recovery of microorganisms in the food after treatment. In addition, the results of studies in buffers or laboratory media cannot directly be extrapolated to real food situations. For instance, milk and cream protect microorganisms against pressure (Carlez et al., 1993a, 1993b; Patterson et al., 1995; Gervilla et al., 1999). Arroyo et al. (1997) subjected lettuce and tomatoes to pressure and found a decrease in the number of CFUs of about 2 log cycles at a pressure of 350 MPa for 10 minutes. Shigehisha et al. (1991) subjected pork slurries to pressure and found that pressures as high as 600 MPa were necessary to inactivate *Micrococcus luteus, Staphylococcus aureus,* and *Streptococcus faecalis*. Proteins in general seem to protect microorganisms against pressure inactivation. Also glucose seems to have a protective effect apart from water activity (Simpson & Gilmour, 1997a). Szczawinski et al. (1997) found that *Listeria monocytogenes* survived in Edam cheese after pressure treatment for 15 minutes at 500 MPa. Tanaka and Hatanaka (1992) developed an interesting application of hydrostatic pressure by stopping acidification of yogurt using pressures between 200 and 300 MPa.

Acidity

The change in pH due to temperature is easy to measure in thermal studies but is often ignored. However, although it is known that pressure can alter pH significantly during treatment, this cannot be measured directly. Theoretical calculations can be made to estimate the pH shift in buffers, although each buffer will respond differently. When the activation volume of the buffer is large, the pH change due to pressure is also large (Kitamura & Itoh, 1987). For instance, pressure causes a large pH shift in phosphate buffer, despite the fact that Sörensen phosphate buffer is frequently used in microbial inactivation studies. In addition, buffers have their specific effects on the cell physiology. Pressure-stable buffers such as TRIS (Tris(hydroxymethyl)-aminomethane) and imidazol cause more rapid inactivation at the same pH than more physiological buffers such as PIPES (Piperazin-z-ethanesulfonic acid) and ACES (N-(z-acetamido)Z-aminoethane-sulfonic acid). (Hellemons & Smelt, 1997). Thus, the appropriate buffer depends on the purpose of the investigation (i.e., whether the study focuses on the specific effect of pH or on real food situations). For instance, organic acids are relevant for many foods; however, because they have several different physiological effects, the interpretation of the results can be difficult. This is particularly true for citric acid. Organic acids are antimicrobial due to the effects of pH and of undissociated acid molecules on microbial cells, but no specific effect of organic acids during pressure treatment have been observed. This might be due to the fact that pressure favors ionization of organic acids.

Yeasts and molds are relatively resistant to low pH, and a pH of 4.0 has little effect on these microorganisms when they are subjected to heat or pressure. Vegetative cells of bacteria become more sensitive to pressure and heat in low pH conditions (Ritz et al., 1998). In studying effects of physical treatments, one should distinguish between the effects during treatment and the effects of the treatment on subsequent recovery. Bacteria surviving heat or pressure treatments become more sensitive to suboptimal pH after treatment. For instance Garcia-Graells et al. (1998) observed that a pressure-resistant mutant of *Escherichia coli* was able to survive a pressure treatment of 500 MPa in fruit juice; however, unlike untreated cells, these cells gradually died in the fruit juice. Thus, low pH not only enhances inactivation directly during treatment but also can inhibit outgrowth of cells that have been subjected to heat or pressure.

Water Activity

A low water activity protects proteins and whole organisms against heat and pressure. There is a plethora of data showing the protective effect of low water activity against heat. There are also numerous reports showing the protective effects of low water activity against pressure (Oxen & Knorr, 1993; Palou et al., 1997c). Apart from water activity's general osmotic effect on the cell, it also has specific effects. Salts confer less protection than carbohydrates against thermal or pressure damage. At the same water activity, thermotolerance and barotolerance is lower in glycerol than in solutions of monosaccharides and disaccharides. Trehalose seems to be particularly effective (Smelt et al., 1998). There are a number of explanations for the role of trehalose as a particularly effective saccharide against environmental stress. According to Shinsuke et al. (1996), macromolecules are stabilized against heat and pressure and barotolerance and thermotolerance seem to be linearly related to the number of equatorial OH groups. Although low water activity values protect cells against environmental stress during treatment, the surviving microorganisms that have been injured by heat or pressure are more difficult to recover at water activity values suboptimal for growth. For instance, Patterson et al. (1995) showed that recovery of pressure-treated cells was much lower when 2% salt was added to the medium. In conclusion, the net effect of lower water activity is not always easy to predict.

Other Antimicrobial Compounds

Sorbic acid, which acts as an organic acid but also interferes with the microbial membrane, is more active in combination with pressure. Palou et al. (1997a) showed that sorbate, regardless of whether it was combined with reduced water activity, could be used as an extra "hurdle" to obtain pressure-treated foods. Microbes are particularly sensitive to nisin during or after pressure treatment. Kalchayanand et al. (1994) reported that sublethal injury sensitized not only *Listeria monocytogenes*, but also *Escherichia coli* and *Salmonella typhimurium* to nisin when pressure treated. Apparently Gram-negative bacteria such as *Escherichia coli* and *Salmonella*, which are normally resistant to nisin, can be sensitized to the agent when pressurized (Kalchayanand et al.). This might be explained by the specific action of nisin. Nisin interacts with the cell membrane, and it could possibly penetrate to the inner-cell membrane. The specific synergistic action of nisin and high pressure might occur via local immobilization of phospholipids resulting in enhanced susceptibility to pressure (Russell et al., 1995; ter Steeg et al., 1999). During pressure treatment, *Escherichia coli* was found to be sensitive to lysozyme, nisin, and ethylenediamintetraacetic acid (EDTA) separately and even more so in a combination of these compounds (Hauben et al., 1997b). Propyl hexo benzoate and sodium ascorbate did not influence pressure resistance of *Listeria monocytogenes.*

Ogawa et al. (1990) observed that the addition of 5 to 10 g allyl-isothiocyanate increased the reduction of *Salmonella* at 200 MPa by 5 log cycles. Papineau et al. (1991) showed that chitosan salts had an additive effect on pressure inactivation. Kalchayanand et al. (1998a, 1998b) investigated the combined effect of hydrostatic pressure, temperature, and pediocin AcH (3000 units/mL). They found that the presence of pediocin during pressure resulted in an additional inactivation of 2 log cycles. In many studies on the effect of antimicrobials, the agent was present only during treatment. In real food situations, the agent is usually still present after treatment, which can contribute to an additional effect. Ponce et al. (1998a, 1998b) studied the combined effect of nisin and high pressure on the destruction of *Listeria monocytogenes* and *Escherichia coli* in liquid whole egg. An additional effect of nisin was found after a relatively severe pressure treatment (400 MPa for 15 minutes at 20°C).

DESIGN OF SAFE PASTEURIZATION CONDITIONS

Setting Criteria for Inactivation

A problem arises in setting inactivation criteria for any new decontamination technique. In principle, the requirements should be equivalent to those for heat treatment. Many products are processed far beyond the minimum inactivation criteria for heat treatment, but lower thermal treatment does not always contribute to a better product. Often, the minimum heat treatment is not known. On the other hand, a high pressure treatment equivalent to the current heat treatment might be difficult to achieve, even though in the context of the risk assessment a microbiologically safe pressure treatment is possible. Ideally, the required lethality (in numbers of log cycles) should be established by a complete risk assessment, taking into account the initial contamination, the processing conditions, and the storage and distribution conditions. However, because the risk of infectious microorganisms in the final product must be very small (10^{-6} to 10^{-8}), this low risk can be guaranteed only by raw materials with an extremely low contamination or by a process that results in sufficient kill of the microorganism(s). Availability of robust and accurate kinetic models is a prerequisite for quantitative risk assessment. Because temperature effects due to combinations of adiabatic heat and heat dissipation play a role in pressure treatments, models should be available both for pressure and temperature. This means that the classical two-dimensional time-temperature inactivation model must be replaced by three-dimensional time-temperature-pressure models.

Inactivation Kinetics at Constant Heat or Constant Pressure

Traditional thermal processes are often built on experience rather than systematic investigation of the inactivation kinetics of the relevant organism(s). In the formulation of thermal processes, simple models are generally used. Thermal inactivation of microorganisms is generally believed to be log-linear, analogous to radioactive decay and inactivation of many pure enzymes. Decimal reduction times (*D* values) (the time needed to inactivate 90% of the population) or inactivation rate constants (*k* values) are used. This approach has proven its value for many years in the formulation of heat processes. The log-linear inactivation model assumes a single hit, and it is difficult to see that this assumption is consistent with the observation that most organisms pass a transient stage of sublethal damage before they die. To build a model consistent with these observations, Moats (1971) assumed that it is necessary to inactivate a number of critical sites before the cell dies. These models always result in a "shoulder." In addition, "tails" have often been observed, and various explanations can be given for them.

Many deviations from log-normality have been observed that cannot be explained by the assumption of mere artifacts. Consequently, *D* values should be used with caution when small safety limits are required and particularly when extrapolation is made. If the number of critical sites to be investigated before inactivation is different per cell, the assumption leads to the idea that the population of microorganism is heterogeneous. Peleg (1996) and Augustin et al. (1998) worked out this idea and they assumed that the distribution of the resistance of the cell in a population follows a log-logistic or a log-normal distribution. Peleg and Cole (1998), who investigated a large number of curves, also concluded that some curves that were considered to be semilogarithmic were in fact slightly curved. This curvature can have a significant effect on inactivation when determined by extrapolation. If this is true, it can be expected that models based on heterogeneous populations of microorganisms are not limited to inactivation by heat. Indeed, Peleg and Cole could describe inactivation by heat and by pressure by assuming that resistance was represented by a Weibull distribution. The idea of a Weibull distribution in pressure inactivation was also put forward by Heinz and Knorr (1996). Their model has some resemblance to the model of Moats (1971) because it assumes that there is a transient stage before the organisms are killed. The distribution models do not take into account specific mechanisms of inactivation apart from the assumption of heterogeneity.

Inactivation by pressure is often characterized by an initial drop that can partly be explained by the pressure come-up time, which is often ignored in pressure studies (Palou et al., 1997b). Whereas inactivation by temperature can often be satisfactorily described as log-linear, deviations from log-linearity seem to occur more frequently in inactivation by pressure (Metrick et al., 1989; Ludwig et al., 1992; Patterson et al., 1995). However, log-linear inactivation is also found in pressure inactivation (Chen & Tseng, 1997; Mussa & Ramaswamy, 1997; Mussa et al., 1999).

Temperature Relationships

Although we have limited information on the mechanism of heat and pressure inactivation, it is still unclear how these elements can be built into a kinetic model. In the traditional model describing heat inactivation, the relation between inactivation rate and temperature is given by

$$\log(D) = a - \frac{1}{z}T \tag{1}$$

where D is the decimal reduction time in minutes or seconds, a the intercept, T the temperature (usually in °C) and $-1/z$ the slope; z is usually given in °C. The assumption of linearity between log (D) and temperature seems even more questionable than the log-linear decrease, and it seems to hold for only a limited temperature range. However, as for D values, the z concept has proven its value for formulation of many heat processes. When distribution models are used, temperature relation can no longer be described by z values, because temperature relations of the distribution parameters should be established. Augustin et al. (1998) could describe thermal inactivation of *Listeria monocytogenes* by a log-logistic distribution. They concluded that the standard deviation (in log time) was not significantly dependent on the temperature, but that the mean (log time) was linearly dependent on the temperature. Thus, they could calculate a figure analogous to a z value. Another complication in the D and z concept is that slow heating has less lethal effect that can be expected from the total integrated heating (Mackey & Derrick, 1987a). This might be explained by the development of a stress response as outlined above.

Pressure Relationships

Problems similar to those with temperature arise with pressure relations. In pressure inactivation, an empirical descriptor analogous to z is sometimes used, which is designated as z_P. The equation becomes

$$\log(D) = a - \frac{1}{z_P}P \tag{2}$$

where D is the decimal reduction time in minutes or seconds, a the intercept, P the pressure (mostly in MPa), and $-1/z_P$ the slope; z_P is usually given in MPa. However, it is questionable how widely the z_P can be used. In the distribution models other parameters should be formulated. It seems that the effect of decompression is often overestimated for vegetative cells. The only clear evidence of killing cells by decompression is with gas vacuolated cells such as *Micrococcus aquaticus*. Chastain and Yayanos (1991) found that barophilic bacteria were fully viable directly after decompression, but they gradually died after longer exposure to atmospheric conditions. Experiments with nonvacuolated bacteria reveal virtually complete survival after rapid decompression (Hemmingsen & Hemmingsen, 1978). Although Palou et al. (1998a) claimed that *Zygosaccharomyces bailii* was more easily inactivated by oscillatory high pressure treatment, their results show little or no difference between oscillatory and continuous pressure treatment. Aleman et al. (1996) reported that pulse high pressure treatments in pineapple was more effective than static pulse. However, careful examination of pressure and temperature during processing might show that this effect is due to the integrated pressure-temperature-time history.

Combined Effect of Temperature and Pressure

Models to quantify the combined effects of pressure and temperature are still in a purely empirical state. Sonoike et al. (1992) proposed a quadratic model describing the combined effect of pressure and temperature on inactivation of *Lactobacillus casei* and *Escherichia coli* by assuming log-linear inactivation. Chen and Tseng (1997) studied the combined effect of temperature and pressure on the inactivation of yeasts. Although they found reduction in numbers when higher pressures were applied, the activation energy proved to be independent of pressure. The activation energy corresponded to a z_P value of about 165 MPa. Some deviant results were obtained, probably due to experimental conditions. Hellemons and Smelt (1998) developed a similar model for the combined action of pressure and temperature on the inactivation of *Listeria monocytogenes* by assuming inactivation according to log-normal inactivation. The model is consistent with the model of Augustin et al. (1998). Although the model is purely empirical, it is consistent with the mechanisms of enzyme inactivation. However, further investigations are necessary to confirm this.

Consequences for the Use of Time-Temperature-Pressure Integrators

Time-temperature integrators are microorganisms, enzymes, or other molecules that show at denaturation the same activation energy and hence the same z value. They are preferably more resistant to the treatment than the target organism. Thus, from inactivation data of the time-temperature integrator, the lethality in log cycles of the target organism can be calculated. Enzyme inactivation by heat and by pressure can be considered often as log-linear; since log-linear models are often applicable in thermal processing, they are ideal as time-temperature integrators. If inactivation of target organisms is not log-linear, the integrators should be used with care. Thus, it is necessary to establish exactly how time-temperature-pressure integrators can be applied for high pressure processes. This topic will be addressed in additional detail elsewhere in this volume.

Sublethal Damage

Although from a mechanistic point of view there is little difference between sublethal and lethal damage, sublethal damage brings another dimension into inactivation kinetics. To establish the relevance of sublethal damage, it is useful to take into account the type of microorganism to be investigated as outlined in Table 3–2.

Sublethal inactivation is particularly relevant to toxigenic and spoilage organisms, because it often results in a longer lag time (the time before an organism resumes growth) or in an increased sensitivity to low pH or antimicrobial substances. Quantification of inactivation that includes sublethal damage has not often been described. A model describing both sublethal damage and lethal dam-

Table 3–2 Defining Target for Pressure Inactivation

Group of Microorganisms	*Health Implications*	*Requirements*
Spoilage organisms Yeasts Molds Lactic acid bacteria	No health risk	Inactivation; prevention of growth; delay of growth
Toxigenic pathogenic bacteria *Staphylococcus aureus* *Clostridium botulinum* Mycotoxigenic molds	Microorganisms themselves are not infectious, but they are able to form toxins in the food	Inactivation; prevention of growth
Infectious pathogens *Listeria monocytogenes* *Salmonella* *Campylobacter* *Yersinia enterocolitica* *Aeromonas hydrophilia*	Microorganism itself is infectious	Inactivation

age for *Lactobacillus plantarum* was presented by Smelt et al. (1998). The model assumes that the physiological state of a population is heterogeneously distributed, which can result in a prolonged lag time. As the cells are subjected to a more damaging treatment, the distribution of the physiological state changes and finally more cells are unable to multiply; there is an intermediate stage in which the cells are still capable of performing some physiological functions. A schematic outline of the model is given in Figure 3–1. However, more investigations are required before this model can be used for practical conditions.

CONCLUSION

High pressure pasteurization is now becoming a mature decontamination technique, and a number of products that use this technique have been on the market for a couple of years. If the inactivation kinetics are better understood, the amount of pressure can be reduced. The use of combined treatments might further reduce the required pressure, but that technique has not yet been fully explored. In addition, it is important to consider the problem of unusual pressure resistance in some pathogens.

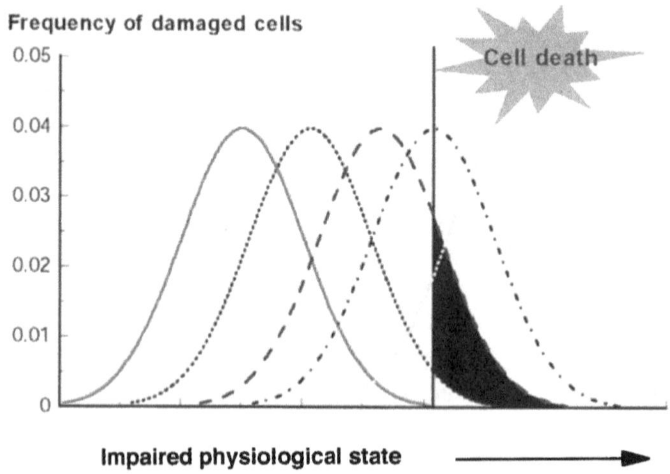

Figure 3–1 Graphical presentation of a model describing the impaired physiological state. The lag time or the increased sensitivity to pH can be a reflection of the impaired physiological state.

REFERENCES

Abe, F., & Horikoshi, K. (1995). Hydrostatic pressure promotes the acidification of vacuoles in *Saccharomyces cerevisiae. FEMS Microbiol. Lett., 130,* 307–312.

Albright, L.J. (1975). The influence of high pressure upon biochemical activities of heterotrophic bacteria. *Can. J. Microbiol., 21,* 1405–1412.

Aleman, G.D., Ting, S.C., Mordre, A.C., Hawes, M., Walker, M., Farkas, D.F., & Torres, J.A. (1996). Pulse ultrahigh pressure treatments for pasteurisation for pasteurisation of pineapple juice. *J. Food Sci., 61,* 388–390.

Allen, E.E., Facciotti, D., & Bartlett, D.H. (1999). Monounsaturated but not polyunsaturated fatty acids are required for growth of the deep-sea bacterium *Photobacterium profundum* SS9 at high pressure and low temperature. *Appl. Environ. Microbiol., 65,* 1710–1720.

Appleyard, J., & Gaze, J. (1993). The effect of exposure to sublethal temperature on the final heat resistance of *Listeria monocytogenes* and *Clostridium botulinum.* Technical memorandum No. 683, CCFRS, Chipping Campden, Golos, UK.

Arroyo, G., Sanz, P.D., & Prostamo, G. (1997). Effect of high pressure on the reduction of microbial populations in vegetables. *J. Appl. Microbiol., 82,* 735–742.

Augustin, J.C., Carlier, V., & Rozier, J. (1998). Mathematical modelling of the heat resistance of *Listeria monocytogenes. J. Appl. Microbiol., 84,* 185–191.

Balny, C., & Masson, P. (1993). Effects of high pressure on proteins. *Food Rev. Int., 9,* 611–628.

Bartlett, D.H., Kato, C., & Horikoshi, K. (1995). High pressure influences on gene and protein expression. *Res. Microbiol., 146,* 697–706.

Benito, A., Ventoura, G., Casadei, M., Robinson, T., & Mackey, B. (1999). Variation in resistance of natural isolates of *Escherichia coli* O157 to high hydrostatic pressure, mild heat and other stresses. *Appl. Environ. Microbiol., 65,* 1564–1569.

Berlin, D.L., Herson, D.S., Hicks, D.T., & Hoover, D.G. (1999). Response of pathogenic *Vibrio* species to high hydrostatic pressure. *Appl. Environ. Microbiol., 65,* 2776–2780.

Bridgeman, P.W. (1958). *The Physics of High Pressure.* London: G. Bell & Sons.

Carlez, A., Rosec, J.P., Richard, N., & Cheftel, J.-C. (1993). Bacterial growth during chilled storage of pressure treated minced meat. *Lebensm. Wiss. Technol., 27,* 48–54.

Chastain, R., & Yayanos, A. (1991). Ultrastructural changes in an obligately barophilic marine bacterium after decompression. *Appl. Environ. Microbiol., 57,* 1489–1497.

Cheftel, J.C. (1995). Review: High-pressure, microbial inactivation and food preservation. *Food Sci. Technol. Int., 1,* 75–90.

Chen, C.S., & Tseng, C.W. (1997). Effect of hydrostatic pressure on the temperature dependence of *Saccharomyces cerevisiae* and *Zygosaccharomyces rouxii. Process Biochem., 32,* 337–343.

Chilton, P., Isaacs. N.S., Mackey, B., & Stenning, R. (1997). The effects of high hydrostatic pressure on bacteria. In K. Heremans (Ed.), *High Pressure Research in the Biosciences and Biotechnology* (pp. 225–228). Leuven: Leuven University Press.

Cioni, P., & Strambini, G.B. (1997). Pressure-induced dissociation of yeast glyceraldehyde-3-phosphate dehydrogenase: Heterogeneous kinetics and perturbations of subunit structure. *Biochemistry, 36,* 8586–8593.

Coote, P.J., Cole, M.B., & Jones, M.V. (1991). Induction of increased thermotolerance in *Saccharomyces cerevisiae* may be triggered by a mechanism involving intracellular pH. *J. Gen. Microbiol., 137,* 1701–1708.

Crowe, L.M., Mouradian, R., & Crowe, J.H. (1984). Effects of carbohydrates on membranes at low water activities. *Biochim Biophys. Acta, 769,* 141–150.

Degraeve, P., Delorme, P., & Lemay, P. (1996). Pressure-induced inactivation of *E. coli* β-galactosidase: Influence of pH and temperature. *Biochim. Biophys. Acta, 1292(1),* 61–68.

DeLong, E.F., & Yayanos, A.A. (1985). Adaptation of the membrane lipids in deep-sea bacteria to changes in hydrostatic pressure. *Science, 228,* 1101–1103.

Earnshaw, R.G., Appleyard, J., & Hurst, R.M. (1995). Understanding physical inactivation processes opportunities using heat, ultrasound and pressure. *Int. J. Food Microbiol., 1995,* 197–219.

Fedio, W.M., & Jackson, H. (1989). Effect of tempering on the heat resistance of *Listeria monocytogenes. Lett. Appl. Microbiol., 9,* 157–160.

Fernandes, P.M.B., Panek. A.D., & Kurtenbach, E. (1997). Effect of hydrostatic pressure on a mutant of *Saccharomyces cerevisiae* deleted in the trehalose 6 phosphate synthase gene. *FEMS Microbiol. Lett., 152,* 17–21.

Garcia-Graells, C., Hauben, K., & Michels, C.W. (1998). High pressure inactivation and sublethal injury of pressure resistant *Escherichia coli* mutants in fruit juices. *Appl. Environ. Microbiol., 64,* 1566–1568.

Gervilla, R., Sendra, E., Ferragut, V., & Guamis, B. (1999). Sensitivity of *Staphylococcus aureus* and *Lactobacillus helveticus* in ovine milk subjected to high hydrostatic pressure. *J. Dairy Sci., 82,* 1099–1107.

Gross, M., & Jaenicke, R. (1990). Pressure-induced dissociation of tight couple ribosomes. *FEBS Lett., 267,* 239–241.

Gross, M., Lehle, K., Jaenicke, R., & Nierhaus, K.H. (1993). Pressure induced dissociation of ribosomes and elongation cycle intermediates: Stabilising conditions and identification of the most sensitive functional state. *Eur. J. Biochem., 218,* 463–468.

Hamada, K., Nakatomi, Y., & Shimada, S. (1992). Direct induction of tetraploids or homozygous diploids in the industrial yeast *Saccharomyces cerevisiae* by hydrostatic pressure. *Curr. Genet., 22,* 371–376.

Hashizume, C., Kimura, K., & Hayashi, R. (1995). Kinetic analysis of yeast inactivation by high pressure treatment at low temperatures. *Biosci. Biotechnol. Biochem., 59,* 1455–1458.

Hauben K.J.A, Bernaerts, K., & Michiels, C.W. (1997). Protective effects of calcium on inactivation on *Escherichia coli* by high hydrostatic pressure. *J. Appl. Microbiol., 85,* 678–684.

Hawley, S.A. (1971). High pressure techniques. *Meth. Enzymol., 49,* 14–25.

Hayakawa, K., Ueno, V., Kawamura, S., Kato, T., & Hayashi, R. (1998). Micro-organism inactivation using high pressure generation in sealed vessels under subzero temperature. *Appl. Microbiol. Biotechnol., 50,* 415–418.

Hei, D.J., & Clark, D.S. (1994). Pressure stabilisation of proteins from extreme halophiles. *Appl. Environ. Microbiol., 60,* 932–939.

Heinz, V., & Knorr, D. (1996). High pressure inactivation kinetics of *Bacillus subtilis* cells by a three-state model considering distributed resistance mechanisms. *Food Biotechnol., 10,* 149–161.

Hellemons, J.C., & Smelt, J.P.PM. (1997). Annual report for EU project FAIR CT96-1175.

Hellemons, J.C., & Smelt, J.P.PM. (1998). Annual report for EU project FAIR CT96-1175.

Hemmingsen, B.B., & Hemmingsen, E.A. (1978). Tolerance of bacteria to extreme gas supersaturation. *Biochem. Biophys. Res. Commun., 85,* 1379–1384.

Hite, B.H. (1899). The effect of pressure in the preservation of milk. *Bull. West Virg. Univ. Agric. Exp. Station, 58,* 15–35.

Hoover, D.G., Metrick, C., Papineau, D.F., Farkas, D.F., & Knorr, D. (1989). Biological effects of high pressure on microorganisms. *Food Technol., 43,* 99–107.

Iwahashi, H., Kaul, S.C., Obuchi, K., & Komatsu, Y. (1991). Induction of barotolerance by heat shock treatment in yeast. *FEMS Lett., 80,* 325–328.

Iwahashi, H., Obushi, K., Fuji, S., & Komatsu, Y. (1997). Barotolerance is dependent on trehalose and heat shock protein 104 but is essemtially different from thermotolerance in *Saccharomyces cerevisiae. Lett. Appl. Microbiol., 25,* 43–47.

Kalchayanand, N., Sikes, T., Dunne, C.P., & Ray, B. (1994). Hydrostatic pressure and electroporation have increased bactericidal efficiency in combination with bacteiocins. *Appl. Environ. Microbiol., 60,* 4174–4177.

Kalchayanand, N., Sikes, A., Dunne, C.P., & Ray, B. (1998a). Interaction between hydrostatic pressure, time and temperature of pressurization and pediocin AcH on activation of foodborne bacteria and electroporation have increased bactericidal efficiency in combination with bacteiocins. *J. Food Prot., 61,* 425–431.

Kalchayanand, N., Sikes, A., Dunne, C.P., & Ray, B. (1998b). Factors influencing death and injury of foodborne pathogens by hydrostatic pressure-pasteurisation. *Food Microbiol., 15,* 207–214.

Kamimura, K., Fuse, H., Takimura, O., & Yamaoka, Y. (1993). Effect of growth pressure and temperature on fatty acid composition of a barotolerant deep-sea bacterium. *Appl. Environ. Microbiol., 59,* 924–936.

Kitamura, Y., & Itoh, T. (1987). Reaction volume of protonic buffering agents. Prediction of pressure dependence of pH and pOH. *J. Solution Chem., 16,* 715–725.

Kobori, H., Sato, M., Tameike, A., Hamada, K., Shimada, S., & Osumi, M. (1995). Ultrastructural effects of pressure stress to the nucleus in *Saccharomyces cerevisiae*: a study by immunoelectron microscopy using frozen thin sections. *FEMS Microbiol. Lett., 132(3),* 253–258.

Lanciotti, R., Gardini, F., Sinigaglia, M., & Guerzoni, M.E. (1996). Effects of growth conditions on the resistance of some pathogenic and spoilage species to high pressure homogenization. *Lett. Appl. Microbiol., 22,* 165–168.

Landau, J. (1967). Induction, transcription and translation in *Escherichia coli*: A hydrostatic pressure study. *Biochim. Biophys. Acta, 149,* 506–512.

Leadly, C.E., & Williams, A. (1997). High pressure processing of foods and drink—An overview of recent developments and future potential. *New Technologies Bulletin No. 14,* March, Campden and Chorleywood Food Research Association, Chipping Campden, Gloucester, GL55 6LD, U.K.

Lema, M.W., Brown, A., Butler C.A., & Hoffman, P.S. (1988). Heat-shock response in *Legionella pneumophila. Can. J. Microbiol., 34,* 1148–1153.

Linton, R.H., Pierson, M.D., & Bishop, J.R. (1990). Increase in heat resistance of *Listeria monocytogenes* Scott A., by subsequent heat shock. *J. Appl. Bacteriol., 61,* 389–393.

Ludwig, H., Bieler, C., Hallbauer, K., & Scigilla, W. (1992). Inactivation of microorganisms by hydrostatic pressure. In C. Balny, R. Hayashi, K. Heremans, & P. Masson (Eds.), *High Pressure and Biotechnology*, Colloque INSERM (Vol. 224, pp. 25–32). Montrouge: John Libbey Eurotext.

MacDonald, A.G. (1992). Effects of high hydrostatic pressure on natural and artificial membranes. In C. Balny, R. Hayashi, K. Heremans, & P. Masson (Eds.), *High Pressure and Biotechnology*, Colloque INSERM (Vol 224, pp. 67–75). Montrouge: John Libbey Eurotext.

Mackey, B.M., & Derrick, C.M. (1986). Elevation of the heat resistance of *Salmonella typhimurium* by sublethal heat shock. *J. Appl. Bacteriol., 61,* 389–393.

Mackey, B.M., & Derrick, C.M. (1987a). Changes in the heat resistance of *Salmonella typhimurium* during heating at rising temperatures. *Lett. Appl. Microbiol., 4,* 13–16.

Mackey, B.M., & Derrick, C.M. (1987b). The effect of prior heat shock on the thermoresistance of *Salmonella thompson* in foods. *Lett. Appl. Microbiol., 5,* 115–118.

Mackey, B.M., Miles, C.A., Parsons, S.E., & Seymour, D.A. (1991). Thermal denaturation of whole cells and cell components of *Escherichia coli* by differential scanning calorimetry. *J. Gen. Microbiol., 137,* 2361–2374.

Mackey, B.M., Forestiere, K., Isaacs, N.S., & Brooker, B. (1994). The effect of high hydrostatic pressure on *Salmonella thompson* and *Listeria monocytogenes* examined by electron microscopy. *Lett. Appl. Bacteriol., 19,* 429–432.

Mackey, B.M., Forestiere, K., & Isaacs, N. (1995). Factors affecting the resistance of *Listeria monocytogenes* to high hydrostatic pressure. *Food Biotechnol., 9,* 1–11.

Makita, T. (1992). Application of high pressure and thermophysical properties of water to biotechnology. *Fluid Phase Equilibria, 76,* 87–95.

Marquis, R.E., & Bender, G.R. (1980). Isolation of a variant of *Streptococcus faecalis* with enhanced barotolerance. *Can. J. Microbiol., 26,* 371–376.

Masson, P. (1992). Pressure resistance and proteins. In C. Balny, R. Hayashi, K. Heremans, & P. Masson (Eds.), *High Pressure and Biotechnology*, Colloque INSERM (Vol 224, pp. 89–99). Montrouge: John Libbey Eurotext.

Metrick, C., Hoover, D.G., & Farkas, D.F. (1989). Effects of high hydrostatic pressure on heat resistant and heat sensitive strains of *Salmonella. J. Food Sci., 54,* 1547–1564.

Moats, W.A. (1971). Kinetics of thermal death of bacteria. *J. Bacteriol., 105,* 165–171.

Mussa, D.H., & Ramaswamy, H.S. (1997). Ultra high pressure processing of milk: Kinetics of microbial destruction and changes in physicochemical characteristics. *Lebensmittel Wiss. Technol., 30,* 551–557.

Mussa, D.H., Ramaswamy, H.S., & Smith, J.P. (1999). High pressure destruction kinetics of *Listeria monocytogenes* in milk. *J. Food Prot., 62,* 40–45.

Nakashima, K., Horikoshi, K., & Mizuno, T. (1995). Effect of hydrostatic pressure on the synthesis of outer membrane proteins in *Escherichia coli. Biosci. Biotechnol. Biochem., 59,* 130–132.

Niven, G.W., Miles, C.A., & Mackey, B.M. (1999). The effects of hydrostatic pressure on ribosome confirmation in *Escherichia coli*: An in vivo study using differential scanning calorimetry. *Microbiology, 145,* 419–425.

Ogawa, H., Fukuhisa, K., Kubo, Y., & Fukumoto, H. (1990). Pressure inactivation of yeasts, molds, and pectinesterases in satsuma mandarin juice: Effects of juice concentration, pH, and organic acids, and comparison with heat sanitation. *Agric. Biol. Chem., 54,* 1219–1225.

Osumi, M., Yamada, N., Sato, M., Kobori, H., Shimada, S., & Hayashi, R. (1992). Pressure effects on yeast cells ultrastructure: Change in the ultrastructure and cytoskeleton of the dimorphic yeast *Candida tropicalis*. In C. Balny, R. Hayashi, K. Heremans, P. Masson (Eds), *High Pressure and Biotechnology,* Colloque INSERM (Vol 224, pp. 9–18). Montrouge: John Libbey Eurotext.

Oxen, P., & Knorr, D. (1993). Baroprotective effects of high solute concentrations against inactivation of Rhodotorula rubra. *Lebensm. Wiss. Technol., 26,* 220–223.

Palou, E., Lopez-Malo, A., Barbosa-Canovas, G.V., Welti-Chanes, J., & Swanson, B.G. (1997a). High hydrostatic pressure as a hurdle for *Saccharomyces bailii* inactivation. *J. Food Sci., 62,* 855–857.

Palou, E., Lopez-Malo, A., Barbosa-Canovas, G.V., Welti-Chanes, J., & Swanson, B.G. (1997b). Kinetic analysis of *Zygosaccharomyces bailii* inactivation by high hydrostatic pressure. *Lebensm. Wiss. Technol., 30,* 703–708.

Palou, E., Lopez-Malo, A., Barbosa-Canovas, G.V., Welti-Chanes, J., & Swanson, B.G. (1997c). Combined effect of high hydrostatic pressure and water activity on *Zygosaccharomyces bailii* inhibition. *Lett. Appl. Microbiol., 24,* 417–420.

Palou, E., Lopez-Malo, A., Barbosa-Canovas, G.V., Welti-Chanes, J., & Swanson, B.G. (1998). Oscillatory high hydrostatic pressure inactivation of *Zygosaccharomyces* bacilli. *J. Food Prot., 61,* 1213–1215.

Palou, E., Lopez-Malo, A., Barbosa-Canovas, G.V., Welti-Chanes, J., & Swanson, B.G. (1999). High-pressure treatment in food preservation. In M.S. Rahman (Ed.), *Handbook of Food Preservation* (pp. 533–575). New York and Basel: Marcel Dekker.

Papineau, A.H., Hoover, D.L., Knorr, D., & Farkas, D.F. (1991). Antimicrobial effect of water soluble chitosans with hydrostatic pressure. *Food Biotechnol., 5*(1), 45–57.

Patterson, M.F., Quinn, M., Simpson, R., Doyle, M.P., Beuchat, L.R., Montville, T.J., & Gilmour, A. (1995). The sensitivity of vegetative pathogens to high hydrostatic pressure treatment in phosphate buffered saline and foods. *J. Food Prot., 58,* 524–529.

Patterson, M.F., & Kilpatrick, D.J. (1998). The combined effect of high hydrostatic pressure and mild heat on inactivation of pathogens in milk and poultry. *J. Food Prot., 61,* 432–436.

Peleg, M. (1996). A model of microbial growth and decay in a closed habitat base on combined Fermi's and logistic equations. *J. Sci. Food Agric., 71,* 225–227.

Peleg, M., & Cole, M.B. (1998). Reinterpretation of microbial survivor curves. *Crit. Rev. Food Sci., 38*(5), 353–380.

Perrier-Cornet, J.M., Marechal, J.M., & Gervais, P. (1995). A new design intended to relate high-pressure treatment to yeast-cell mass transfer. *J. Biotechnol., 41,* 49–58.

Ponce, E., Pla, R., Sendra, E., Guamis, B., & Mor-Mur, M. (1998a). Combined effect of nisin and high hydrostatic pressure destruction of *Listeria innocua* and *Escherichia coli* in liquid whole egg. *Int. J. Food Microbiol., 43,* 15–19.

Ponce, E. Pla, R., Mor-Mur, M., Gervilla, R., & Guamis, B. (1998b). Inactivation of *Listeria innocua* in liquid whole egg by high hydrostatic pressure. *J. Food Prot., 61,* 119–122.

Rijke, G., & Smelt, J.P.P.M. (1996). [Heat and pressure resistance of *Lactobacillus plantarum*]. Unpublished results

Ritz, M., Courcoux, P., Semenou, M., & Federighi, M. (1998). High hydrostatic pressure inactivation of Salmonella typhimurium; effects of pressure duration, pH and temperature studied by analysis of variance. *Vet. Res., 29,* 547–556.

Russell, N.J., Evans, R.I., ter Steeg, P.F., Hellemons, J.C., & Abee, T. (1995). Membranes as a target for stress adaptation. *J. Food Microbiol., 28,* 255–261.

Sato, M., Nakamura, Y., Nakashima, K.K., Kato, C., & Horikoshi, K. (1996). High pressure repression of the malB operon in *Escherichia coli. FEMS Microbiol. Lett., 135,* 111–116.

Schmid, G., Lemann, H.-D., & Jaenicke, R. (1975). High pressure effects on the activity of glycolytic enzymes. *Biophys. Chem., 3,* 90–98.

Schwarz, J.R., & Landau, J.V. (1972). Inhibition of dell free protein synthesis by hydrostatic pressure. *J. Bacteriol., 112,* 1222–1227.

Szczawinski, J., Szczawinska, M., Stanczak, B., Fonberg-Broczek, M., Arabas, J., & Szcsepek, J. (1997). Effect of high pressure of *Listeria monocytogenes* in ripened, sliced cheese at ambient temperature. In K. Heremans (Ed.), *High Pressure Research in the Biosciences and Biotechnology* (pp. 289–298). Leuven: Leuven University Press.

Shigehisha, T., Ohmori, T., Saito, A., Taji, S., & Hayashi, R. (1991). Effects of high hydrostatic pressure on characteristics of pork slurries and inactivation of microorganisms associated with meat and meat products. *Int. J. Food Microbiol., 12,* 207–216.

Shimada, S., Andou, M., Naito N., Yamada, N., Osumi, M., & Hayashi, R. (1993). Effects of hydrostatic pressure on the ultrastructure and leakage of internal substances in the yeast *Saccharomyces cerevisiae. Appl. Environ. Microbiol., 3,* 924–926.

Shinsuke, F., Obuchi, K., Iwahashi, H., Fujii, T., & Komatsu, Y. (1996). Saccharides that protect yeast against hydrostatic pressure stress correlated to the mean number of equatorial OH groups. *Biosci. Biotech. Biochem., 60,* 476–478.

Simpson, R.K., & Gilmour, A. (1997a). The effects of high hydrostaic pressure on *Listeria monocytogenes* in phosphate buffered saline and model food systems. *J. Appl. Microbiol., 83,* 181–188.

Simpson, R.K., & Gilmour, A. (1997b). The effect of high hydrostatic pressure on the activity of intracellular enzymes of *Listeria monocytogenes. Lett Appl. Microbiol., 25,* 48–53.

Smelt, J.P.P.M. (1998). Recent advances in the microbiology of high pressure proceessing. *Trends Food Sci. Technol., 9,* 152–158.

Smelt, J.P.P.M., Rijke, A.G.F., & Hayhurst, A. (1994). Possible mechanism of high pressure inactivation of microorganisms. *High Pressure Res., 12,* 199–203.

Smelt, J.P.P.M., Wouters, P.C., & Rijke, A.G.F. (1998). Inactivation of microorganisms by high pressure. In D.S. Reid (Ed.), *The Properties of Water in Foods* ISOPOW 6 (pp. 398–413). London: Blackie Academic & Professional.

Smith, W., Pope, D., & Landau, J.V. (1975). Role of bacterial ribosome subunits in barotolerance. *J. Bacteriol., 124,* 582–584.

Somero, R. (1992). Adaptations to high hydrostatic pressure. *Ann. Rev. Physiol., 54,* 557–577.

Sonoike, K., Setoyama, T., Kuma, Y., & Kobayashi, S. (1992). The effect of pressure and temperature on the death rates of *Lactobacillus casei* and *Escherichia coli.* In C. Balny, R. Hayashi, K. Heremans, & P. Masson (Eds.), *High Pressure and Biotechnology*, Colloque INSERM (Vol. 224, pp. 297–301). Montrouge: John Libbey Eurotext.

Styles, M., Hoover, D., & Farkas, D. (1991). Response to *Listeria monocytogenes* and *Vibrio parahaemolyticus* to high hydrostatic pressure. *J. Food Sci., 56,* 1404–1407.

Tanaka, T., & Hatanaka, K. (1992). Application to yoghurt to prevent its after-acidification. *J. Jap. Soc. Food Sci. Technol., 39,* 173–177.

Thakur B.R., & Nelson, P.R. (1998). High pressure and preservation of food. *Food Rev. Int., 14,* 427–447.

ter Steeg, P.F., Hellemons, J.C., & Kok, A.E. (1999). Synergistic action of nisin, sublethal ultra-high pressure and reduced temperature on bacteria and yeasts. *Appl. Environ. 65,* 4148–4154.

Tsuchido, T., Takano, M., & Skibasaki, I. (1984). Effect of temperature elevating process on the subsequent isothermal death of *E. coli* K12. *J. Ferment. Technol., 52,* 788–792.

Welch, T.J., Farewell, A., Neidhart, F.C., & Bartlett, D.H. (1993). Stress response of *Escherichia coli* to elevated hydrostatic pressure. *J. Bacteriol., 175,* 7170–7177.

Wirsin, C.O., Jannasch, H.W., Wakeham, S.G., & Canuel, E.A. (1987). Membrane lipids of a psychrophilic and barophilic deep-sea bacterium. *Curr. Microbiol., 14,* 319–322.

Wouters P.C., Glaasker E., & Smelt, J.P.P.M. (1998). Effects of high pressure on inactivation kinetics and events related to proton efflux in *Lactobacillus plantarum. Appl. Environ. Microbiol., 64,* 509–514.

Yuste, J., Mor-Mur, M., Capellas, M., Guamis, B., & Pla, R. (1998). Microbiological quality of mechanically recovered poultry meat treated with high hydrostatic pressure and nisin. *Food Microbiol., 15,* 407–414.

CHAPTER 4

Effects of High Pressure on Spores

Volker Heinz and Dietrich Knorr

INTRODUCTION

The history of modern food preservation is a history of heat treatment. This technique dates back to the last decade of the 18th century, when the Frenchman Nicholas Appert discovered a method that allowed people to store food for a comparably long period of time without fermenting, smoking, or salting the products or removing water by evaporation (Appert, 1810). He exposed foods packaged in sealed glass bottles to heat, simply by immersing them in boiling water for a defined treatment time. The prolonged shelf life was thought to be mainly due to the prevention of the access of air. The role of microorganisms in the spoilage of food was not yet known. Because of this lack of scientific knowledge, investigators used trial-and-error experiments to develop heat treatment regimes for preservation of various products. The first commercial large-scale production of foods using heat treatment is reported to have started in 1804 (Lopez, 1987). For several decades only minor improvements were introduced to this technology.

In 1866, Louis Pasteur identified the microbiological background of food deterioration (Pasteur, 1866). This discovery also yielded an explanation for the preservation process that already had become known as "canning." Shortly before Pasteur's discovery, investigators had found empirically that raised processing temperatures decrease the rate of spoilage during processing and, hence, reduce the risk of infections or poisonings. The key step forward in thermal preservation was the invention of the retort in 1851, which allowed "pressure cooking." Heat treatment, which was now feasible at temperatures above 100°C, resulted in products with enhanced microbiological safety and considerably prolonged shelf life. Following Pasteur's advice to use a treatment temperature of 121.1°C (250°F) destroyed all spoilage-causing microorganisms. In 1876, Ferdinand Cohn and Robert Koch provided an explanation for the necessity of using such high tem-

The authors are grateful for assistance from Edwin Ananta and Martin Bunzeit during the preparation of this manuscript. Parts of the presented work were carried out within the European Community projects, AIR1.CT92 00296 and FAIR CT96-1175.

peratures when they discovered that the enormous heat resistance of some bacteria (which can survive boiling in water at atmospheric pressure) is related to morphological differentiation of bacterial cells that vary from vegetative multiplication (Cohn, 1876). They observed that alterations in the environment of *Bacillus anthracis* can induce the formation of internal structures with high refractivity, which are finally liberated from the decomposing vegetative cells. After they harvested the so-called spores and transferred them onto nutritious medium, these scientists were able to recover the vegetative state of the organisms. In spite of the far-reaching importance of this discovery, Cohn and Koch became famous for the proof of the infectious nature of *Bacillus anthracis* rather than for the discovery of spore formation. Thus, these findings had no impact on the understanding or the development of food preservation. Although remarkable scientific efforts during the following years yielded knowledge on a huge variety of spore-forming bacteria, it took another 20 years until Van Ermengem (1896) demonstrated the link between food poisoning and the toxin-producing *Clostridium botulinum*.

Because of their extreme resistance against heat and other stress factors such as chemicals, radiation, and desiccation, spores are widely distributed in nature and can be found on all types of raw food material. However, as scientists learned more about the different patterns of microbial life and their transition mechanisms, they recognized that the mere presence of thermotolerant, dormant spores is not the key issue to food deterioration. What is important is the potential of these organisms to reestablish the less resistant but actively metabolizing vegetative state.

The physical and chemical characteristics of the environment can strongly affect the survival of spore formers. In the first decades of the 20th century, scientists demonstrated the importance of pH for heat resistance and germination and suggested a classification of foodstuffs in acid food (pH < 4.5) and low-acid food (pH ≥ 4.5), making use of the finding that spore germination and growth is inhibited in food with an equilibrium pH below 4.5. Various sensitization techniques such as the addition of sodium chloride were developed to lower spores' resistance to heat. These techniques permitted the use of less severe thermal treatment conditions and, hence, reduced detrimental changes in flavor, texture, and nutritional quality of suitable foods.

In 1899, Hite subjected milk to high hydrostatic pressure instead of high temperature to prevent it from turning sour (Hite, 1899). His attempt was inspired by the sensorial shortcomings of heat-sterilized milk that had a "cooked" taste and recent discoveries by researchers in the field of marine ecology that microorganisms are affected by pressure. Hite achieved a 4-log reduction in the microbial count in milk by a 10-minute treatment with approximately 700 MPa at room temperature. Treatment for 30 minutes at 400 MPa successfully preserved grape juice, cider, peaches, and pears without destroying the fresh flavor. The finding that, below 200 MPa, the lethal effect of pressure was strongly reduced was in agreement with the findings of Chlopin and Tammann (1903). They found that bacterial spores were resistant to hydrostatic pressure. This finding was later proven by Larson et al. (1918), who found that even an applied pressure of 1200 MPa was not sufficient to kill *Bacillus subtilis* spores.

In the early 1920s, Bigelow and colleagues introduced a scientifically based method and calculation procedure for the heat sterilization of food (Bigelow et al., 1920; Bigelow, 1921). These authors proved the effectiveness of heat treatment by quantifying the death of bacterial spores that are responsible for product deterioration or foodborne intoxication. Based on these calculations, the food industry developed clear processing rules that accounted for the transient nature of heat transfer. The production of shelf-stable food was (and it still is) the goal of the canning industry, and it became clear that the resistance of spores to hydrostatic pressure impaired the quality-retaining benefits of this alternative treatment.

For much of the 20th century, the food industry focused on the optimization of thermal preservation. In fact, the development of rotary retort systems, high-temperature–short-time treatment, aseptic processing, and novel packaging systems substantially improved the quality characteristics of thermally processed foods. However, heat treatment still has its shortcomings. It is the aim of this chapter to explore ways in which high pressure can be used to control spores in foods in which conventional thermal processing fails to obtain competitive, high-quality products.

PRESSURE AND TEMPERATURE

Thermodynamic properties and phase equilibriums of any system as well as transport properties such as viscosity, thermal conductivity, or diffusivity must be considered in their functional relationship with temperature and pressure. In addition, the equilibrium concentrations of chemical reactions depend on temperature and pressure. From experiments, we know that the formation of ions from neutral molecules (e.g., the self-ionization of water) is promoted by increasing pressure. Applying the Le Châtelier-Braun principle to the resulting drop in pH, it can be assumed that the volume of the ions is less than the volume of the neutral molecule. In general, the impact of pressure (P) and temperature (T) can be derived quantitatively from Gibbs definition of free energy (G):

$$G = H - TS \tag{1}$$

where S and H are the entropy and enthalpy, respectively. Using $H = U + PV$, where U is inner energy and V is volume, the total derivative of equation 1 yields:

$$dG = dU + PdV + VdP - TdS - SdT \tag{2}$$

When the first and second laws of thermodynamics ($\mathrm{d}U = \mathrm{d}q - P\mathrm{d}V$ and $\mathrm{d}q = T\mathrm{d}S$, where dq denotes the change in heat) are applied and when only changes in free energy (ΔG) are considered, this converts to

$$d(\Delta G) = \Delta VdP - \Delta SdT \tag{3}$$

From this equation, it follows that transitions or reactions within physicochemical systems that show a difference in free energy between educts and products are determined by simultaneously occurring changes in volume and entropy. Hence, reactions such as phase transitions or molecular reorientation depend on both temperature and pressure and cannot be treated separately.

Integration of equation 3 using a Taylor series expansion up to second-order terms yields the following equation:

$$\Delta G = \Delta G + \Delta V_0(P - P_0) - \Delta S_0(T - T_0) + \left(\frac{\Delta\beta}{2}\right)(P - P_0)^2 - \left(\frac{\Delta C_p}{2T_0}\right)(T - T_0)^2 + \Delta\alpha(P - P_0)(T - T_0) \quad (4)$$

where the following expressions are derived from thermodynamic relationships: $\alpha = (\mathrm{d}V/\mathrm{d}T)_p$ represents the thermal expansion coefficient, $\beta = (\mathrm{d}V/\mathrm{d}P)_T$ represents compressibility, and $C_p = (\mathrm{d}q/\mathrm{d}T)_p$ represents heat capacity.

This approximation yields a quadratic dependence of the change in free energy from temperature and pressure. Figure 4–1 illustrates that this quadratic, two-variable function converts to an ellipsoidal phase-transition line in the P/T plane for constant ΔG. In Figure 4–1A, the particular situation where $\Delta G = 0$ is considered for the relevant pressure and temperature range ($P \geq 0$ and $T \geq 0$). Usually state A and state B denote molecular or physicochemical states that are reversibly or irreversibly connected by reactive pathways that finally bring about the transition to the energetically favored configuration ($\Delta G < 0$). In a more general approach, this concept may also apply to the inactivation of spores. In spite of our limited knowledge about the mechanistic background of the lethal action, it is clear that the action involves biochemical reactions that are sensitive to changes in pressure and temperature. Hence, it is still a matter of speculation whether only one site or multiple target sites of the sporulated organism is the weakest link when exposed to external stress. From the point of view of preservation technology, these considerations are of less importance. Therefore, it is justifiable to reduce the highly complex strategy of survival of bacterial spores (which is described later in this chapter) to two distinct states: recoverable and not recoverable. It can be assumed that the inactivation of spores can be brought about at various pressure-temperature conditions along the equilibrium line $\Delta G = 0$ (this is analogous to state A and B in Figure 4–1B).

The rate at which the spore turns from the recoverable to the nonrecoverable state cannot be deduced from the formulations above. This is because these thermodynamic formulations are based on the assumption that during the transitions all components involved are equilibrated, which theoretically takes an infinitely long time. In fact, the kinetics of reactive pathways must be treated in a different way.

For process engineering, it is important to know which treatment time is required to obtain the expected results and how the process is influenced by changes in pressure and temperature.

The rate of processes such as chemical reactions can be accelerated as well as delayed by variations in pressure or temperature. This effect strongly depends on the molecules involved and the reaction mechanism under consideration. In order to find a quantitative formulation of this observation, the reaction rate (k) must be related to these parameters (P and T). A reliable equation on the temperature

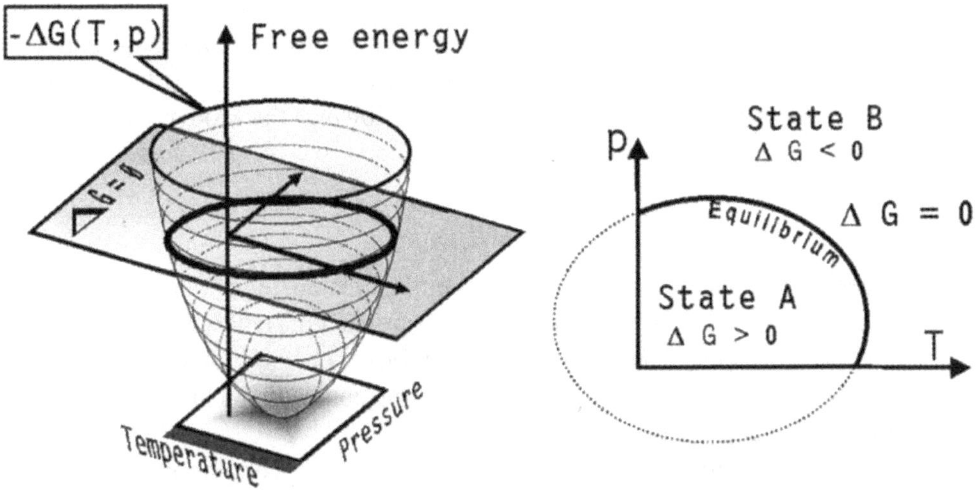

Figure 4–1 Changes in free energy in response to pressure and temperature.

---

dependence of k was first published by Arrhenius in 1889 (equation 5). However, it was not until 1935 that a similar expression for the pressure dependence (equation 6) was derived by Eyring (1935a, 1935b).

$$\left(\frac{\delta \ln(k)}{\delta T}\right)_P = -\frac{E_a}{RT^2} \tag{5}$$

$$\left(\frac{\delta \ln(k)}{\delta P}\right)_T = -\frac{\Delta V^{\neq}}{RT} \tag{6}$$

where R is the universal gas constant ($R = 8.314$ J/K mol), E_a is the activation energy, and $\Delta V^{\neq}$ is the activation volume.

From equation 6, it is evident that a negative $\Delta V^{\neq}$ produces an increased reaction rate k, whereas positive $\Delta V^{\neq}$ indicates that the reaction is retarded. Examples of accelerated reactions can be found among polymerizations, cycloadditions, and solvolytic reactions. Unfortunately, the general application of equation 6 is hindered by the empirical finding that $\Delta V^{\neq}$ is not always constant but may vary with pressure. Another limitation is that the order of reaction might vary at different pressure levels (which is not considered in equation 6).

From equations 5 and 6 it is obvious that experimental designs must include both pressure and temperature as independent variables. Hence, systematic investigations of any pressure-temperature response of a system are based on kinetic experiments at constant pressure and temperature to identify the reaction order or the possible underlying mechanism of action of a more complex response. One way to present results of those experiments is to use a contour plot with lines of constant reaction rate (k) in a P/T diagram similar to Figure 4–1.

The importance of these results is even more relevant when the thermodynamic relation between pressure and temperature must be taken into account. If the boundary of a system under consideration is assumed to be adiabatic and if phase transitions are excluded, an increase in pressure will produce a temperature rise. This observation is expressed in the following thermodynamically derived equation:

$$\left(\frac{\delta T}{\delta P}\right)_{adiabatic} = \frac{\beta T}{\rho C_p} \qquad (7)$$

where ρ denotes the density of the system. Unfortunately, in most cases the compressibility and the specific heat capacity are not constant within the temperature and pressure range, and few experimental data on food materials are available. The temperature response of liquid water to a compression up to 800 MPa is estimated in Figure 4–2.

The temperature increase (ΔT) under adiabatic conditions depends on the maximum pressure reached as well as on the temperature of the system before compression. For example, pressurization of water from ambient pressure up to 800 MPa with an initial temperature of 5° and 100°C results in temperatures of 23°C and 140°C, respectively. This increase in temperature has substantial implications on reaction kinetics during high pressure treatment cycles. Additionally, in real (nonadiabatic) situations, thermal equilibration with the environment occurs, which produces a time and spacially varied temperature field.

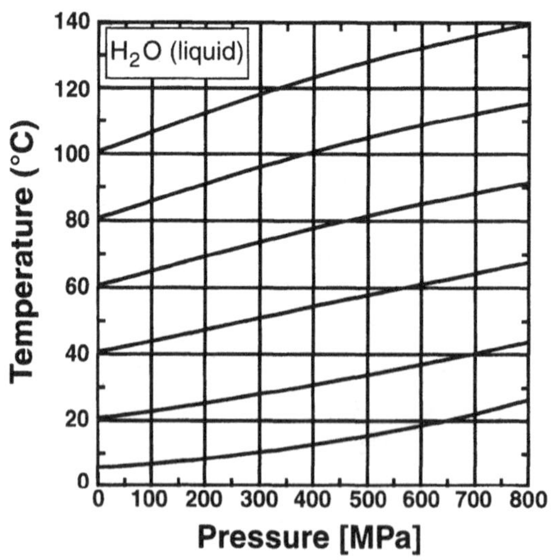

Figure 4–2 Temperature increase of liquid water in response to compression under adiabatic conditions.

MICROBIOLOGICAL ASPECTS

Bacteria are present in nearly every natural habitat. Hence, the presence of bacteria must be considered for all kinds of foods—regardless of whether the food originates from plants or from animals. The presence of certain bacterial strains can be crucial for preservation (e.g., lactic acid fermentation) or deterioration (e.g., protein decomposition). Because bacteria have the ability to reproduce by cell division within generation times of about 30 minutes, optimal growth conditions can trigger an avalanche-like multiplication of the endogenous population.

Some microorganisms respond to starvation or environmental stress not only in the often observed way of producing antibodies or stress proteins (Stevens et al., 1992) but by starting a process of structural differentiation that results in highly compartmentalized cells with an extremely low metabolic activity. The formation of so-called spores is a unique survival strategy of some bacterial genera; the most important of these bacteria for food preservation are *Bacillus* and *Clostridium*. An essential part of the life cycle of such bacteria is the presence of a sensor mechanism that triggers the germination (conversion from the dormant to the vegetative state), if suitable conditions for growth and multiplication are found in the habitat.

In particular, the ability of *Clostridium* to outgrow and to multiply in the absence of oxygen and the ability of some strains to synthesize substances that are highly toxic to humans make this type of bacteria a key problem of the canning industry.

Sporulation and Structure

Approximately 8 hours are required to produce dormant bacterial spores that are extremely resistant to physical treatments such as heat, pressure, or radiation, or to the attack of chemical agents or enzymes. Extensive changes in morphology and physiology of the cells undergoing this transition have been identified. The process is controlled by molecular genetics, which involves more than 125 genes (Stragier & Losick, 1996). In order to classify the progressively changing organisms and to facilitate discussion about the process of sporulation, scientists have proposed dividing the process into seven distinct stages. The most practical determining criteria for this classification are morphological characteristics accessible by microscopic observation. Figure 4–3 presents three distinct pathways of differentiation of spore-forming organisms.

During vegetative growth, bacteria multiply by symmetric division of single organisms into two daughter cells with identical morphological and genetic characteristics (stage 0). For spore-forming bacteria, the occurrence of an unequal cell division can be regarded as the first morphological manifestation of the transition from the vegetative to the dormant state (stage II). At this stage, the formation of a septum divides the cell into two compartments of unequal size but

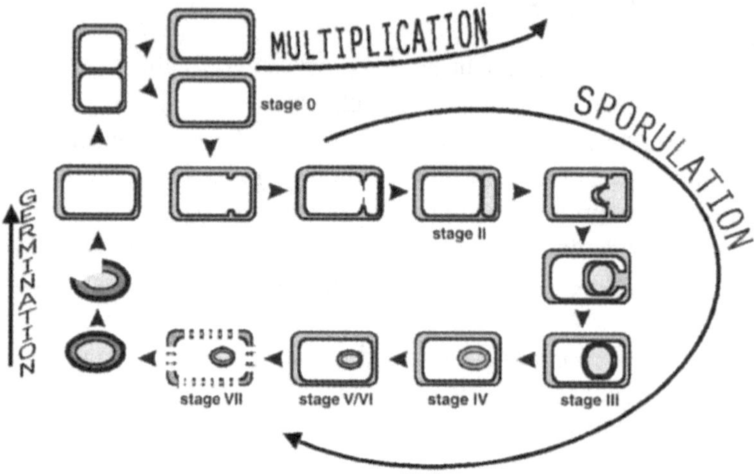

Figure 4–3 Differentiation scheme of spore-forming bacteria. In addition to the multiplication of cells by division, these organisms embark on a second pathway of subsequent sporulation and germination when they face unfavorable environmental conditions for vegetative growth. During sporulation, highly compartmentalized and extremely resistant structures are formed. The dormant spores can convert to actively metabolizing cells by the irreversible process of germination.

with an identical set of chromosomes. The forespore develops within the smaller of the two compartments. The existence and the characteristics of a stage between 0 and II is disputed. However, chromosome replication must be completed before any differentiation steps following stage 0 can occur. Stage I can be defined by the occurrence of gene expression, even though this stage lacks clear morphological differentiation.

From stage II onward, the membrane of the mother cell grows around the forespore compartment, which is already surrounded by an intact membrane. Finally, three membranes enclose the forespore compartment, which will later develop to the central protoplast of the spore. Due to the mechanism of engulfment during stage III, the orientation of the middle membrane is reversed. Because the resulting structure is unique in prokaryotes, it has been speculated that the formation of spores might be related to the diversification during the evolution of cellular life. This process produces more complex cells with enclosed, metabolically active organelles, which might be regarded as the first manifestation of eukaryotic life (Gould, 1983). In any case, the membrane disposition of stage III is considered to be crucial for the subsequent dehydration of the forespore compartment.

The loss of ions such as potassium and the simultaneous decrease in pH by approximately 1 unit that happens during the transition from stage III to stage IV mark the beginning of the water removal. Before this occurs, the accumulation of peptidoglycan between the inner and outer forespore membrane initiates the formation of the cortex and the germ cell wall. In addition, small, acid-soluble proteins (SASPs), which seem to play an important role in spore resistance, are first detected before stage IV is reached. The synthesis of spore coats is most pronounced during the transition from stage IV to stage V. Later, during maturation (the transition from stage V to stage VI), the uptake of considerable quantities of divalent cations such as Ca^{++}, Mg^{++}, or Mn^{++} occurs; simultaneously, the mother cell synthesizes the protoplasmatic accumulation of dipicolinic acid (DPA). Due to the progressive dehydration of the core, the final DPA concentration exceeds its saturation level, and it is assumed that DPA-Ca^{++} complexes are formed. The high degree of dehydration causes the spore to have a bright appearance when observed microscopically in phase contrast illumination. The release of the spore occurs during the transition to stage VII, when lysis of the mother cell is initiated by endogenous lytic enzymes.

After sporulation, the resulting organism is extremely resistant to external attack by physical or chemical means. This resistance is due mainly to the spore's high degree of compartmentalization (Figure 4–4).

The outermost layers, the proteinaceous exosporium and the spore coats, are considered to act as a protective barrier for the internal integuments, which are sensitive to enzymatic degradation and to the action of oxidizing agents. Additionally, the neighboring outer forespore membrane, with a thickness of less than 5 nm, prevents small molecules from access to the spore interior. The much larger underlying peptidoglycan, with a thickness of about 100 nm, consists of cross-linked polymer chains. Compared to similar structures in the cell wall of vegetative cells, the degree of cross-linkage is reduced and more peptide chains remain unlinked. This feature in combination with the presence of free carboxyl groups produces a net negative charge, which has often been assumed to be the structural basis for the function of the cortex during sporulation and for its role in spore resistance.

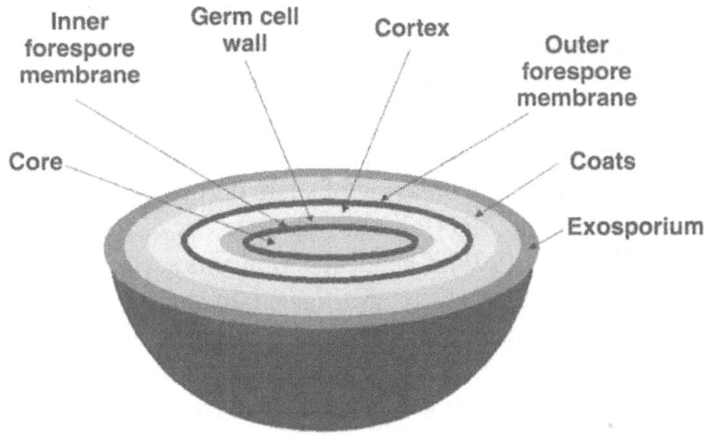

Figure 4–4 Simplified structure of bacterial spore.

In contrast to this highly spore-specific structure, the germ cell wall mainly consists of mother-cell type peptidoglycan, which later forms the basis for vegetative wall synthesis during outgrowth (Buchanan et al., 1994). The inner forespore membrane between core and germ cell wall serves as a highly effective permeability barrier, which prevents influx into the central protoplast of hydrophobic molecules and molecules with a molecular weight higher than 150 (Setlow & Johnson, 1997). Most enzymes, the DNA, ribosomes, as well as considerable amounts of DPA and divalent cations (Ca^{++}) are localized there. The surprisingly low water content of the core (0.4 to 1 g of water per g of dry weight, compared to approximately 4 g in vegetative cells) seems to be a key feature of the extreme resistance of dormant spores. Additionally, the accumulating SASPs are assumed to support the stabilization of vital biochemical components (such as the DNA) of the core.

In spite of the fact that the relation between SASP content and wet heat resistance is not generally proven, there is consensus that the mechanisms of dormancy and resistance are multicomponent and that DPA-binding SASPs as well as immobilization of small molecules by high levels of CaDPA are synergistic factors of the dehydration of the protoplast. It is evident that the mechanisms that produce and maintain this dry state are closely related to the structure and the activity of the surrounding structure of cortical peptidoglycan, which itself is highly hydrated. However, we still lack knowledge about the basic functionality of the cortex.

Germination and Outgrowth

The dormant spores appear to be highly compartmentalized organisms with well-protected components within the dehydrated central region. There are comparably fewer resistant functional units in the outer parts, but these are nevertheless an important part of the survival strategy of spore-forming organisms.

The conversion from the dormant state to the state of active metabolism, which is characterized by growth and multiplication of vegetative cells, is the result of a trigger reaction that is presumably located within the less dehydrated domain of the spore. This sensory system monitors the environment and initiates the germination if it recognizes suitable conditions for vegetative growth.

Germination in this context is defined as "an irreversible series of degradative reactions that result in the breaking of spore dormancy and loss of typical spore properties" (Johnstone, 1994). It can be initiated by different chemical and physical stimuli (Bergère, 1991). Chemical germinants such as L-alanin, inosine, and adenosin, have been extensively investigated. In addition, the application of high hydrostatic pressure has been used for mechanistic studies as well as for the development of sterilization techniques. The latter makes use of the observation that the early stages of germination are marked by loss of resistance to heat, radiation, ultraviolet light, chemicals, and pressure in consequence of the onset of the depolymerization of the cortex and the hydration of the core. At this stage of the germination process biochemical reactions that were suppressed during the state of dormancy are reactivated. The recovery of enzymatic activities produces ATP and triggers the generation of amino acids from SASPs, which are required for the subsequent synthesis of proteins during outgrowth (Setlow & Setlow, 1995).

Genetic analysis (Moir et al., 1994) has helped us to understand the mechanisms involved, although the complete sequence of the germination process is still unknown. However, in *Bacillus megaterium* it has been demonstrated that the trigger reaction produces a series of changes in enzyme activity after a nutrient germinant interacts with the receptor site of the spore (Johnstone, 1994). Conformational reorientation converts subunits of the receptor to proteolytic enzyme activities, which then cleave the pro-form of another enzyme resulting in mobile peptidoglycan-lytic enzyme activities.

The degradative events take place during the first 30 minutes after the initiation of germination. The subsequent process, termed as outgrowth, also requires exogenous nutrient sources. In optimal environmental situations, outgrowth can be terminated after 90 minutes (Setlow, 1983). Because of the uptake of water and nutrients and the onset of biosynthesis of membrane and cell wall components, the spore swells. This swelling finally causes the release from the ruptured spore coat and enables the cell to divide and to multiply. It is questionable whether this intermediate phase between dormancy and vegetative growth is characterized by a reduced stability of vital components. It is possible that a treatment that first germinates the spore and then kills the outgrowing organism could reduce required processing intensity.

Resistance

Spore resistance to physical or chemical treatments may be related to a number of factors acquired at different times during sporulation. Core dehydration, impermeability of coat layers and membranes, and the presence of SASPs are considered to be general components of passive protection mechanisms. The target site of freezing, desiccation, γ-radiation, and ultraviolet light is the DNA. However, the DNA-α/β-type SASP complex formation has a stabilizing effect on the helix structure during freezing and drying, and it changes the photochemistry of DNA so that lethal or mutagenic solutes such as thymine residues are not generated during exposure to ultraviolet light.

Spores are resistant to most chemical attacks because of the effective permeability barrier formed by multilayer coat structures and the inner and outer forespore membrane. However, small molecules that can damage the DNA, including oxidizing and cross-linking agents, may still have access to the core. The saturation of spore DNA with α/β-type SASP is assumed to be responsible for the increased resistance to small molecules such as formaldehyde (Loshon et al., 1999).

In the case of heat treatment, there is evidence that the site of attack is a protein rather than the DNA because general mutagenesis and DNA damage have not been observed (Fairhead et al., 1993). Again, it is assumed that SASPs binding to spore DNA are the key factor for the high degree of thermal stability that is not found with free DNA. Alternatively, Flowers and Adams (1976) have suggested that the impact of heat on the integrity of the membranes is crucial for survival. However, the crystalline state seems to protect at least the inner forespore membrane against thermal attack.

The high degree of mineralization, which is due to the presence of large amounts of divalent cations (Ca^{++}, Mn^{++}, Mg^{-+}) and DPA, seems to have a synergistic effect on stability, although it is unlikely that DPA has an important role in heat resistance. On the other hand, numerous investigations with various *Bacillus* species proved that there is a direct correlation between core water content and thermal stability. It is assumed that the low level of free water stabilizes macromolecules and inhibits molecular motion, which delays or stops chemical reactions.

Despite the body of evidence that protein denaturation takes place during heat treatment in wet environments, the identity of the target site is still unknown. In 1972, Busta and Adams found that the germination system is affected by heat. It was assumed (Stewart et al., 1981) that the reactions that initialize germination might be most sensitive to heat. Consequently, an interruption of germination at this early stage would imply that these injured organisms cannot be recovered on nutrition medium after heat treatment. However, Warth (1980) demonstrated that, under certain circumstances, heat-killed spores can be triggered to germinate. Obviously, it is possible to bypass the destroyed germination system.

To localize the site of heat killing, Belliveau et al. (1992) used differential scanning calorimetry. From these measurements, they concluded that probably a crucial protein is the rate limiting target. By simultaneous detection of the change in refractivity and the release of core material (e.g., DPA) that exhibits a temporal delay compared to the spore killing, these investigators found that membrane disintegration and core hydration are not prerequisites of the inactivation. They suggested that the crucial step involves protein denaturation inside the protoplast.

Instead of protein denaturation Setlow and Setlow (1998) suggested that deamination of labile asparagine residues of proteins may be the key target. However, the identity of these proteins is still unknown.

From Figure 4–1, in which the free energy change is plotted as a function of temperature and pressure, it can be deduced that the proteinous target site of heat inactivation is also sensitive to increased pressure. Spores are much less sensitive than vegetative cells to pressure, as well as to heat. However, the phenomenon of pressure-induced germination (Gould & Sale, 1970), which is observable in a pressure range even below the levels that are necessary for inactivation of vegetative cells, indicates that pressure may have additional effects on the functionality of dormant spores.

MODELING APPROACH

In food microbiology, mathematical modeling is frequently used to quantify the effect of intrinsic or environmental key factors on growth or kill of microbial populations. Quantitative modeling of the fluctuations in number or physiological state of the targeted organisms in response to alterations in the habitat offers a useful tool, particularly for those who are concerned with the stabilization of perishable foods. Moreover, the derived relationships between external stress and the macroscopic response can be used to derive or validate the underlying mechanisms of action on the cellular level.

Bacterial life is characterized by a succession of growth and multiplication. Some genera, as described earlier, have further options of differentiation pathways (sporulation and germination) during their life cycle. Bacteria reproduce rapidly by cell division, which occurs with a maximum repetition rate of about two divisions per hour. Hence, shortly after the onset of multiplication, we must consider populations rather than individual organisms. Thus, it is advisable to observe the dynamics of growth in terms of logarithmic cell concentrations. The same is true for dying cells within a population. Chick (1910) first described the reduction of the logarithmic number of bacterial cells in response to chemical agents or heat to be a linear function of time. Earlier investigators had defined the "thermal death time" as a criterion for the susceptibility of a contaminated sample to heat. This approach, which used the temperature at which a suspension of microorganisms is killed in 10 minutes, was less applicable for the derivation of a mathematical relationship between the time of exposure and microbial inactivation. The slope (k) obtained from linear regression analyses of the survival count relative to the initial number of microorganisms (N/N_0) plotted logarithmically to the basis of e versus time provided a powerful tool that still is one of the fundamentals of descriptive microbiology. Using these notations, the mathematical formulation of microbial inactivation at constant temperature is given by:

$$\ln\left(\frac{N}{N_0}\right) = -kt \quad \text{or} \quad \log\left(\frac{N}{N_0}\right) = -\frac{k}{2.303}t \tag{8}$$

Apart from this empirical finding, the same result can be obtained on integrating the following linear first-order differential equation:

$$\frac{dN}{dt} = -kN \tag{9}$$

However, equation 9 indicates that the decrease in survivor count (dN/dt) with time (the rate of inactivation or, as it is sometimes called, the frequency of death) is proportional to the number of remaining survivors (N). This principle is strictly valid in radioactive decay and in chemical reaction kinetics, where the decomposition rate of pure atoms or molecules depends on the actual concentration. The transposition of this rule to living organisms, however, is complicated by cells' biological diversity (such as fluctuations in their age, metabolism, and resistance). Hence, the goal of food microbiology to work with high concentrations of genotypically and phenotypically identical organisms, which would be adequate for the derivation of quantitative kinetic considerations, is almost impossible to reach. In addition, it is questionable whether our understanding of microbial life and death obtained from experiments is complete (Kilsby, 1999). Further difficulties arise from the fact that the molecular target that may be regarded as the weakest spot for heat or other physical or chemical treatments is mostly unknown. In many cases, the impact of physical treatments such as heat or pressure is not localized at single but at multiple targets of vital metabolic reactions within the cell. Hence, a general inactivation model should consider the response to external stress of relevant reaction pathways in a dynamic system approach (Hills & Mackey, 1995). However, the development of such a model requires additional biochemical information even on the subcellular level that is still difficult to obtain experimentally. As long as this challenging task has not been accomplished, the use of reduced models in descriptive microbiology will be justified. However, microbiologists still debate how far the microbial response to environmental changes can be simplified to make the deduced mathematical formulation a useful tool for reliable predictions of decontamination in biotechnology and process engineering.

Scientists have always attempted to understand the mechanism of cell death beyond the quantitative description of inactivation. However, because only data on survival counts are available, the interpretation of inactivation kinetics has been a subject of speculative discussion since the beginning of the 20th century. Because of the limited accuracy of microbiological enumeration methods of the surviving organisms, it is often unclear whether the experimentally found shape of the inactivation curve is a true macroscopic picture of the lethal events in the single cells or whether it is masked by experimental error. As an example, the early experimental data of Esty and Meyer (1922) are presented in Figure 4–5. Without further experimental evidence, it is difficult to decide whether a straight line or a curve with initial "shoulder" and pronounced "tail" yields the correct interpolation of the data. Recently, this lack of knowledge has triggered a debate on the reliability of current sterilization practice in general (Casolari, 1994; Peleg & Cole, 1998; Peleg, 1999).

However, careful inspection of the data of Esty and Meyer (1922) suggests that, in addition to the linear relationship between logarithmic inactivation and treatment time (equation 8), the sigmoidal curve that resembles a cumulative distributional function yields similar statistical significance of the regression analysis. As a simple example, the two-parameter Weibull distribution (Weibull, 1951) has been applied to the data in Figure 4–5 after logarithmic transformation (A, B: parameters):

$$\log\left(\frac{N}{N_0}\right) = e^{\frac{-t^A}{B}} - 1 \qquad (10)$$

Other distributional functions that yield essentially the same result were recently reviewed by Xiong et al. (1999). Generally, the application of equation 10 (or similar distributions) instead of equation 8 for regressive data analysis yields a completely different interpretation with regard to the mechanistic background. From the frequency of death that is obtained after differentiation of

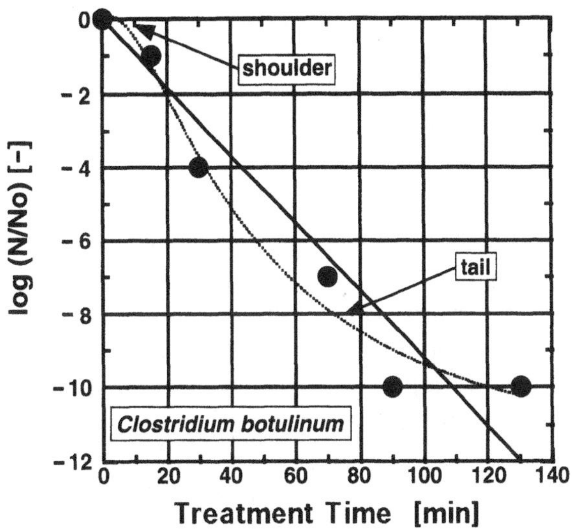

Figure 4–5 Spore inactivation kinetics of *Clostridium botulinum* 62A in phosphate buffer during heat treatment at 100°C.

equation 10 (on substituting the left-hand side with $N^{\neq}$), it is evident that here the rate of inactivation does not depend on the actual number of survivors (like in equation 9) but solely on the treatment time t and vice versa:

$$\frac{dN^{\neq}}{dt} = -t^{A} \cdot \frac{A}{(t \cdot B)} \cdot e^{\frac{t^{A}}{B}} \tag{11}$$

In 1918, the dissimilarity of these approaches was identified and the derived inactivation theories were grouped into "mechanistic" and "vitalistic" models (Lee & Gilbert, 1918) The mechanistic theories, similar to chemical reaction kinetics, explain the progressive heat inactivation by the discrete and random action of energy-transferring collisions, which are more likely to happen when the number of target sites (the number of microorganisms) is higher. This conception served very well for more detailed investigations of spore inactivation that examined intermediate steps such as activation and germination, and it resulted in the development of a system of first-order differential equations (Rodriguez et al., 1988; Abraham et al., 1990). In addition, one-step inactivation models often yield satisfying results, especially in situations when additional parameters affecting the inactivation of spores must be included (Fernández et al., 1995; Gonzalez et al., 1995).

The vitalistic interpretation considers the biological diversity of resistance to external stress within a microbial population to be the dominant effect on the gradual death of the cells. The validity of this alternative explanation that the survival curve is the cumulative form of a temporal distribution of lethal events was recently demonstrated by a reexamination of various published microbial survival curves (Peleg & Cole, 1998). Fernández et al. (1999) showed that the prediction of thermal inactivation of *Bacillus cereus* spores was closer to the experimental data when the Weibull distribution (equation 10) was used instead of the log-linear model (equation 8). Similar results were obtained when the logistic distribution was applied to thermal inactivation data of *Clostridium botulinum* 213B with logarithmically transformed treatment times (Anderson et al., 1996).

The fact that some mechanistic as well as vitalistic theories are convincing gives rise to the question of whether the application of hybrid models is beneficial. The starting point could be the assumption of diverse and continuously distributed microbial resistance or the probability of a lethal collision that is proportional to the number of survivors. The latter approach was used by Körmendy et al. (1999), who introduced distributed rate constants to the first-order inactivation model. Using this approach, equation 8 takes the following form:

$$\frac{N}{N_0} = \int_0^\infty f(k)e^{-kt}\,dk \qquad (12)$$

where $f(k)$ is the density function of the assumed continuous distribution of the rate constant (k). This means that the microbial population is assumed to consist of fractions with different sensitivity to heat or other types of external stress.

Another hybrid model postulated a hypothetical intermediate state (B) that represents the physiological state of a cell that is susceptible to continuously applied treatment intensity (Heinz & Knorr, 1996). It is assumed that active or passive homeostatic resistance mechanisms prevent the intact cell (state A) from immediate destruction when exposed to external stress. The rate of conversion from the unaffected state (A) to state (B) is expressed by a time-dependent density function such as the Weibull distribution $W_D(t)$. Only the meta-stable fraction (B) follows a first-order inactivation kinetics with the rate constant (k) to the not recoverable state (C). Hence, the formation of B is given by:

$$\frac{dB}{dt} = W_b(t) - kB \qquad (13)$$

Equation 13 must be solved for the concentration $A + B$, as both fractions are assumed to be recoverable under optimal growth conditions.

Following the definition of Whiting (1995), the previously described quantifications of microbial inactivation can be classified as primary-level models. A mathematical function describes the temporal change in survivor count under well-defined and constant boundary conditions with one or a set of factors like the rate constant or the distribution parameters. Secondary-level models are required when changing boundary conditions affect the lethality of the treatment. In those situations, it is necessary to find a functional relationship between the characteristic parameters of the primary-level models and the varying external factors.

The earliest secondary-level model was developed in the canning industry. During the past 200 years, processing rules have evolved that are based on empirical findings from a time when mechanisms of food deterioration and poisoning were not understood. Increasing knowledge about the nature of microbial life led to an improved understanding of the mechanisms of heat preservation. Ball (1924) was the first to convert this knowledge into a clear quantitative model of thermal spore destruction. The outline of his approach was the hypothesis of logarithmic-linear inactivation kinetics. His goal was to develop a calculation procedure that could predict the lethal effect of a thermal treatment on variable temperature levels (for example, during retorting of cylindrical cans). A solution of this problem can be obtained by integrating equation 5, which yields the more common formulation of Arrhenius's law:

$$k = Ae^{\frac{-E_a}{RT}} \tag{14}$$

where the Arrhenius factor (A) has been introduced as the integration constant. In addition to this relation between treatment temperature (T) (in Kelvin) and the rate constant (k), a variety of other exponential or square root laws have been proposed that have sometimes resulted in a better fit of the experimental data (reviewed in McMeekin et al., 1993). Using the general expression $k(T)$, equation 7 can be rewritten as:

$$\frac{dN}{dt} = -k(T)N \tag{15}$$

and, if $T = f(t)$ (the temperature is a function of time),

$$\frac{dN}{dt} = -k(t)N \tag{16}$$

The integration within the time limits $t^{\neq}$ and 0 yields:

$$\frac{N}{N_0} = \int_0^{t^{\neq}} k(t)dt \tag{17}$$

From equation 17, the relative logarithmic reduction in microbial count can be obtained for any time-dependent temperature profile $T = f(t)$ if the functional relationship $k = f(T)$ is known and under constant boundary conditions (i.e., all environmental changes that affect the lethality or the sensitivity to heat are excluded).

In the canning industry, a similar but more obvious approach is used. The decimal reduction time (D) is introduced as the treatment time at a constant temperature that is required to reduce the microbial count by 90%. There is a simple reciprocal correlation between D and the rate constant (k):

$$k = \frac{2.303}{D} \tag{18}$$

Hence, an increased lethality of thermal treatments using higher temperatures is expressed by higher rate constants (k) or by shorter decimal reduction times (D). A linear functional relationship is obtained if the decimal reduction time (D) is plotted logarithmically versus the treatment temperature (T). In contrast to the approach of Arrhenius, where 0 K has been chosen as the temperature reference point, this method allows the definition of T_{ref} within a range that is relevant for practical purposes (Kyereme et al., 1999):

$$D = D_{ref} 10^{\left(\frac{T_{ref}-T}{z}\right)} \qquad (19)$$

From equation 19, it follows that D at any temperature (T) can be derived if the decimal reduction time (D_{ref}) at the reference temperature level (T_{ref}) and the proportionality constant (z) is known. For treatment temperatures being a function of time $T = f(t)$ an integral parameter can be derived from this equation:

$$F = \int_0^{t^{\#}} 10^{\left(\frac{T(t)-T_{ref}}{z}\right)} dt \qquad (20)$$

The F value is a useful tool to compare the thermal effect of treatments where the temperature of the product is not constant. In food canning, 121.1°C (250°F) is commonly used as a reference temperature. Spores of thermophilic *Bacillus stearothermophilus* have decimal reduction time $D_{121.1}$ of 3 minutes, whereas spores of the toxin-forming *Clostridium botulinum* have a decimal reduction time of only 0.2 minutes. Consequently, a 12-log reduction can be achieved for these organisms by treatment times at 121.1°C of 36 minutes and 2.4 minutes, respectively. However, these values can only be used as approximations, because the true inactivation times strongly depend on the media in which the spores are suspended.

The advantage of equation 20 is that it allows us to evaluate the lethal effect on relevant organisms of treatments at variable temperature. Processes that have equal F values are identical in microbial destruction, independent of their specific temperature-time profiles. For example, the reduction of *Clostridium botulinum* spores by 12 log-cycles can be achieved by a 2.4-minute treatment at constant 121.1°C or by any temperature-time profile that yields an F value of 2.4. The accuracy of the prediction is determined by z, which is strain and medium specific and must be checked experimentally. As a rule of thumb z = 10°C can be used for preliminary calculations of spore inactivation. Surprisingly, it was not until 1950 in the United States and 1960 in Europe that the calculation procedures described above became relevant for the food-processing industry.

With the help of the F-value concept, it is possible to account for the temporal fluctuations in temperature that may occur during thermal processing of food. However, this concept does not consider the spatial inhomogeneity of the temperature within packaged food or in fluid flow. Unfortunately, the transport of heat is not instantaneous, but, heat conduction is governed by Fourier´s law:

$$\frac{\delta T}{\delta t} = a \nabla^2 T \qquad (21)$$

The Laplace operator (∇^2) indicates that a heat flux in a certain direction can only occur if a temperature gradient exists in the same direction. In addition, the heat flux is delayed by the heat diffusivity (a), which is a thermophysical property of the product under consideration.

From the simple example of heating a cylindrical can filled with an immiscible product (e.g., corned beef) by immersing it in a hot liquid in an autoclave, it is evident that the geometrical center reaches the autoclave temperature later than the boundary layer. Applying the F-value concept to this sterilization task, equation 20 requires the temporal temperature profile of the coldest volume element within the can (the "critical point"). Because of the minimum thermal effect to meet sterility at the critical point, it is necessary to overprocess the boundary layer, which may have detrimental effects on food quality. As the heat diffusivity (a) is given by the product itself, the local heating rate (left-hand side of equation 21) can only be improved by an increased temperature gradient at the boundary layer (i.e., by applying higher autoclave temperatures, which again may thermally damage the outermost product layers).

In conventional canning, the problem is minimized by reducing the geometrical dimensions of the package or, in the case of miscible products, inducing convectional heat transfer by rotation. Alternative methods of increasing the heating rate in the core of bulky products (such as microwave or resistive heating) also have shortcomings due to limited penetration depth or to electromagnetic shielding by metallic packaging materials.

The heat of compression can be used to raise the temperature of packaged products homogeneously to a level where extensive spore inactivation occurs. The implementation of a pressurization step has the additional advantage of requiring a lower lethal temperature level than heat treatment at ambient pressure. In addition, the temperature increase during compression is largely reversible when the pressure is released.

SPORES UNDER PRESSURE

Previously, it has been stated that changes in free energy (ΔG) can be brought about by variations in temperature and pressure. Although the application of this thermodynamic approach to biological systems is limited because of the high degree of compartmentalization and the complex biochemistry, some general aspects of this approach may be useful to demonstrate the combined action of pressure and temperature. If the mechanism of spore inactivation is thought to be a simple transition between a dormant recoverable state and an inactivated state, equation 3 may be applied. The volume is less important than the entropy, as long as ΔV is small. Hence, the conventional practice in thermo-bacteriology and the food canning industry focuses on the impact of temperature on destruction rates. In an aqueous environment, this simplification holds true if the pressures applied are less than approximately 100 MPa. However, in excess of this threshold level, the lethal effect of pressure and temperature cannot be treated separately. Chlopin and Tammann (1903) proved this theoretical result experimentally when they studied the effect of high pressure up to 300 MPa on various microbial species and found that the lethality of pressure is higher when the treatment is made at a higher temperature. The same authors stated that spore-forming bacteria are among the most resistant organisms investigated. Other early findings on spore inactivation by pressure are reported by Larson et al. (1918). These authors used high pressure equipment that reached pressures up to 1800 MPa. With these treatments, sterility could not be achieved at room temperature. Some years later, Johnson and Zobell (1948) found that pressure can prevent spores from heat inactivation, which is not consistent with previous findings. However, the pressure level they worked with was well below the levels used by Chlopin and Tammann (1903). An overview of publications regarding spore inactivation by high pressure is given in Table 4–1. It provides information on the investigated organisms, the pressure and temperature range, the media, and the methodology used.

The inconsistency of the findings of Johnson and Zobell (1948) were later explained to be due to a germinative effect of high pressure. Clouston and Wills (1969, 1970) and others (Gould, 1970; Gould, et al., 1970; Gould & Sale, 1970) convincingly showed that comparably low pressures (below 200 MPa) can trigger spore germination that is detectable by common indicators. This finding has received a great deal of attention, and microbiologists as well as process engineers have been fascinated by the idea that hydrostatic pressure can impair spores' high resistance to physical and chemical treatments. Table 4–1 notes whether the reviewed publications investigated spore inactivation or germination or both.

Although there is consensus that pressures as low as 50 MPa can make *Bacillus* spores germinate, the maximum pressure level seems to depend strongly on the strain under investigation. Of course, the temperature at which the pressure treatment is performed has an enormous effect—at least on the kinetics of germination. Heinz and Knorr (1998) used in situ scanning of the optical density of *Bacillus subtilis* during pressure treatment and found that the maximum germination rate was in the range between 100 and 200 MPa (Figure 4–6).

Temperature level strongly affects the rate at which the optical density changes. Because parts of the germination process (such as the efflux of DPA) are basically mass-transport phenomena, increasing temperature reduces the viscosity of the suspending fluid and increases the equilibration rate of highly concentrated core material after the breakdown of the permeability barrier. For *Bacillus subtilis*, this explanation holds true if the temperature does not exceed 40°C. Surprisingly, the rate of pressure-induced germination detected by in situ optical density measurements is reduced at higher temperatures. This lack of homogeneity was also noticed when the release of DPA or Ca^{++} was measured immediately after pressure treatment. Using kinetic data of core material efflux, a model has been put forward that also considers that the moment of onset of the irreversible germination as well as the degradation rate vary among the individual organisms within a spore population (Coote et al., 1995). Following an early approach of McCormick (1965), the Weibull distribution (equation 10) is used to describe quantitatively the time-dependent scattering of the initiation of mass

Table 4–1 High Pressure Effect on Germination and Inactivation of Bacterial Spores: A Literature Overview

Reference	*Organism*		*Treatment*		*Type*	*Methodology*	*Medium*	*Comments*
	Bacillus	*Clostridium*	*Pressure (MPa)*	*Temperature (ºC)*				
Chlopin & Tammann 1903	ant	—	0.1–300	0–40	I	m	Nutrient broth	Spore formers are highly resistant; lethal pressure effect increases at higher temperature
Larson et al., 1918	sub	—	300–1200	Room	I	m	Infusorial earth	Still survivors after 14h at 1200 MPa
Basset et al., 1932	—	—	—	—	—	—	—	Rapid inactivation of vegetative form; survival of spores
Johnson & Zobell, 1948	sub	—	0.1–60	25–94	I	m	Buffer pH 7, water	1 log at 60MPa/94°C/30min; 4 log at 0.1MPa/94°C/30min
Timson & Short, 1965	sub		0.1–800	−30–100	I	m	Milk	Spore can survive phase transitions of water
Clouston & Wills, 1969	pum		0.1–170	25	I/G	m/b	Water, buffer pH 6.8	2 log at 170MPa/25°C/270min; 4.5 log with simultaneous irradiation
Clouston & Wills, 1970	pum	—	80–100	25	I/G	m	Buffer pH 6.8	Less than 1 log at 80MPa/25°C/100min
Gould & Sale, 1970	sub, cer, co, pum	—	25–100	20–80	G	m	Water	At 50 and 100MPa, germination optimum at 50°C
Sale et al., 1970	co, cer, sub	sp	0.1–800	25–75	I/G	m/b	Buffer pH 8	6 log after (70°C/30min) + (300MPa/75°C/120min)
Murrell & Wills, 1977	sub, pum, cer	—	50–70	25–44	I/G	m	Buffer pH 6.8	1 log at 63MPa/41°C/40min; germination optimum at 63MPa/50°C
Bender & Marquis, 1982	meg	—	30–100	24–60	G	m	Buffer pH 6.8	>30 MPa/45°C can induce germination
Mallidis & Drizou, 1991	st	—	1–30	117–128	I	m	Water	5 log at 10MPa/123°C/3 min
Seyderhelm & Knorr, 1992	st	—	0.1–600	20–90	I	m	Water	6 log at 350MPa/80°C/30min
Hayakawa et al., 1994a	st	—	0.1–800	20–70	I	m	Water	6 log at 600MPa/70°C/6×5min oscillatory treatment
Hayakawa et al., 1994b	st	—	0.1–800	20–70	I	m	Water	6 log at 600MPa/70°C/6×5min oscillatory treatment
Nishi et al., 1994	lic, sub, pol	—	50–200	25–60	I/G	m	Milk, nutrient broth	6 log after (200MPa/60°C/10min) + (incub.37°C/60min) + (65°C/30min)
Okazaki et al., 1994	sub	—	0.1–400	25–111	I	m	Buffer pH 7	4 log at 400 MPa/100°C/10 min

continues

Table 4–1 continued

Reference	Organism		Treatment		Type	Methodology	Medium	Comments
	Bacillus	*Clostridium*	*Pressure (MPa)*	*Temperature (ºC)*				
Sojka & Ludwig, 1994	sub	—	20–500	40–80	I/G	m	Physiol. NaCl	>8 log at 50°C/6×30min/cycling between 150 and 500MPa
Crawford et al., 1996	—	sp	410–820	80	I	m	Chicken	2 log at 680MPa/80°C/20min and 6 log with subsequent irradiation (3kGy)
Gola et al., 1996	cer, li, st, co	sp	300–900	—	I	m	Buffer pH 7	st:5 log at 700MPa/70°C/5min; li:6 log at 700MPa/70°C/3min; sp:2 log at 900MPa/80°C/10min
							Truffle cream	sp(2log): 900MPa/80°C/10min
Ludwig et al., 1996	sub		0.1–500	4–70	I/G	m/b	Physiol. NaCl	4.5 log at 500 MPa/70°C/15min 8 log by oscillation: 70°C/7×(1min/500MPa + 1min/0.1MPa)
Nakayama et al., 1996	sub, st, lic, meg, co		200–1000	5–10	I	m	Water	Less than 1 log inactivation at 1000MPa/10°C/40 min
Okazaki et al., 1996	sub, co	sp	0.1–400	35–110	I	m	Buffer pH 7	sub:6 log at 400MPa/65°C/15min; co:6 log at 400MPa/110°C/18min; sp:6 log at 400MPa/110°C/18min
Kakugawa et al., 1996	st	—	0.1–400	50–120	I	m	Buffer pH 7	5 log at 150MPa/90°C/30min
Roberts & Hoover, 1996	co	—	400	25–70	I	m	Buffer pH 4–7	6 log at 400MPa/70°C/30min/ pH4/+nisin
Heinz, 1997	sub	—	0.1–600	5–80	I/G	m/b	Ringer solution	6 log at 150 MPa/70°C/30min; depending on the pressure-temperature level, germination is either initiated or inhibited
Hölters et al., 1997	sub/st	—	0.1–500	30–70	G	m/b	Physiol. NaCl	Germination at 60MPa/30°C or 30MPa/50°C
Marquis, 1997	meg	—	100	25	I	m	Aqueous H_2O_2 solutions	0 log at 100MPa/25°C/3h; 5.5 log at 100MPa/25°C/3h in the presence of 1% H_2O_2
Sojka & Ludwig, 1997	sub	—	0.1–520	30–50	G	m/b	Physiol. NaCl	>50 MPa/>30°C can induce germination
Wuytack et al., 1997	sub	—	100	40	G	m/b	Water	Rate of pressure-induced germination in mutant spores is decreased

Hayakawa et al., 1998	st	—	30–200	5–100	I	m	Buffer pH 7	6 log at 200MPa/95°C/60min + rapid decompression (1.5 ms) 1 log at 30MPa/95°C/720min + rapid decompression (1.5 ms)
Heinz & Knorr, 1998	sub	—	50–300	10–70	I/G	m/b	Water	6 log at 150 MPa/70°C/30 min
Herdegen & Vogel, 1998	sub	—	0.1–700	40–80	I	m	Peptone solution pH 7	5 log at (500MPa/20°C/15min) +(0.1MPa/37°C/30min) +(600MPa/20°C/15min)
Mills et al., 1998	—	sp	60–600	20–80	I	m	Water	2 log at 0.1MPa/80°C/10 min + 400 MPa/60°C/70min
Raso et al., 1998	cer	—	250–690	25–60	I/G	m	Buffer pH 7	8 log at 690MPa/60°C/1min (sporulation at 37°C) 6 log at 690MPa/60°C/15min (sporulation at 20°C)
Rovere et al., 1998	cer, li, st, co	bot, sp	700–900	50–70	I	m	Beef broth, buffer pH7	At 700 MPa/70°C/5min: cer, 5log; li, 6log; st, 5log; co, 1log At 800 MPa/88°C/9min: bot,3log,
Wuytack et al., 1998	sub	—	100–600	40–55	I/G	m/b	Buffer pH 7	3 log at (600MPa/40°C/60min) + (0.1MPa/55°C/10min) 4.5 log at (100MPa/40°C/60min) + (0.1MPa/55°C/10min)
Hölters et al., 1999	—	pas	0.1–500	30–60	I/G	m/b	Physiol. NaCl	2 log at 420MPa/60°C/75min 4 log by oscillation: 60°C/6×(10min/420MPa + 5min/0.1MPa)
Rovere et al., 1999	—	sp	600–1200	90–110	I	m	Meat broth	D(90°C) at 400/600/800/ 1200MPa: 76.9/16.7/5.3/4.38min D(110°C) at 0.1/400/600/ 800MPa: 13.3/6.1/1.3/0.7
Wuytack, 1999	sub	—	0.1–600	20–80	I/G	m/b	Various aqueous media	6 log at 500MPa/60°C/30min, at >55°C; not all spores germinated prior to inactivation
Reddy et al., 1999	—	bot	689–827	35–60	I	m	Buffer pH 7	5 log at 827MPa/40°C/10min 5 log at 227MPa/50°C/5min
Wuytack et al., 2000	sub	—	100–600	40	G	m	Water	Different pathway for germination at 100 and 600MPa

Organisms: sub, *B.subtilis;* st, *B.stearothermophilus;* cer, *B.cereus;* li, *B.licheniformis;* meg, *B.megaterium;* pum, *B.pumilis;* co, *B.coagulans;* pol, *B.polymyxa*; ant, *B.anthracis;* sp, *C. sporogenes;* pas, *C. pasteurianum;* bot, *C. botulinum.*

Type (of experiment): I, inactivation; G, germination.

Methodology used: m, microbiological; b, biochemical.

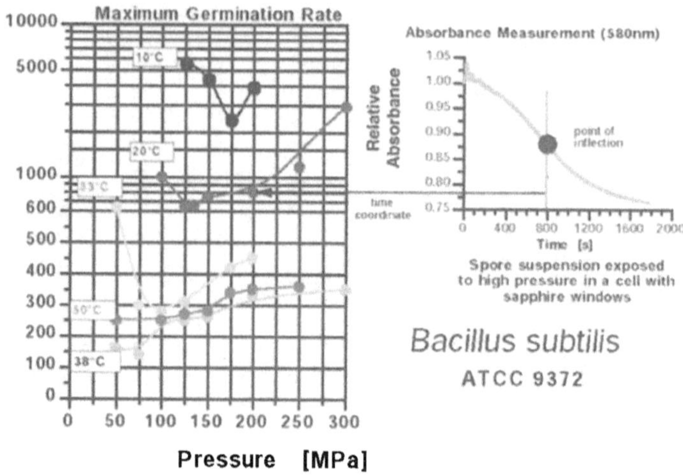

Figure 4–6 In situ scanning of the optical density (relative absorbance at 580 nm) of suspended spores of *Bacillus subtilis* ATCC 9372. (A) Pressure and temperature dependence of germination. (B) The ordinate of the graph represents the time coordinate where the highest germination rates were detected (point of inflection).

transport. The subsequent release of DPA is modeled as the accumulation of individual equilibration reactions at single spores using Fick's First Law of Diffusion. When this model was applied to a pressure treatment at 150 MPa, the modulus of the distributed onset times is strongly reduced when the temperature is elevated up to 40°C and remains at a constant level at higher temperatures. The lack of homogeneity of the efflux rate of DPA is even more evident. It is assumed that the combined action of pressure and temperature can trigger the de-immobilization of cortex-degrading enzyme activities and cause inactivation of the same enzymes by structural unfolding. Consequently, the extent of pressure-induced cortex degradation depends on the ratio between these competing reactions. Hence, the finding that temperatures higher than 40°C delay the release of core material of *Bacillus subtilis* may be due to the predominance of the unfolding reaction, which stops the degradation at a very early stage or completely inhibits the activity of the germination-specific lytic enzymes.

With the aid of transmission X-ray microscopy, Mönch et al. (1999) investigated the same strain of *Bacillus subtilis* after high pressure treatment in a wet environment without further sample preparation. At 150 MPa and below 50°C, where microbiological cultivation methods clearly indicated germination of the spores, X-ray microscopy imaging detected that the spore protoplast no longer absorbed the X-ray radiation (Figure 4–7).

It is assumed that the high levels of DPA, proteins, and electrolytes were released during pressure-induced germination. However, a temperature greater than 70°C at the same pressure level killed the spores without any indication of

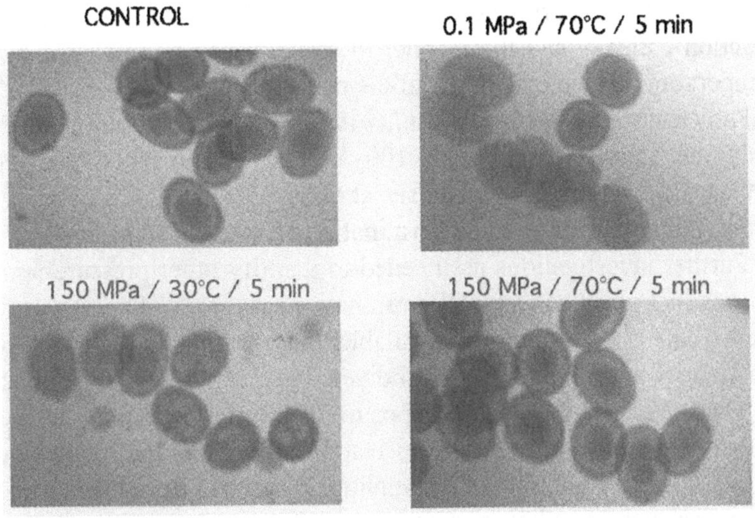

Figure 4–7 X-ray micrographs from spores of *Bacillus subtilis*. After a 150 MPa treatment at 30°C, the germinative action of high pressure is detectable by a loss of absorbance of the central protoplast. At a temperature of 70°C at the same pressure, no such change was observed. However, the same heat intensity at ambient pressure reduced the absorbance.

germination. X-ray microscopy image analysis yielded no significant difference between the spore structure of treated and untreated samples, which proves that the cortex lytic enzymes were irreversibly damaged at these treatment conditions. If the same heat intensity is applied at ambient pressure, the reduced X-ray absorption shows that the spores underwent at least partial cortex degradation. Hence, the germination apparatus appears to be unaffected.

These observations can be explained by applying the concept of free energy changes to the activity of cortex lytic enzymes. It can be assumed that, similar to the changes depicted in Figure 4–1, the transition from the actively catalyzing state to the inactivated state occurs along an elliptical line in the *P/T*-plane which, in real situations, may also include irregularly skewed convex sections. If another transition line is overlaid that marks the initialization of the enzymatic depolymerization of the cortex peptidoglycan, pressure-temperature conditions can be identified where germination is either favored, delayed, or totally inhibited. This conclusion can also apply to the more probable scenario that the pathway of pressure-induced germination comprises many more catalyzed reaction steps requiring a variety of enzymes, each with an individual *P/T*-behavior.

Wuytack et al. (2000) recently found that high pressure triggers germination of *Bacillus subtilis* in a similar way as it is observed in the presence of nutrients by activating the receptors for alanine and asparagine. Further, experiments with chemical inhibitors have proved that pressure triggers a cascade of biochemical reactions rather than breaking mass transfer barriers and bringing about the hydration of the core solely by physicochemical means. The complexity of the reaction pathway and the manner in which it can be impacted by pressure and temperature is also evident from the observation that both a 100 MPa and a 600 MPa pressure treatment of *Bacillus subtilis* at 40°C can initiate germination, but only the spores treated with 100 MPa finally reached the vegetative state (Wuytack et al., 1998). Obviously, at 600 MPa and 40°C a competing reaction is predominant and arrests the germination at an early stage.

Further investigations are needed to identify other pressure-temperature conditions that produce identical germinative responses of spores. However, even if a complete phase diagram is available, there would still be a lack of information on kinetics and on the biological diversity within the spore population.

There is no doubt that the exposure of a spore population to high hydrostatic pressure produces a nonuniform response of individual organisms in terms of triggering delay and rate of conversion. But, apart from other irreversible events in nature such as radioactive decay or diffusion phenomena, there is experimental evidence that the rate of onset of germination is independent of the number of ungerminated spores within a population. Hence, equation 9 and its integrated version (equation 8—the log-linear model) are not applicable. Instead, some researchers have suggested the use of nonautonomous kinetics on germination data. This approach assumes the number of germination spores per unit time to be solely dependent on duration of the treatment or the time elapsed since the germinating agent was added. Examples of distributional functions used to quantify germination are the Poisson distribution (Woese et al., 1968), the Weibull distribution (McCormick, 1965), and the logistic distribution (Coote et al., 1995; Billon et al., 1997). A common feature of these functions is the concave shape at longer treatment times. However, these distributional functions are just the mathematical expression of the often-observed phenomenon that fractions of spore populations start to germinate much later than the majority of the organisms.

Unfortunately, pressure-induced germination cannot overcome this problem. Kinetic analysis of the germination of *Bacillus subtilis* using the previously introduced distributional-diffusive model (Heinz & Knorr, 1998) yielded that the required pressurization time to germinate the complete population is extensively long (Figure 4–6). Raising the temperature can improve mass transport–related reactions during germination, but simultaneously cortex-degrading enzymes and the triggering system are progressively destroyed.

Various authors have stated that, at temperatures below approximately 60°C, extensive inactivation of spores cannot be achieved (Table 4–1). Below this temperature and at a pressure level of 400 MPa, Mills et al. (1998) and Roberts and Hoover (1996) reduced spores by no more than 3 log-cycles (*Clostridium sporogenes* and *Bacillus coagulans*, respectively). Although these authors did not report on germination, it is probable that pressure-induced germination during the initial phase of pressure treatment sensitized the spores and finally killed them during the continued compression. Sojka and Ludwig (1994, 1997) suggested using a two-step process, possibly with an increased pressure intensity during the second step to improve the killing of the germinated spores. However, due to the diverse ability to germinate within a population, caution must be applied. For practical situations, it can be concluded that a two-step process that uses mild pressure-temperature treatments to germinate and then to inactivate spores is not sufficient to obtain commercial sterility of food (Wuytack, 1999). Whether the application of pressure levels higher than those investigated until now can improve the lethality of the treatment is questionable from the viewpoint of large-scale pressure vessel design. In addition, Heinz and Knorr (unpublished results) investigated spore kill of *Bacillus stearothermophilus* at ambient termperatures with pressures up to 1200 MPa and did not find more than 2 log-cycles of inactivation.

However, this scenario changes completely when the temperature is raised well above the level of optimum pressure-induced germination. In excess of 40°C, *Bacillus subtilis* spores already exhibit a clear lack of homogeneity in germination-related mass transfer when exposed to 150 MPa (Heinz, 1997). X-ray microscopic analysis reveals that this result is due to an inhibition of enzymatic cortex degradation (Mönch et al., 1999). This finding has also been reported from thermal inactivation of *Bacillus megaterium* at temperatures higher than 100°C (Belliveau et al., 1992).

The synergistic effect of pressure and temperature on spore killing was first reported by Mallidis and Drizou (1991). They found that simultaneous pressure and heat treatment can reduce the heat resistance of *Bacillus stearothermophilus*. Seyderhelm and Knorr (1992) confirmed this result and obtained a 6 log-cycle reduction of the same strain at 80°C and 350 MPa. Further systematic investigations on spore inactivation at higher temperatures followed (Table 4–1). Despite the experimental differences of these works, we have attempted to compare the effect of pressure on the heat resistances of the spore-forming bacteria investigated. In Figure 4–8, the thermal resistance of spores from different organisms at 400 MPa are compared. Additionally the behavior at ambient pressure is shown for *Bacillus stearothermophilus* (Navani et al., 1970) and *Clostridium botulinum* (Stumbo, 1948).

Because of the lack of published data on the kinetics of combined pressure-temperature inactivation, the logarithmic-linear approximation of the survival curves appears to be the only feasible way to create a database for further considerations on the potential of this treatment for commercial food sterilization. However, the inspection of the available data (e.g., Okazaki et al., 1994) suggests that the diversity of spore populations with regard to pressure-temperature inactivation is far less pronounced than for pressure-temperature germination.

Because of the empirical nature of the primary model (Whiting, 1995), which quantifies the inactivation kinetics, the secondary model for the pressure and temperature dependence of the kinetic parameter k is simplified. Instead of using the theoretical approach (equations 5, 6, and 14), this model plots the decimal reduction time D (equation 18) versus temperature (Figure 4–8). Obviously, D has a much smaller increase when pressure is applied simultaneously to the ther-

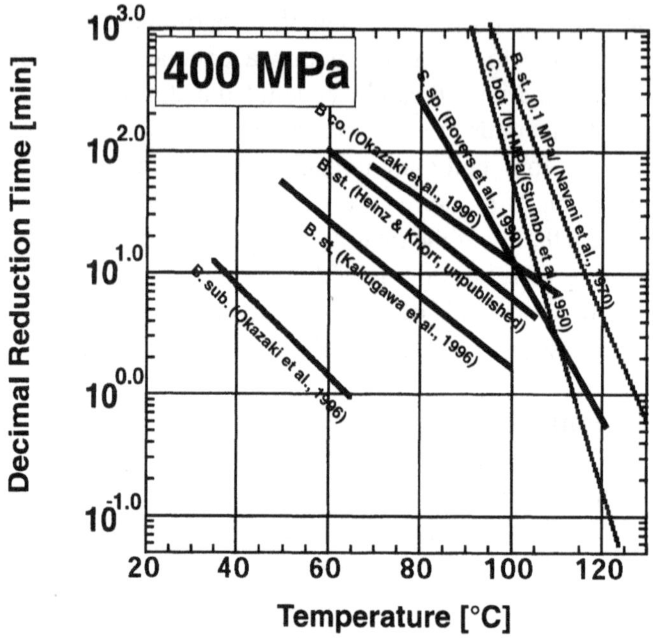

Figure 4–8 Examples of thermal resistances of spore-forming bacteria at 400 MPa (B. sub., *Bacillus subtilis*; B. st, *Bacillus stearothermophilus*; B. co., *Bacillus coagulans*; C. sp., *Clostridium sporogenes*) compared to ambient pressure (dotted lines: B. st, *Bacillus stearothermophilus*; C. bot, *Clostridium botulinum*).

mal treatment. This has important implications on the design of sterilization processes. For example 1 log-cycle of *Bacillus stearothermophilus* can alternatively be inactivated with a 10-minute treatment at 115°C at ambient pressure (Navani et al., 1970) or at 93°C at 400 MPa (Heinz & Knorr, unpublished data, 1998). Other *Bacillus* species and *Clostridium* species behave in a similar way. The convergence of the *D* values of the conventional thermal process with those that are obtained after simultaneous pressure-temperature treatment has already been observed by Mallidis and Drizou (1991). These authors stated that "the relative effect of pressure in reducing heat resistance was diminished with increase in the test temperature." However, their findings were not based on data at pressures in excess of 200 MPa.

In Figure 4–9, the effect of the level of the simultaneous pressure treatment on the thermal resistance is demonstrated with *Bacillus stearothermophilus* and *Clostridium sporogenes*. Especially for *Bacillus stearothermophilus,* a remarkable change in the slope of the temperature function of the *D* value is obtained when the pressure is elevated from 0.1 to 400 MPa. The impact of a further pressure increase is less pronounced. At temperatures higher than 80°C, the change in *D* with pressure is small compared to impact of temperature.

At 100°C and 400 MPa, a 12-minute treatment can inactivate 1 log-cycle of *Clostridium sporogenes* spores. A reduction of the treatment time can be obtained by doubling the pressure level or by increasing the temperature by about 12°C.

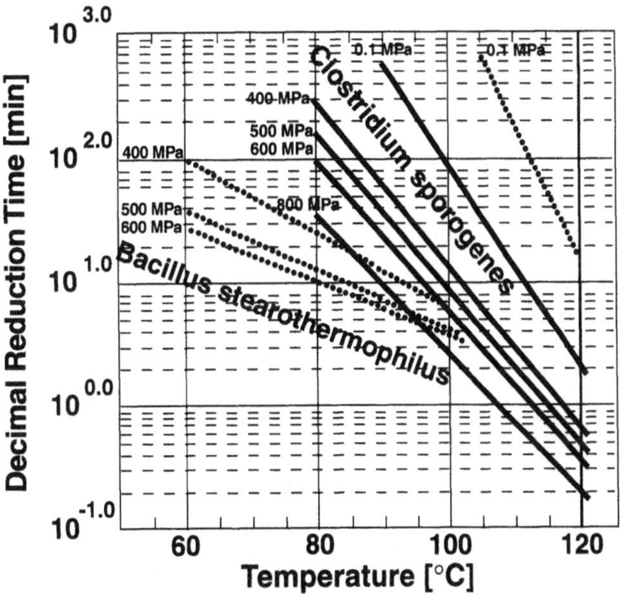

Figure 4–9 Thermal resistances of *Clostridium sporogenes* in beef broth and *Bacillus stearothermophilus* in mashed broccoli with pressure applied simultaneously.

However, the benefit of a combined pressure-temperature treatment for an efficient spore kill is progressively impaired with increasing temperature.

CONCLUSION

It may be helpful to rearrange the previously presented life-cycle scheme of spore-forming organisms (Figure 4–2) with regard to the influence of pressure and temperature. Figure 4–10 includes the population dynamics in the diagram. The extent and the rate of spore germination is largely determined by the diverse sensitivities of individual organisms.

Single vegetative cells grow and multiply by cell division. The probability that the genetic information of the genus survives starvation, hostile environment, or external stress is maximized when the number of organisms of the vegetative population is highest. In addition, spore-forming organisms can convert fractions of the population to a physiological state that is characterized by a high degree of resistance. Individual organisms that have undergone sporulation remain on a comparably high stability level, even if they face hostile environmental conditions that are lethal for vegetative organisms. In Figure 4–10, this phenomenon is symbolized by the collection of the spores within a stability sink and a drop of externally stressed vegetative cells below the minimum level of stability that allows recovery. However, the stable resting positions of these dormant spores become labile if the spores are committed to germinate. Nutritional as well as physical triggering can force the spores to undergo the transition to the vegeta-

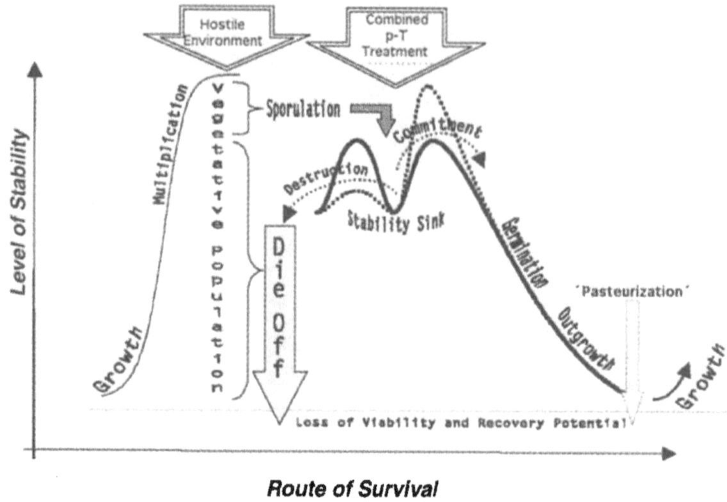

Figure 4–10 Simplified mechanistic scheme of the response of spore-forming bacteria to pressure-temperature treatment at different levels of survival probability.

tive state of growth and multiplication, which increases the susceptibility of the organisms to external stress. When the state of lowest resistance is reached, rather mild treatment conditions (similar to pasteurization) are sufficient to cause these organisms to lose viability and recovery potential.

However, the pressure-induced germination at temperatures below approximately 50°C is characterized by a high degree of statistical fluctuations within the population. In Figure 4–10, the elevated barrier on the right-hand side of the stability sink (the dotted line) depicts this finding. The reduced barrier on the left-hand side expresses that, under certain pressure-temperature conditions (in the temperature range above 70°C for *Bacillus stearothermophilus*), the spores' resistance to heat is lower. The late-germinating dormant spores in response to a pressure-temperature treatment are explained as a fraction of the population that remains within the stability sink, because both barriers are too high to overcome. However, especially in the intermediate temperature range (approximately 50°C to 70°C) it is not clear whether this "superdormancy" is irreversible.

Hence, the major benefit of high hydrostatic pressure treatment for food preservation is the reduction of spores' resistance to heat. The two-step process (also known as "Tyndallization"), which first germinates and then kills the spores under rather mild conditions, has to be rejected as a method of sterilization in food manufacturing because of the diversity in spores' germinative response. On the other hand, it is questionable whether the advantage of reduced decimal reduction times and thus shorter treatment cycles justifies the investment in high pressure processing systems. In particular, the fact that the synergistic effect of pressure is impaired at higher temperatures strongly limits potential applications.

Despite the cautious conclusive remarks, the use of hydrostatic pressure has undeniable benefits over conventional canning. Figure 4–11, which is derived from the previous sections of this chapter, illustrates this concept.

As an example of the reduction in spore resistance to heat, the data of Rovere et al. (1999) are taken to compare the inactivation under 800 MPa with the generally accepted *botulinum* inactivation at ambient pressure (Stumbo, 1948). As the decimal reduction times of both treatments converge at about 121°C (Figure 4–11A), this level has been chosen as the reference temperature (T_{ref}) in equation 20 for the calculation of the *F* value. In the same equation, *z* is derived from the slope and yields 10°C for the conventional process and 18°C for the pressure-assisted process. An *F* value of 2.4 minutes is equivalent to a 12 log-cycle reduction of *Clostridium botulinum* spores and is regarded as the minimum thermal effect that must be applied to every part of the product.

If products are immiscible (e.g., corned beef, tuna, or highly viscous pastes), the temperature increase of the center is delayed by the limitations in heat conduction of the outer layers resulting from Fourier's law (equation 20). For such products, the required minimum processing time is a function of the thermophysical properties of the product, the geometry, and the temperature of the heating medium. The simulation of the time course of the center temperature of a

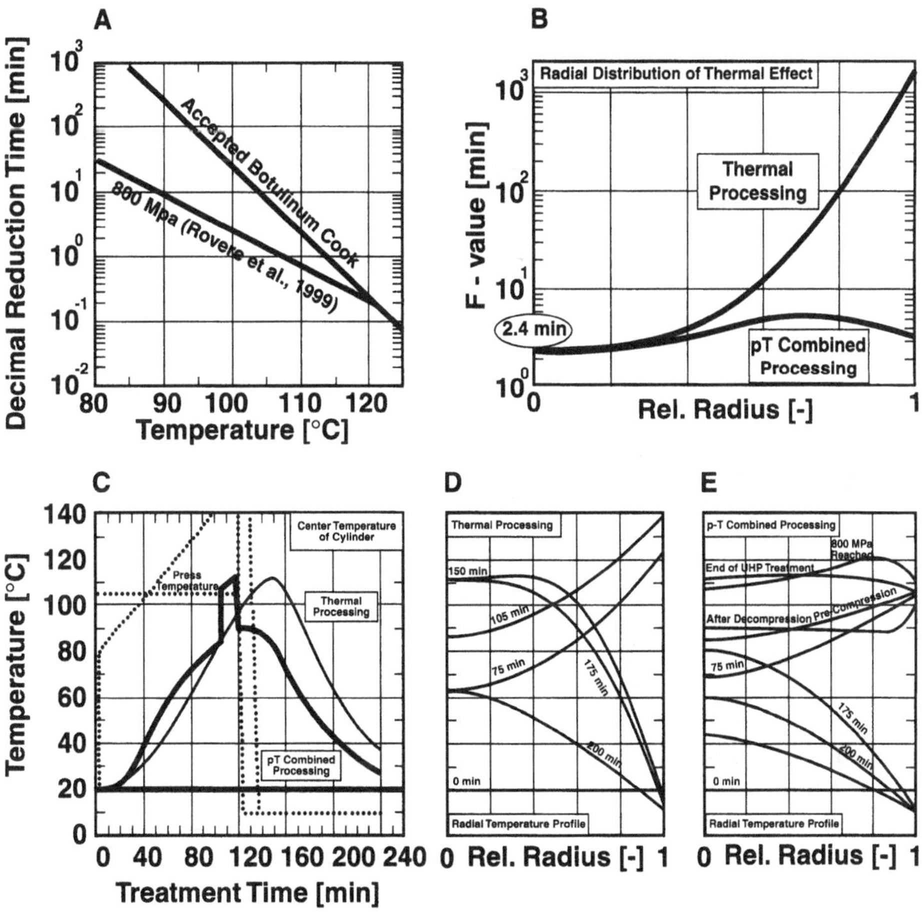

Figure 4–11 Comparison of the uniformity of the thermal effect of conventional treatment and combined pressure-temperature sterilization of immiscible products.

cylindrical package (diameter, 10cm) in Figure 4–11C assumes that, in the case of conventional thermal treatment, the autoclave temperature is raised from 80°C to 140°C and then kept constant for 30 minutes. For cooling, the temperature of the immersing medium is reduced to 10°C. The temperature response of the center (thin line) is a typical profile often obtained in canning. The calculation of its *F* value yields 2.4 minutes. Alternatively, by pressure-assisted heating (thick line) the same *F* value is obtained. The temperature increase after 90 minutes of conventional heating (temperature of the heating medium, 105°C) is produced by the heat of compression during the initial phase of a pressure treatment of 800 MPa (equation 7 and Figure 4–2). After a holding time of 10 minutes, the pressure is released again, which simultaneously reduces the temperature at the center. The subsequent cooling is identical to that of the conventional process.

The major difference between these treatments is more evident if the simulation results for the radial temperature profiles derived from equation 20 are compared (Figure 4–11D and 4–11E). Whereas the temperature of the boundary layer (relative radius, 1) exceeds 120°C after 75 minutes, the pressure-assisted heating remains below this value throughout the whole process. The reason is that the heat of compression increases the temperature homogeneously throughout the product. Hence, the overprocessing of the boundary layer, which is expressed by a local *F* value of more than 1000 minutes (Figure 4–11B), can be avoided.

Consequently, pressure-assisted heating is advisable for the sterilization of immiscible foods without any limitation in packaging scale. The suggested process can homogeneously heat up and cool down the product, and it allows the accurate control of the treatment intensity required for the sterilization.

REFERENCES

Abraham, G., Debray, E., Candau, Y., & Piar, G. (1990). Mathematical model of thermal destruction of *Bacillus stearothermophilus* spores. *Appl. Environ. Microbiol., 56,* 3073–3080.

Anderson, W.A., McClure, P.J., Baird-Parker, A.C., & Cole, M.B. (1996). The application of a log-logistic model to describe the thermal inactivation of *Clostridium botulinum* 213B at temperatures below 121.1°C. *J. Appl. Bacteriol., 80,* 283–290.

Appert, N. (Ed.). (1810). *L'art de conserver pendant plusiers annees toutes les substances animales et vegetales.* Paris: Patris and Co.

Arrhenius, S. (1889). Über die Reaktiosgeschwindigkeit bei der Inversion von Rohrzucker durch Säuren. *Z. Physik. Chem., 4,* 226.

Ball, C.O. (1924). Determining, by methods of calculation, the time necessary to process canned foods. *Bull. Nation. Res. Counc., 7,* 9–76.

Basset, J., Macheboeuf, M.A., & Roux, M.E. (1932). Etude sur les effets biologiques des ultrapressions. Resistance des bacteries, des diastases et des toxines aux pressions élevées. *Compt. Rend. Hebd. Seances Acad. Sci. Paris, 196,* 1431–1440.

Belliveau, B.H., Beaman, T.C., Pankratz, H.S., & Gerhardt, P. (1992). Heat killing of bacterial spores analyzed by differential scanning calorimetry. *J. Bacteriol., 174,* 4463–4474.

Bender, G.R., & Marquis, R.E. (1982). Sensitivity of various salt forms of *Bacillus megaterium* spores to the germinating action of hydrostatic pressure. *Can. J. Microbiol., 28,* 643–649.

Bergère, J.-L. (1991). Spore formation and germination of *Bacillus cereus*: The spore cycle. *Bull. IDF, 275*, 9–14.

Bigelow, W.D. (1921). The logarithmic nature of thermal death time curves. *J. Infect. Dis., 29*, 528–536.

Bigelow, W.D., Bohart, G.S., Richardson, A.C., & Ball, C.O. (1920). Heat penetration in processing canned foods. *Nation. Canners Assoc. Bull., 16-L*, 128.

Billon, C.P.-P., McKirgan, C.J., McClure, P.J., & Adair, C. (1997). The effect of temperature on the germination of single spores of *Clostridium botulinum* 62A. *J. Appl. Microbiol., 82*, 48–56.

Buchanan, C.E., Henriques, A.O., & Piggot, P.J. (1994). Cell wall changes during bacterial endospore formation. In J.-M. Ghuysen & R. Hakenbeck (Eds.), *Bacterial Cell Wall* (pp. 167–186). Amsterdam: Elsevier Science B.V.

Busta, F.F., & Adams, D.M. (1972). Identification of a germination system involved in the heat injury of *Bacillus subtilis* spores. *Appl. Microbiol., 24*, 412–417.

Casolari, A. (1994). About basic parameters of food sterilization technology. *Food Microbiol., 11*, 75–84.

Chick, H. (1910). The process of disinfection by chemical agencies and hot water. *J. Hygiene, 10*, 237–286.

Chlopin, G.W., & Tammann, G. (1903). Über den Einfluß hoher Drücke auf Mikroorganismen. *Z. Hygiene Infektionskrankheiten, 45*, 171–204.

Clouston, J.G., & Wills, P.A. (1969). Initiation of germination and inactivation of *Bacillus pumilus* spores by hydrostatic pressure. *J. Bacteriol., 97*, 684–690.

Clouston, J.G., & Wills, P.A. (1970). Kinetics of initiation of germination of *Bacillus pumilus* spores by hydrostatic pressure. *J. Bacteriol, 103*, 140–143.

Cohn, F. (1876). *Beiträge zur Biologie der Pflanzen*, 253.

Coote, P.J., Billon, C.M.-P., Pennell, S., McClure, P.J., Ferdinando, D.P., & Cole, M.B. (1995). The use of confocal laser microscopy (CLSM) to study the germination of individual spores of *Bacillus cereus. J. Microbiol. Meth., 21*, 193–208.

Crawford, Y.J., Murano, E.A., Olson, D.G., & Shenoy, K. (1996). Use of high hydrostatic pressure and irradiation to eliminate *Clostridium sporogenes* in chicken breast. *J. Food Prot., 59*, 711–715.

Esty, J.R., & Meyer, K.F. (1922). The heat resistance of the spores of *B. Botulinus* and allied anaerobes. *J. Infect. Dis., 31*, 650–663.

Eyring, H. (1935a). The activated complex in chemical reactions. *J. Chem. Phys., 3*, 107–115.

Eyring, H. (1935b). The activated complex and the absolute rate of chemical reactions. *Chem. Rev., 17*, 65–77.

Fairhead, H., Setlow, B., & Setlow, P. (1993). Prevention of DNA damage in spores and in vitro by small, acid-soluble proteins from *Bacillus* species. *J. Bacteriol., 175*, 1367–1374.

Fernández, A., Salmerón, C., Fernández, P.Z., & Martinez, A. (1999). Application of a frequency distribution model to describe the thermal inactivation of two strains of *Bacillus cereus. Trends Food Sci. Technol., 10*, 158–162.

Fernández, P.S., Gomez, F.J., Ocio, M.J., Rodrigo, M., Sanchez, T., & Martinez, A. (1995). *D* values of *Bacillus stearothermophilus* spores as a function of pH and recovery medium acidulant. *J. Food Prot., 58*, 628–632.

Flowers, R.S., & Adams, D.M. (1976). Spore membrane(s) as the site of damage within heated *Clostridium perfringens* spores. *J. Bacteriol., 125*, 429–434.

Gola, S., Foman, C., Carpi, G., Maggi, A., Cassarà, A., & Rovere, P. (1996). Inactivation of bacterial spores in phosphate buffer and in vegetable cream treated with high pressures. In R. Hayashi & C. Balny (Eds.), *High Pressure Bioscience and Biotechnology* (pp. 253–259). Amsterdam: Elsevier Science B. V.

Gonzalez, I., Lopez, M., Mazas, M., Gonzales, J., & Bernardo, A. (1995). The effect of recovery conditions on the apparent heat resistance of *Bacillus cereus* spores. *J. Appl. Bacteriol., 78,* 548–554.

Gould, G.W. (1970). Germination and the problem of dormancy. *J. Appl. Bacteriol., 33,* 34–49.

Gould, G.W. (1983). Mechanisms of resistance and dormancy. In A. Hurst & G.W. Goulds (Eds.), *The Bacterial Spore* (pp. 173–209). London: Academic Press.

Gould, G.W., & Sale, A.J.H. (1970). Initiation of germination of bacterial spores by hydrostatic pressure. *J. Gen. Microbiol., 60,* 335.

Gould, G.W., Stubbs, J.M., & King, W.L. (1970). Structure and composition of resistant layers in bacterial spores coats. *J. Gen. Microbiol., 60,* 347–355.

Hayakawa, I., Furukawa, S., Midzunaga, A., Horiuchi, H., Nakashima, T., Fujio, Y., Yano, Y., Ishikura, T., & Sasaki, K. (1998). Mechanism of inactivation of heat-tolerant spores of *Bacillus stearothermophilus* IFO 12550 by rapid decompression. *J. Food Sci., 63,* 371–374.

Hayakawa, K., Kanno, T., Tomito, M., & Fujio, Y. (1994a). Application of high pressure for spore inactivation and protein denaturation. *J. Food Sci., 59,* 159–163.

Hayakawa, K., Kanno, T., Yoshiyama, K., & Fujio, Y. (1994b). Oscillatory compared with continuous high pressure sterilization of *Bacillus stearothermophilus* spores. *J. Food Sci., 59,* 164–167.

Heinz, V. (1997). Wirkung hoher hydrostatischer Drücke auf das Absterbe- und Keimungsverhalten sporenbildender Bakterien am Beispiel von *Bacillus subtilis* ATCC 9372. Ph.D. Thesis. TU Berlin, Germany.

Heinz, V., & Knorr, D. (1996). High pressure inactivation kinetics of *Bacillus subtilis* cells by a three-state-model considering distribution resistance mechanisms. *Food Biotechnol., 10,* 149–161.

Heinz, V., & Knorr, D. (1998). High pressure germination and inactivation kinetics of bacterial spores. In N.S. Isaacs (Ed.), *High Pressure Food Science, Bioscience and Chemistry* (pp. 435–441). Cambridge: The Royal Society of Chemistry.

Heinz, V., & Knorr, D. (1998). [Pressure/temperature stability of *Bacillus* spores]. Unpublished data.

Herdegen, V., & Vogel, R.F. (1998). Strategies for high pressure inactivation of endospore-forming bacteria. In N.S. Isaacs (Ed.), *High Pressure Food Science, Bioscience and Chemistry* (pp. 394–402). Cambridge: The Royal Society of Chemistry.

Hills, B.P., & Mackey, B.M. (1995). Multi-compartment kinetic models for injury, resuscitation, induced lag and growth in bacterial cell populations. *Food Microbiol., 12,* 333–346.

Hite, B.H. (1899). The effect of pressure in the preservation of milk—A preliminary report. *Bull. West Virg. Univ. Agric. Exp. Station, 58,* 15–35.

Hölters, C., Sojka, B., & Ludwig, H. (1997). Pressure-induced germination of bacterial spores from *Bacillus subtilis* and *Bacillus stearothermophilus*. In K. Heremans (Ed.), *High Pressure Research in the Biosciences and Biotechnology* (pp. 257–260). Leuven: Leuven University Press.

Hölters, C., van Almsick, G., & Ludwig, H. (1999). High pressure inactivation of anaerobic spores from *Clostridium pasteurianum*. In H. Ludwig (Ed.), *Advances in High Pressure Bioscience and Biotechnology* (pp. 65–68). Heidelberg-Berlin: Springer-Verlag.

Johnson, F.H., & Zobell, C.E. (1948). The retardation of thermal disinfection of *Bacillus subtilis* spores by hydrostatic pressure. *J. Bacteriol., 57,* 353–358.

Johnstone, K. (1994). The trigger mechanism of spore germination: Current concepts. *J. Appl. Bacteriol., 76* (Symp. Suppl.), 17S–24S.

Kakugawa, K., Okazaki, T., Yamauchi, S., Morimoti, K., Yoneda, T., & Suzuki, K. (1996). Thermal inactivating behavior of *Bacillus stearothermophilus* under pressure. In R. Hayashi & C. Balny (Eds.), *High Pressure Bioscience and Biotechnology* (pp. 171–174). Amsterdam: Elsevier Science B. V.

Kilsby, D.C. (1999). Food microbiology: The challenges for the future. *Int. J. Food Microbiol., 50,* 59–63.

Körmendy, I., Körmendy, L., & Ferenczy, A. (1999). Thermal inactivation kinetics of mixed microbial populations. A. hypothesis paper. *J. Food Eng., 38,* 439–453.

Kyereme, M., Swartzel, K.R., & Farkas, B.E. (1999). New line intersection procedure for the equivalent point method of thermal evaluation. *J. Food Sci., 64,* 565–570.

Larson, W.P., Hartzell, T.B., & Diehl, H.S. (1918). The effect of high pressures of bacteria. *J. Infect. Dis., 22,* 271–279.

Lee, E.L., & Gilbert, C.A. (1918). On the application of the mass law to the process of disinfection—Being a contribution to the "mechanistic theory" as opposed to the "vitalistic theory." *J. Phys. Chem., 22,* 348–372.

Lopez, A. (1987). *Basic Information on Canning* (pp. 375). Baltimore: The Canning Trade.

Loshon, C.A., Genest, P.C., Setlow, B., & Setlow, P. (1999). Formaldehyde kills spores of *Bacillus subtilis* by DNA damage and small, acid-soluble spore proteins of the a/b-type protect spores against DNA damage. *J. Appl. Microbiol., 87,* 8–14.

Ludwig, H., van Almsick, G., & Sojka, B. (1996). High pressure inactivation of microorganisms. In R. Hayashi & C. Balny (Eds.), *High Pressure Bioscience and Biotechnology* (pp. 237–244). Amsterdam: Elsevier Science B. V.

Mallidis, C.G., & Drizou, D. (1991). Effect of simultaneous application of heat and pressure on the survival of bacterial spores. *J. Appl. Bacteriol., 71,* 285–288.

Marquis, R.E. (1997). Oxidative killing of bacterial spores under pressure. In K. Heremans (Ed.), *High Pressure Research in the Biosciences and Biotechnology* (pp. 229–232). Leuven: Leuven University Press.

McCormick, N.G. (1965). Kinetics of spore germination. *J. Bacteriol., 89,* 1180–1185.

McMeekin, T.A., Olley, J.N., Ross, T., & Ratkowsky, D.A. (1993). *Predictive Microbiology—Theory and Application.* Taunton: Research Studies Press.

Mills, G., Earnshaw, R., & Patterson, M.F. (1998). Effects of high hydrostatic pressure on *Clostridium sporogenes* spores. *Lett. Appl. Microbiol., 26,* 227–230.

Moir, A., Kemp, E.H., Robinson, C. & Corfe, B.M. (1994). The genetic analysis of bacterial spore germination. *J. Appl. Bacteriol., 77* (Symp. Suppl.), 9S–16S.

Mönch, S., Heinz, V., Guttmann, P., & Knorr, D. (1999). X-ray microscopy in food sciences. *Proceedings 6th International Conference on X-ray Microscopy—XRM 99,* University of Berkeley.

Murrell, W.G., & Wills, P.A. (1977). Initiation of *Bacillus* spore germination by hydrostatic pressure: Effect of temperature. *J. Bacteriol., 129,* 1272–1280.

Nakayama, A., Yano, Y., Kobayashi, S., Ishikawa, M., & Sakai, K. (1996). Comparison of pressure resistances of spores of six *Bacillus* strains with their heat resistances. *Appl. Environ. Microbiol., 62,* 3897–3900.

Navani, S.K., Scholfield, J., & Kibby, M.R. (1970). A digital computer program for the statistical analysis of heat resistance data applied to *Bacillus stearothermophilus* spores. *J. Appl. Bacteriol., 33,* 609–620.

Nishi, K., Kato, R., & Tomita, M. (1994). Activation of *Bacillus* spp. spores by hydrostatic pressure. *Nippon Shokukin Kogyo Gakkaishi, 41,* 542–549.

Okazaki, T., Kakugawa, K., Yamauchi, S., Yoneda, T., & Suzuki, K. (1996). Combined effects of temperature and pressure on inactivation of heat-resistant bacteria. In R. Hayashi & C. Balny (Eds.), *High Pressure Bioscience and Biotechnology* (pp. 415–422). Amsterdam: Elsevier Science B. V.

Okazaki, T., Yoneda, T., & Suzuki, K. (1994). Combined effects of temperature and pressure on sterilization of *Bacillus subtilis* spores. *Nippon Shokukin Kogyo Gakkaishi, 41,* 536–541.

Pasteur, L. (1866). Etudes sur le vin. *Des Seances del L'Academie Des Sciences, 63,* 509–510.

Peleg, M. (1999). On calculating sterility in thermal and non-thermal preservation methods. *Food Res. Int., 32,* 271–278.

Peleg, M., & Cole, M.B. (1998). Reinterpretation of microbial survival curves. *Crit. Rev. Food Sci., 38,* 353–380.

Raso, J., Barbosa-Canovas, G., & Swanson, B.G. (1998). Sporulation temperature affects initiation of germination and inactivation by high hydrostatic pressure of *Bacillus cereus. J. Appl. Microbiol., 85,* 17–24.

Reddy, N.R., Solomon, H.M., Fingerhut, G.A., Rhodehamel, E.J., Balasubramaniam, V.M., & Palaniappan, S. (1999). Inactivation of *Clostridium botulinum* type E spores by high pressure processing. *J. Food Saf., 19,* 277–288.

Roberts, C.M., & Hoover, D.G. (1996). Sensitivity of *Bacillus coagulans* spores to combinations of high hydrostatic pressure, heat, acidity and nisin. *J. Appl. Bacteriol., 81,* 363–368.

Rodriguez, A.C., Smerage, G.H., Teixeira, A.A., & Busta, F.F. (1988). Kinetic effects of lethal temperatures on population dynamics of bacterial spores. *Trans. ASAE, 31,* 1594–1601.

Rovere, P., Gola, S., Maggi, A., Scaramuzza, N., & Miglioli, L. (1998). Studies on bacterial spores by combined high pressure-heat treatments: Possibility to sterilize low acid foods. In N.S. Isaacs (Ed.), *High Pressure Food Science, Bioscience and Chemistry* (pp. 354–363). Cambridge: The Royal Society of Chemistry.

Rovere, P., Lonnerborg, N.G., Gola, S., Miglioli, L., Scaramuzza, N., & Squarcina, N. (1999). Advances in bacterial spores inactivation in thermal treatments under pressure. In H. Ludwig (Ed.), *Advances in High Pressure Bioscience and Biotechnology* (pp. 114–210). Heidelberg-Berlin: Springer Verlag.

Sale, A.J.H., Gould, G.W., & Hamilton, W.A. (1970). Inactivation of bacterial spores by hydrostatic pressure. *J. Gen. Microbiol., 60,* 323–334.

Setlow, B., & Setlow, P. (1995). Small, acid-soluble proteins bound to DNA protect *Bacillus subtilis* spores from killing by dry heat. *Appl. Environ. Microbiol., 61,* 2787–2790.

Setlow, B., & Setlow, P. (1998). Heat killing of *Bacillus subtilis* spores in water is not due to oxidative damage. *Appl. Environ. Microbiol., 64,* 4109–4112.

Setlow, P. (1983). Germination and outgrowth. In A. Hurst & G.W. Gould (Eds.), *The Bacterial Spore.* London: Academic Press.

Setlow, P., & Johnson, E.A. (1997). Spores and their significance. In M.P. Doyle, L.R. Beuchat, & T.J. Montville (Eds.), *Food Microbiology Fundamentals and Frontiers* (pp. 35–60). Washington, DC: ASM Press.

Seyderhelm, I., & Knorr, D. (1992). *Reduction of* Bacillus stearothermophilus spores by combined high pressure and temperature treatments. *Z. Lebensmitteltechnol. Verfahrenstechn., 43,* EFS 17–20.

Sojka, B., & Ludwig, H. (1994). Pressure-induced germination and inactivation of *Bacillus subtilis* spores. *Pharm. Ind., 56,* 660–663.

Sojka, B., & Ludwig, H. (1997). Release of dipicolinic acid and amino acids during high pressure treatment of *Bacillus subtilis* spores. *Pharm. Ind., 59,* 355–359.

Stevens, C.M., Daniel, R., Illing, N., & Errington, J. (1992). Characterization of a sporulation gene *spoIVA* involved in spore coat morphogenesis in *Bacillus subtilis. J. Bacteriol., 174,* 586–594.

Stewart, G.S.A.B., Johnstone, K., Hagelberg, E., & Ellar, D.J. (1981). Commitment of bacterial spores to germinate—A measure of the trigger reaction. *Biochem. J., 198,* 101–106.

Stragier, P., & Losick, R. (1996). Molecular genetics of sporulation in *Bacillus subtilis. Annu. Rev. Genet., 30,* 297–341.

Stumbo, C.R. (1948). Bacteriological considerations relating to process evaluation. *Food Technol., 2,* 115–132.

Timson, W.Y. & Short, A.Y. (1965). Resistance of microorganisms to hydrostatic pressure. *Biotechnology and Bioengineering, 7,* 139–159.

Van Ermengem, E. (1896). Centralblatt Bakteriologie Abt. I, *19,* 443–451.

Warth, A.D. (1980). Heat stability of *Bacillus* enzymes within spores and in extracts. *J. Bacteriol., 47,* 321–350.

Weibull, W. (1951). A statistical distribution function of wide applicability. *J. Appl. Mech., 18,* 293–297.

Whiting, R.C. (1995). Microbial modelling in foods. *Food Sci. Nutr., 35,* 467–495.

Woese, C.R., Vary, J.C., & Halvorson, H.O. (1968). A kinetic model for bacterial spore germination. *Proceedings of the Academy of Sciences USA, 59,* 869–879.

Wuytack, E., Peirsman, K., & Michiels, C. (1997). Properties of *Bacillus subtilis* mutant affected in spore germination under high hydrostatic pressure. In K. Heremans (Ed.), *High Pressure Research in the Biosciences and Biotechnology* (pp. 261–268). Leuven: Leuven University Press.

Wuytack, E.Y. (1999). Pressure-induced germination and inactivation of Bacillus subtilis spores. Ph.D. thesis, Faculty of Agricultural and Applied Biological Sciences, Katholieke Universiteit Leuven.

Wuytack, E.Y., Boven, S., & Michiels, C.W. (1998). Comparative study of pressure-induced germination of *Bacillus subtilis* spores at low and high pressures. *Appl. Environ. Microbiol., 64,* 3220–3224.

Wuytack, E.Y., Soons, J., Poschet, F., & Michiels, C.W. (2000). Comparative study of pressure- and nutrient-induced germination of *Bacillus subtilis* spores. *Appl. Environ. Microbiol., 66,* 257–261.

Xiong, R., Xie, G., Edmondson, A.S., Linton, R.H., & Sheard, M.A. (1999). Comparison of the Baranyi model with the modified Gompertz equation for modelling thermal inactivation of *Listeria monocytogenes* Scott A. *Food Microbiol., 16,* 269–279.

CHAPTER 5

Effects of High Pressure on Enzymes Related to Food Quality

Linda Ludikhuyze, Ann Van Loey, Indrawati, Siegfried Denys, and Marc E.G. Hendrickx

INTRODUCTION

An important issue in the application of high pressure technology in the food preservation/processing industry is regulatory approval, which focuses upon microbiological and toxicological safety of food products. High pressure preservation processes should reduce the microbial load to the same level achieved by traditional technologies, while delivering higher-quality products. Combination of high pressure with other treatments (e.g., mild temperature elevation, refrigerated storage, acidification) during processing and storage is a likely route the food industry will take, as the inactivation of bacterial spores and some pressure-resistant enzymes at room temperature cannot be achieved by pressure alone. Several authors suggest that the most safe and economically feasible use of high pressure is in combination with moderately elevated temperature (Gould & Sale, 1970; Knorr, 1993). From the point of view of those who define hazard analysis of critical control points (HACCP) guidelines (risk analysis method carried out to guarantee safety and quality of food products throughout the production chain) or "novel food" regulations (directives that have to be taken into account when introducing new products, new ingredients, or new processing techniques on the market), and also from an engineering point of view, methods to determine the impact of pressure processes on food safety and quality are indispensable. A variety of methods could be evaluated for this purpose: (1) the in situ method, in which the change of an intrinsic food component is measured before and after treatment; (2) the physical mathematical method, in which the actual temperature-pressure-time profile is combined with the kinetics of the attribute under consideration; or (3) the use of process history indicators, which are devices that mimic the change of the attribute under consideration when submitted to the

The authors would like to thank the National Fund for Scientific Research, Vlaams Instituut voor de bevordering van het Wetenschappelijk-Technologisch Onderzoek in de Industrie, the Katholieke Universiteit Leuven Research Council, and the European Commission (project FAIR-CT96-1175) for financial support.

same temperature-pressure-time profile. All these methods rely largely on kinetic models. In this context, quantitative kinetic studies on the combined effect of pressure and temperature on several aspects of food safety and quality (pathogenic/spoilage microorganisms, enzymes related to food quality, nutritional and sensorial quality characteristics) are required. From an engineering point of view, such kinetic data are very important for evaluation of process uniformity, for process validation, and for process optimization.

The purpose of this chapter is to discuss in detail the combined effect of pressure and temperature on enzymes and to give an overview of the available kinetic information of pressure-temperature inactivation of enzymes related to food quality. In addition, the chapter discusses how this information can be integrated with kinetic data on pressure-temperature inactivation of microorganisms and the value of kinetic data in the context of designing and optimizing high pressure processes.

MECHANISMS AND KINETICS OF PRESSURE INACTIVATION OF ENZYMES

Protein/Enzyme Structure

Proteins are linear polymers of amino acids, for which three or four levels of structural organization can be discerned (Table 5–1). In addition to being dependent upon stabilizing and destabilizing interactions within the polypeptide backbone, the native protein structure is dependent on interactions between the surrounding solvent and molecules present within or at the surface of the protein. The net volume of a protein in solution consists of three contributions: (1) the compositional volume, which refers to the constitutive volume of the atoms;

Table 5–1 Structural Hierarchy of Protein/Enzyme Molecules

Structural Level	*Description*	*Bonds/Interactions Involved*
Primary	Amino acid sequence and location of any disulfides along the covalent polypeptide chain	Covalent bonds (including disulfide bonds)
Secondary	Local spatial arrangement of amino acid residues by means of α-helixes and β-pleated sheets	Hydrogen bonds
Tertiary	Arrangement of the secondary structure segments in spatially separated domains, which results in a compact, three-dimensional form	Hydrogen bonds, hydrophobic interactions, electrostatic interactions, van der Waals interactions
Quaternary	Overall organization of subunits to form an oligomer, without regard to the internal geometry of the subunits	Hydrogen bonds, hydrophobic interactions, electrostatic interactions, disulfide bonds

(2) the void volume, which results from an imperfect packing of the constituting atoms; and (3) the solvation volume, which refers to the volume decrease due to solvation of peptide bonds and amino acid residues (Heremans, 1982; Balny & Masson, 1993).

Enzymes are a special group of protein molecules characterized by two striking properties—their enormous catalytic power and their specificity. Quite often, the rate of an enzyme-catalyzed reaction is 10^6 to 10^{20} times that of an uncatalyzed one. Enzymes are highly specific, both in the type of reaction catalyzed and the choice of substrate. Generally, the specificity is determined by the rate at which the enzyme catalyzes similar reactions or by its ability to distinguish between closely related substrates. A "good" substrate may be cleaved 10^4 times faster than a "poor" substrate. Some amino acid side chains are important in determining enzyme specificity or in accelerating the reaction rate. Enzymes thus contain two important regions or sites: one that recognizes the substrate and one that catalyzes the reaction once the substrate has been bound. These two regions are called collectively the active site, which takes up only a relatively small part of the total protein volume. The active site is a three-dimensional entity formed by amino acid groups from different parts of the linear polypeptide chain, brought into proximity in the folded enzyme structure (Richardson & Hyslop, 1985; Darnell et al., 1992).

The delicate balance of stabilizing and destabilizing interactions emanating from various intramolecular forces, as well as interactions with the surrounding solvent, makes proteins and hence enzymes exhibit only small intrinsic stability. Changing temperature, pressure, or microenvironmental conditions may disturb this subtle balance and bring about changes in the overall enzyme conformation or local changes at or near the active site, which may result in a loss of enzyme activity (Pace, 1990; Mozhaev et al., 1994; Barbosa-Canovas et al., 1997). Thermal inactivation generally takes place in the temperature range 40°C to 80°C (Damodaran, 1988). Pressure inactivation of enzymes can occur at pressures exceeding about 200 to 300 MPa (Cheftel, 1991).

Effect of High Pressure on Enzymes

In view of the specificity of enzymatic reactions, enzymes may be affected by pressure in several ways (Cheftel, 1992): (1) Pressurization at room temperature may bring about reversible or irreversible, partial or complete enzyme inactivation resulting from conformational changes in the protein structure. These changes depend on the type of enzyme, microenvironmental conditions, pressure, temperature, and processing time. (2) Enzymatic reactions may be enhanced or inhibited by pressure, depending on the volume change (positive or negative) associated with the reaction. Pressure-induced changes in the catalytic rate may be due to changes in the enzyme-substrate interaction, changes in the reaction mechanisms, or the effect of a particular rate-limiting step on the overall catalytic rate. (3) A macromolecular substrate (protein, starch) may become more sensitive to enzymatic action once it has been unfolded or gelatinized by pressurization. (4) Provided that the cell membrane or the membrane of intracellular organelles is altered, intracellular enzymes may be released from extracellular fluids or cell cytoplasm, hereby facilitating enzyme-substrate interactions.

Mechanisms of Pressure Inactivation of Enzymes

As explained in detail in Chapter 2, the effect of pressure on chemical or biochemical systems is described by the thermodynamical parameter ΔV (equation 1), the change of partial molar volume between initial and final state at constant temperature. It is governed by the principle of Le Chatelier, which predicts that application of pressure shifts an equilibrium to the state that occupies the smallest volume. Hence, pressure favors reactions accompanied by a volume decrease and vice versa (Heremans, 1982; Gross & Jaenicke, 1994).

$$\left(\frac{\delta \Delta G}{\delta P}\right)_T = -RT\left(\frac{\delta \ln K}{\delta P}\right)_T = \Delta V \qquad (1)$$

In general, covalent bonding is not affected by pressure processing, except for oxidation of sulfhydryl groups and thiol-disulfide interchange reactions (Funtenberger et al., 1997). Therefore, the primary structure of enzymes is only minimally affected by pressure (Heremans, 1993; Mozhaev et al., 1994).

Hydrogen bonds are assumed to be rather pressure insensitive. Hydrogen bond formation results in a decrease in molar volume as a result of shortening of interatomic distances. However, the bonds are strongly directed and may create a fairly open structure, partly compensating for the volume decrease, and resulting in a change in global volume in the range of $+1$ to -3 cm^3/mol (Morild, 1981; Balny & Masson, 1993). Hence, hydrogen bonds are slightly stabilized or destabilized by pressure, depending on the system under consideration (Gross & Jaenicke, 1994). Consequently, changes in the secondary structure take place only at very high pressure, usually above 700 MPa.

The well-organized tertiary structure, on the other hand, is lost during pressure treatment (Heremans, 1993; Mozhaev et al., 1994). Modification of electrostatic and hydrophobic interactions, which are the major forces maintaining the tertiary structure, is accompanied by large hydration changes, which are assumed to be primary sources of volume decrease associated with pressure denaturation of enzymes (Mozhaev et al., 1994; Barbosa-Canovas et al., 1997). The rupture of electrostatic interactions is accompanied by exposure of charged groups, which has the tendency to reorient water molecules and lead to a loss of the loosely packed water structure and consequently to a volume reduction. The volume change resulting from the development of one full electric charge seems to be from -30 to -10 cm^3/mol (Weale, 1967; Morild, 1981). The disruption of hydrophobic bonds leads to the exposure of nonpolar amino acid residues to the aqueous solution. The loosely packed water structure is disturbed, and a structured and densely packed hydrophobic hydration layer is formed. Volume changes associated with disruption of hydrophobic bonds are from -20 to -10 cm^3/mol (Weale, 1967; Morild, 1981).

From the above, it can be concluded that pressure inactivation of monomeric enzymes is predominantly caused by disturbance of hydrophobic and electrostatic interactions. Along with the tertiary structure, the quaternary structure of enzymes is also affected by pressure. Oligomeric enzymes dissociate into individual subunits at about 150 to 200 MPa. Imperfect packing of atoms at the subunit interface, along with disruption of hydrophobic and electrostatic interactions in the intersubunit area may account for the large volume changes (Penniston, 1971; Mozhaev et al., 1994). At higher pressure, unfolding or reassociation (aggregation) of the dissociated subunits may take place (Silva & Weber, 1993).

Reversibility versus Irreversibility

Reversibility can be defined as the spontaneous reversion to the native state within a reasonable experimental time upon removal of the denaturing agent (Ahern & Klibanov, 1985). It is generally believed that reversible changes take place with low pressure, while higher pressure induces irreversible inactivation (Miyagawa et al., 1964). The rate and extent of reactivation are largely dependent on the extent of inactivation, the type of enzyme, several microenvironmental factors, and possible recompression at low pressure. Reassociation of individual subunits of oligomeric proteins generally occurs so slowly that the process is considered to be irreversible. In addition, dissociated subunits may refold in the wrong way or reassociation may begin before refolding of the individual subunits has been accomplished, in each case producing a nonfunctional enzyme form (Silva & Weber, 1993).

Kinetics of Enzyme Inactivation

In general, the decrease of enzyme activity (A) as a function of processing time (t) can be written as

$$\left(\frac{dA}{dt}\right) = -kA^n \tag{2}$$

Mathematical models describing inactivation kinetics of enzymes are displayed in Table 5–2. Often, enzyme inactivation can be described by a first-order kinetic model (Richardson & Hyslop, 1985; Lencki et al., 1992). This is remarkable, since enzyme inactivation is a very complex phenomenon involving several events such as disruption/formation of different interactions, decomposition of amino acids, and aggregation/dissociation. It is therefore suggested that, in the case of apparent first-order kinetics, one of these reactions predominates over the others. If several reactions occur more or less at the same rate, complex non–first-order kinetics are expected (Lencki et al., 1992). Thus, first-order kinetics do not necessarily imply a monomolecular reaction, but they can have a wide variety of phenomenological meanings.

Next to the general n^{th}-order equation (equation 5), several mathematical expressions have been suggested for modeling non–first-order inactivation behavior (Table 5–2). A two-isozyme model (equation 6) refers to the existence

Table 5–2 Mathematical Models Describing Inactivation Kinetics of Enzymes

Inactivation Model	*Mathematical Expression*	*Equation Number*
First-order models	$A = A_0 \exp(-k_{obs}t)$: Arrhenius terminology (Arrhenius, 1889)	(3)
$N \xrightarrow{k} I$	$A = A_0 10^{\left(\frac{-t}{D_{obs}}\right)}$: TDT terminology (Bigelow, 1921)	(4)
General nth-order model	$A = (A_0^{1-n} + (n-1)k_{obs}t)^{\frac{1}{1-n}}$	(5)
$N \longleftrightarrow U \xrightarrow{k} I$		
Distinct isozyme model		
$N_s \longleftrightarrow U_s \xrightarrow{k_s} I_s$	$A = A_{0s} \exp(-k_{obs,s}t) + A_{0l} \exp(-k_{obs,l}t)$	(6)
$N_l \longleftrightarrow U_l \xrightarrow{k_l} I_l$		
Consecutive-step model	$A = \left(A_1 - A_2\left(\frac{k_{obs,1}}{k_{obs,1} - k_{obs,2}}\right)\right)\exp(-k_{obs,1}t)$	(7)
$N \longleftrightarrow U \xrightarrow{k_1} I_1 \xrightarrow{k_2} I_2$	$+\left(A_2\left(\frac{k_{obs,1}}{k_{obs,1} - k_{obs,2}}\right)\right)\exp(-k_{obs,2}t)$	
Fractional conversion model		
$N + RF \longleftrightarrow U + RF \xrightarrow{k} I$	$A = A_{rf} + (A_0 - A_{rf})\exp(-k_{obs}t)$	(8)

of two isozymes/fractions, behaving differently when subjected to inactivating conditions. This results in an initial fast inactivation period, followed by a much slower decay. A consecutive-step model can be used to describe two types of enzyme inactivation. The first type is an inactivation process proceeding as a succession of two irreversible reaction steps. (1) The native enzyme converts to an intermediate form, the activity of which may be dependent on the inactivating condition applied. (2) On its turn, this intermediate form is converted to the inactive form. The second type of inactivation involves enzymes that exist as two different forms (e.g., free enzyme and enzyme interacting with impurities or substrates) that are in equilibrium with each other. The transformation of each enzyme form to the inactive form is characterized by a distinct inactivation rate constant. Clearly, the position of the equilibrium between the two forms may be dependent on temperature, pressure, or microenvironmental conditions.

The fractional conversion model (equation 8) refers to a first-order inactivation process in which a resistant enzyme fraction, not affected after prolonged treatment at the preset inactivating condition, is taken into account.

Table 5–3 summarizes the most common temperature and pressure coefficient models. The Arrhenius equation (equation 9) and the Eyring equation (equation 10) describe respectively the temperature and pressure sensitivity of the inacti-

Table 5–3 Survey of Temperature (*T*) and Pressure (*P*) Coefficient Models

Coefficient Model	*Applicability*	*Mathematical Expression*	*Equation Number*
Arrhenius equation	All *n*, constant *P*	$k = k_{obs,Tref} \exp\left(\frac{-E_a}{R}\left(\frac{1}{T} - \frac{1}{T_{ref}}\right)\right)$	(9)
Eyring equation	All *n*, constant *T*	$k = k_{obs,Pref} \exp\left(\frac{-V_a}{RT}(P - P_{ref})\right)$	(10)
Themal death time (TDT) model	$n = 1$, constant *P*	$D = D_{obs,Tref}\, 10^{\left(\frac{T_{ref}-T}{z_T}\right)}$	(11)
Pressure death time (PDT) model	$n = 1$, constant *T*	$D = D_{obs,Pref}\, 10^{\left(\frac{P_{ref}-P}{z_P}\right)}$	(12)

vation rate constant both for nth-order and first-order inactivation processes (Arrhenius, 1889; Morild, 1981). Temperature and pressure dependence of the inactivation rate constants are expressed as the activation energy (E_a) and activation volume (V_a), respectively. In the food and pharmaceutical area, the decimal reduction time ($D = 2.303/k$) is often used instead of the reaction rate constant to describe enzyme or microbial inactivation of a first-order reaction. In this case, the thermal death time (TDT) model (Bigelow, 1921) and its analogue, the pressure death time (PDT) model (Mussa & Ramaswamy, 1997), can be used to characterize temperature and pressure dependence of the decimal reduction time, respectively as z_T and z_P values. The z_T and z_P values are defined as the temperature or pressure increase necessary to reduce the decimal reduction time by a factor 10.

Hitherto, no generally applicable model to describe the combined pressure-temperature dependence of the inactivation rate constants has been reported in the literature. Available kinetic models for individual enzyme systems will be discussed in detail later in this chapter .

THE EFFECT OF HIGH PRESSURE ON ENZYMES RELATED TO FOOD QUALITY

Based on a screening study of the pressure resistance of enzymes related to food quality in buffer solutions, Seyderhelm et al. (1996) ranked the enzymes according to increasing pressure stability in the following order: lipoxygenase, lactoperoxidase, pectinmethylesterase, lipase, phosphatase, catalase, polyphenoloxidase, and peroxidase. Nevertheless, because pressure stability of enzymes is strongly dependent on several intrinsic (e.g., medium composition) and extrinsic (e.g., temperature) factors, one may encounter contradictory results when surveying the literature.

Polyphenoloxidase

Polyphenoloxidase (PPO) activity (EC 1.14.18.1) causes enzymatic browning of damaged fruits and vegetables, which is believed to be one of the main causes of quality deterioration (brown coloration and concomitant changes in appearance and organoleptic properties) during postharvest handling, storage, and processing. Because it represents a major problem in the food industry (Lambrecht, 1995; Ferrar & Walker, 1996), inactivation of PPO is highly desirable.

Although researchers have carried out several studies on the effect of pressure on PPO from a variety of origins (Table 5–4), the available information is often fragmentary, and only few detailed kinetic data (such as data from a broad pressure-temperature domain) have been determined. To date, researchers have found that, upon pressurization, PPO may display either inactivation or activation (enhanced catalytic activity). From Table 5–4 it can be concluded that the pressure level needed for PPO inactivation at ambient temperature is strongly dependent on its origin and the pH of the medium. In addition, microenvironmental conditions such as the presence of salts, sugars, and other additives have been proven to influence the pressure stability of PPO.

A detailed kinetic study on pressure inactivation of PPO from different origins has been carried out by Weemaes et al. (1998a). Because of the extreme pressure stability of PPO from some species, these researchers attempted to combine some methods (such as different intrinsic and extrinsic stress factors) to reduce the pressure stability of PPO. In this context, they examined whether changes in pH, addition of antibrowning agents, and temperature increase could be used to alter the pressure stability of mushroom and avocado PPO. At room temperature, pressure inactivation of mushroom and avocado PPO at their optimal pH (6.5 and 7, respectively) could be accurately described by a first-order kinetic model (Figure 5–1).

For both mushroom PPO and avocado PPO, the threshold pressure for inactivation decreased when the pH was lowered below the optimal pH value (Figure 5–2). Moreover, acidification brought about an increase in the pressure sensitivity of the inactivation rate constant. That is, the absolute value of the activation volume increased with decreasing pH (Figure 5–2). Reducing pH to 4 resulted in a change in the pressure inactivation behavior of avocado PPO, and the pressure inactivation could no longer be described by a first-order model. In this case, a fractional conversion model seemed appropriate to describe the inactivation course (Figure 5–3). This change in inactivation kinetics was suggested to be due to a distinct pressure stability of two avocado PPO isozymes at this low pH (Matser et al., 1998; Weemaes et al., 1998b).

Whereas the effect of pH on avocado and mushroom PPO was quite similar, the enzymes were differently affected by the presence of antibrowning agents (benzoate, ethylenediaminetetraacetic acid [EDTA], glutathione, sodium chloride [NaCl], and 4-hexylresorcinol). The pressure stability of mushroom PPO at pH 6.5 was dramatically reduced in the presence of 4-hexylresorcinol and not

Table 5–4 Effect of Pressure at Ambient Temperature on PPO from Various Origins

Origin	*Variety*	*Medium*	*Pressure Treatment*	*Activation/ Inactivation*	*Reference*
Apple	NM	PB (pH 4.5, 0.5 M)	150–300 MPa/25°C/1 min	Activation	Anese et al., 1995
			≥500 MPa/25°C/1 min	Inactivation	Anese et al., 1995
	NM	PB (pH 5.4, 0.5 M	<300 MPa/25°C/1 min	Activation	Anese et al., 1995
			>700 MPa/25°C/1 min	Inactivation	Anese et al., 1995
	NM	PB (pH 7, 0.5 M)	≥100 MPa/25°C/1 min	Inactivation	Anese et al., 1995
	Golden	Crude extract (pH 7)	500 MPa/25°C/10 min	Activation	Jolibert et al., 1994
		Crude extract (pH 3–6)	500 MPa/25°C/10 min	Inactivation	Jolibert et al., 1994
		Cell-free extract (pH 3.9)	>250 MPa/25°C/10 min	Inactivation	Jolibert et al., 1994
		Puree	≥1000 MPa/20°C/5 min	Inactivation	Rovere et al., 1996
		PB (pH 6, 0.1 M)	≥600 MPa/25°C/5–60 min	Inactivation	Weemaes et al., 1998a
	Granny Smith	Cell-free extract (pH 3.5)	≥125 MPa/25°C/10 min	Inactivation	Jolibert et al., 1994
	Jonagolden	Cell-free extract (pH 3.9)	≥250 MPa/25°C/10 min	Inactivation	Jolibert et al., 1994
	Red Chief	Cell-free extract (pH 4.1)	>250 MPa/25°C/10 min	Inactivation	Jolibert et al., 1994
Apricot	NM	Cell-free extract (pH 3.9)	≥125 MPa/25°C/15 min	Inactivation	Jolibert et al., 1994
Avocado	NM	PB (pH 7, 0.1 M)	≥800 MPa/25°C/>60 min	Inactivation	Weemaes et al., 1998b
Grape	Trebbiano	Must (pH 3.2)	>600 MPa/15°C/10 min	Inactivation	Castellari et al., 1997
		Synthetic must (pH 2.9)	900 MPa/15°C/2 min	Inactivation	Castellari et al., 1997
	NM	McIlvaine (pH 6)	≥700 MPa/25°C/5–30 min	Inactivation	Weemaes et al., 1998a
Guava	Chung shan	Puree (pH 3.8)	≥600 MPa/25°C/15 min	Inactivation	Yen and Lin, 1996
Onion	Stuttgarter Riese	TB (pH 6.5, 0.01 M)	≤700 MPa/25°C/10 min	Activation	Butz et al., 1994
Pear	Bartlett	Slices	400 MPa/25°C/10 min	Activation	Asaka and Hayashi, 1991
		Cell-free extract	200–600 MPa/25°C/10 min	Activation	Asaka and Hayashi, 1991
	La France	PB (pH 7, 0.02 M)	400–700 MPa/20°C/≥3 min	Activation	Asaka et al., 1994
	Durondeau	PB (pH 7, 0.1 M)	900 MPa/25°C/200 min	Inactivation	Weemaes et al., 1998a
Plum	Mirabelle	Cell-free extract (pH 4.6)	≤500 MPa/25°C/15 min	No inactivation	Jolibert et al., 1994
	Red Beauty	PB (pH 7, 0.1 M)	900 MPa/25°C/180 min	No inactivation	Weemaes et al., 1998a

continues

Table 5–4 continued

Origin	*Variety*	*Medium*	*Pressure Treatment*	*Activation/ Inactivation*	*Reference*
Potato	NM	Cubes	400 MPa/25°C/15 min	No inactivation	Knorr, 1993
	NM	Crude extract	≥200 MPa/25°C/10 min	Inactivation	Gomes and Ledward, 1996
	Desirée	Cell culture	100–200 MPa/25°C/5–10 min	Activation	Dörnenburg and Knorr, 1997
	Desirée	Water surrounded cubes	400 MPa/20°C/15 min	Inactivation	Esthiaghi and Knorr, 1993
Strawberry	NM	Crude extract (pH 2–5)	500 MPa/25°C/10 min	Inactivation	Jolibert et al., 1994
		Crude extract (pH 6)	500 MPa/25°C/10 min	Activation	Jolibert et al., 1994
		Cell-free extract (pH 3.8)	≥375 MPa/25°C/15 min	Inactivation	Jolibert et al., 1994
	Pajaro	Puree (pH 3.9)	50–285 MPa/20°C/15 min	Inactivation	Cano et al., 1997
Tomato	Pera	Puree (pH 4.1)	<200 MPa/20°C/15 min	Activation	Hernández and Cano, 1998
			200–500 MPa/ 25 °C/15 min	Inactivation	Hernández and Cano, 1998
Mushroom	NM	PB (pH 6.5, 0.1 M)	≥750 MPa/25°C/20 min	Inactivation	Weemaes et al., 1997
	NM	PB (pH 6.5, 0.05 M)	>100 MPa/25°C/>1 min	Inactivation	Gomes and Ledward, 1996
	NM	Crude extract	200 MPa/25°C/10 min	Inactivation	Gomes and Ledward, 1996
			400 MPa/25°C/10 min	Activation	Gomes and Ledward, 1996
			≥600 MPa/25°C/10 min	Inactivation	Gomes and Ledward, 1996
	NM	NM	300–700 MPa/20°C/>10 min	Activation	Aoyama et al., 1994
	NM	Intact tissue	600 MPa/25°C/5 min	Activation	Matser et al., 1998
			800 MPa/25°C/5 min	Inactivation	Matser et al., 1998

NM = not mentioned; PB = phosphate buffer; TB = Tris buffer

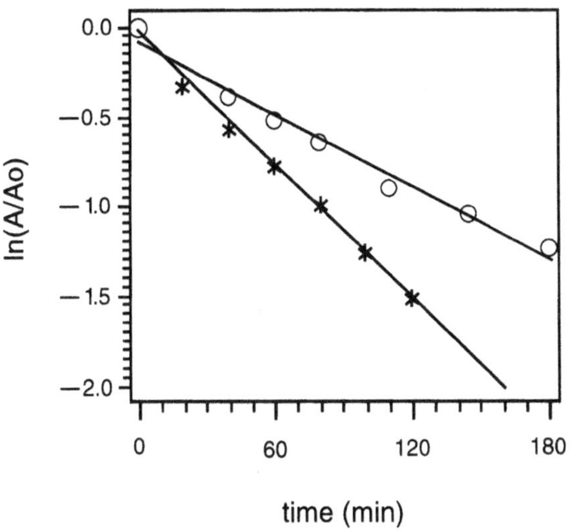

Figure 5–1 First-order pressure inactivation (850 MPa) of avocado (*) and mushroom (○) PPO at 25°C.

affected or only slightly affected in the presence of EDTA, glutathione, NaCl, or benzoate. For avocado PPO at pH 5, pressure stability was reduced by addition of EDTA, and the effect was completely attributable to the pH decrease brought about by EDTA. However, upon addition of benzoate, NaCl, or glutathione, a higher pressure was required to induce inactivation, indicating an increased pressure stability. The effect of 4-hexylresorcinol was dependent on the pressure. At pressures below 700 MPa, the pressure stability was decreased; at higher pressures, it was increased (Weemaes et al., 1999a, 1999b).

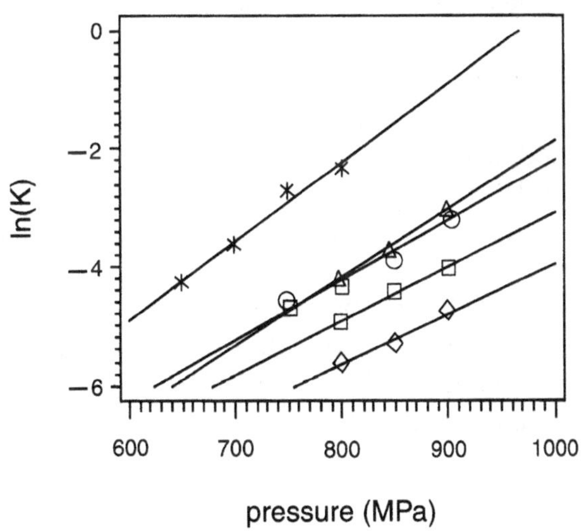

Figure 5–2 Pressure dependence of the inactivation rate constant of avocado PPO at pH 5 (*), pH 6 [McIllvaine buffer (○), phosphate buffer (Δ)], pH 7(□) and pH 8 (◊).

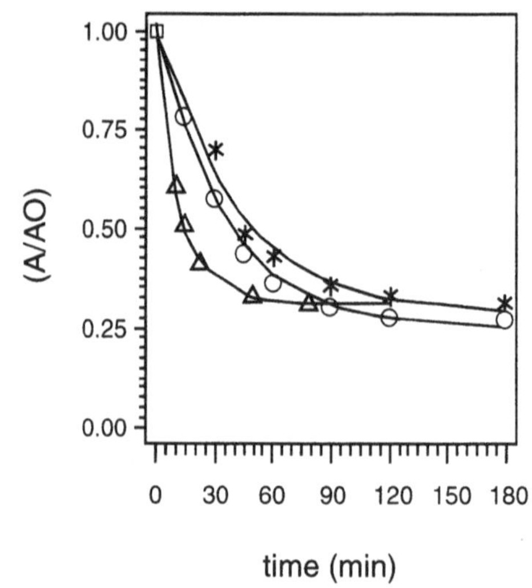

Figure 5–3 Pressure inactivation of avocado PPO in McIlvaine buffer (pH 4) at 25°C and 550 (∗), 600 (○), or 700 (Δ) MPa.

Finally, the combined effect of pressure and temperature was studied in depth for avocado PPO dissolved in phosphate buffer (0.1M) at pH 7. More than 100 combinations of constant pressure (0.1 to 900 MPa) and temperature (25°C to 77.5°C) were studied as a function of time and characterized by first-order inactivation rate constants. An antagonistic effect of pressure and temperature was noted at high temperature (>62.5°C) and low pressure (<250 MPa), indicating that low pressure exerts a protective effect against thermal inactivation. This antagonistic effect can be seen in the pressure-temperature kinetic diagram for inactivation of avocado PPO (Figure 5–4), which is a two-dimensional diagram representing combinations of constant pressure and temperature resulting in the same inactivation rate constant. Temperature dependence (E_a) of the inactivation rate constant was determined at different constant pressures and found to decrease from 320 kJ/mol at 0.1 MPa to 60 kJ/mol at 900 MPa. On the other hand, pressure dependence could not be described by the Eyring equation over the entire pressure range studied. A screening study of the combined pressure-temperature inactivation of plum, pear, and mushroom PPO revealed that not all types of PPO show identical pressure-temperature behavior. In the case of pear PPO, pressure and temperature appeared to act antagonistically in the low-temperature–high-pressure domain, a tendency that was not observed for avocado PPO (Weemaes et al., 1998c).

Pectinmethylesterase

Pectinmethylesterase (PME) (EC 3.1.1.11) catalyzes the demethylation of carboxylate groups on pectic substances, allowing them to engage in cross-linking

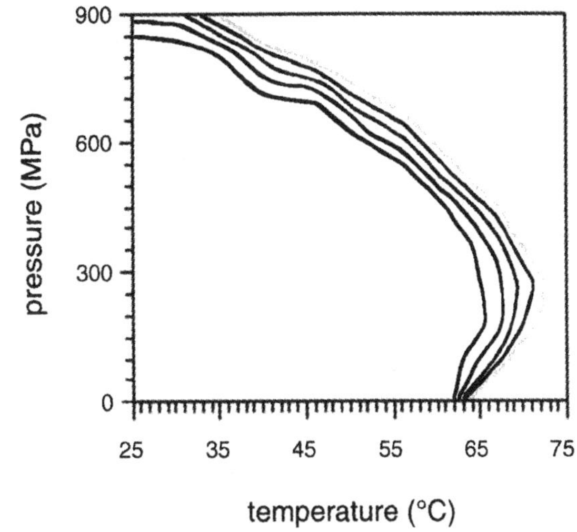

Figure 5–4 Pressure-temperature kinetic diagram for inactivation of avocado PPO in phosphate buffer at pH 7 (lower line: k_{obs} = 0.005 min^{-1}; upper line: k_{obs} = 0.015 min^{-1}).

reactions via calcium. PME is of major importance for the food industry because it induces cloud destabilization of orange juices (Seymour et al., 1991) and gelation of concentrates and loss of consistency in tomato products (Giovane et al., 1990). Threshold pressures for inactivation of PME from different sources have been reported to vary from 150 to 1200 MPa, depending on its origin and the medium in which inactivation is carried out (Table 5–5).

A detailed kinetic study on pressure activation and inactivation of PME from oranges, which examined the influence of environmental factors such as pH, Ca^{++}, sucrose, and enzyme concentration, was performed by Van den Broeck et al. (2000b). These authors studied pressure inactivation from 400 to 900 MPa at various conditions. Under all conditions studied, they observed a pressure-resistant fraction of PME, indicating that the kinetics follow a fractional conversion model (equation 8) (Figure 5–5). Pressure stability of orange PME was generally higher in more concentrated solutions. However, in the case of a very low inactivation, enzyme concentration had no significant influence on the inactivation rate constant. Acidification largely reduced the pressure stability of orange PME. As compared to distilled water, the threshold pressure at pH 3.7 (citric acid buffer) was reduced from 600 to 400 MPa. In addition, pressure sensitivity of the inactivation rate constant was slightly reduced. The effect of Ca^{++} on the inactivation rate was strongly dependent on its concentration (Table 5–6). At relatively low molarities (0.5 to 1.0 M), the inactivation rate constant increased; at higher molarity (1.5 M), a decreased inactivation was noted. In the presence of sucrose, pressure stability of PME and pressure sensitivity of the inactivation rate constants increased (Table 5–6).

Table 5–5 Pressure Effects at Room Temperature on PME from Various Origins

PME Origin	*Medium*	*Pressure Treatment*	*Activation/ Inactivation*	*Reference*
Apple	Puree	⩾1000 MPa/5 min	Inactivation	Rovere et al., 1996
Citrus	Juice	⩾300 MPa/10 min	Inactivation	Cited in Mertens and Knorr, 1992
	Juice	600 MPa/10 min	Inactivation	Ogawa et al., 1990
Grapefruit	Juice	⩾600 MPa/1 s	Inactivation	Goodner et al., 1998
Guava	Puree (Chung shan) (pH 3.8)	⩾600 MPa/15 min	Inactivation	Yen and Lin, 1996
Strawberry	Whole tissue	⩾1200 MPa/NM	Inactivation	Berg, 1996
Tomato	Puree (Nema) (pH 4.0)	⩾800 MPa/5 min	Inactivation	Crelier et al., 1995
	Aqueous solution	<400 MPa/NM	Activation	Crelier et al., 1995
	Aqueous solution	600 MPa/NM	Inactivation	Crelier et al., 1995
	Juice	500 MPa/15 min	No inactivation	Crelier et al., 1998
	Distilled water	<300 MPa/60 min	Activation	Van den Broeck et al., 2000a
	Distilled water	⩾850 MPa/20 min	Inactivation	Van den Broeck et al., 2000a
	Citric acid buffer (pH 3.5, 0.005 M)	⩾800 MPa/20 min	Inactivation	Van den Broeck et al., 2000a
	Puree (Pera) (pH 4);	150 MPa/15 min	Inactivation	Hernández and Cano, 1998
Orange	Freshly squeezed juice	⩾600 MPa/0.5 min	Inactivation	Irwe and Olsson, 1994
	Freshly squeezed juice (Hamlin)	⩾600 MPa/0.5 min	Inactivation	Parish and Olsson, 1994
	Juice (pH 3.2)	⩾270 MPa/30 min	Inactivation	Basak and Ramaswamy, 1996
	Juice (pH 3.7)	⩾300 MPa/60 min	Inactivation	Basak and Ramaswamy, 1996
	Juice	600 MPa/10 min	Inactivation	Seyderhelm et al., 1996
	Tris buffer (pH 7)	800 MPa/5 min	Inactivation	Hendrickx et al., 1998
	Juice (Salustiana) (pH 3.7)	50–200 MPa/15 min	Inactivation	Cano et al., 1997
		>200 MPa/15 min	Activation	Cano et al., 1997
	Distilled water	⩾750 MPa/10 min	Inactivation	Van den Broeck, 2000
	Citric acid buffer (pH 3.7)	⩾400 MPa/10 min	Inactivation	Van den Broeck, 2000
	Juice	⩾700 MPa/1s	Inactivation	Goodner et al., 1998
	Juice (pH 3.7)	⩾500 MPa/0.5 min	Inactivation	Parish, 1998
	Juice (Satsuma mandarin)	⩾300 MPa/10 min	Inactivation	Ogawa et al., 1990
	Water	⩾300 MPa/10 min	Inactivation	Ogawa et al., 1990
	Juice	600 MPa/5 min	Inactivation	Takahashi et al., 1993

NM = not mentioned

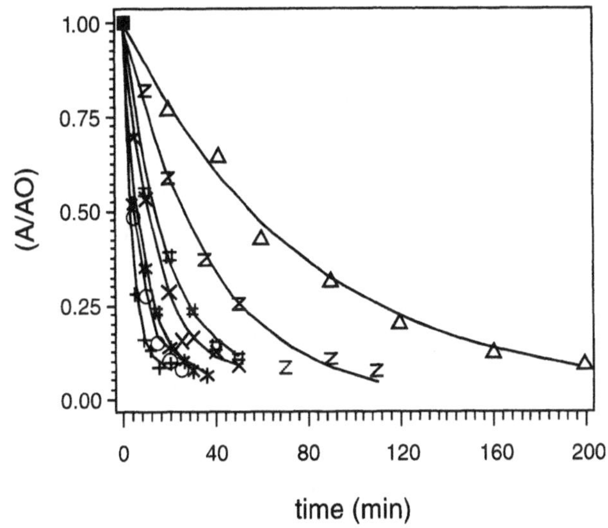

Figure 5–5 Inactivation of commercial orange PME at different combinations of pressure (MPa) and temperature (°C): (#) 0.1/57; (○) 0.1/60; (Δ) 100/57; (z) 150/60; (x) 400/60; (+) 800/35; (∗) 850/20.

A detailed kinetic study on inactivation of orange PME by combined pressure (0.1 to 900 MPa) and temperature (15°C to 65°C) has been performed (Van den Broeck et al., 2000b), revealing an antagonistic effect at low pressure (<200 to 300 MPa) and high temperature (>55°C) (Figure 5–6).

PME from tomato, on the other hand, was extremely pressure resistant, and a very pronounced antagonistic effect between pressure and high temperature was noted (Van den Broeck et al., 2000a). With pressure of 100 to 500 MPa at 60°C (a temperature that induces thermal inactivation at atmospheric pressure), inactivation of tomato PME was completely inhibited. Even at 900 MPa and 60°C, the inactivation rate was slower than for the same treatment at atmospheric pressure (Figure 5–7).

In addition to studying to the effect of high pressure on pure tomato PME in solution, the same investigators also studied the effect of high pressure in the presence of its substrate (pectin). It was noted that tomato PME retains its activ-

Table 5–6 The Effect of $CaCl_2$ and Sucrose on Pressure Inactivation of Commercial Orange PME at 25°C

	Treatment				
Characteristic	*None*	*$CaCl_2$ (0.5 M)*	*$CaCl_2$ (1.0 M)*	*$CaCl_2$ (1.5 M)*	*Sucrose (20%)*
k_{ref} (800 MPa)	0.119 ± 0.003	0.161 ± 0.010	0.149 ± 0.0007	0.083 ± 0.005	0.047 ± 0.002
V_a(cm³/mol)	−27.22 ± 0.96	−46.05 ± 3.99	−32.79 ± 2.09	−31.5 ± 2.27	−23.51 ± 0.95
% stable PME	4.9 ± 0.6	5.5 ± 1.3	4.1 ± 0.9	5.0 ± 0.9	8.3 ± 1.4

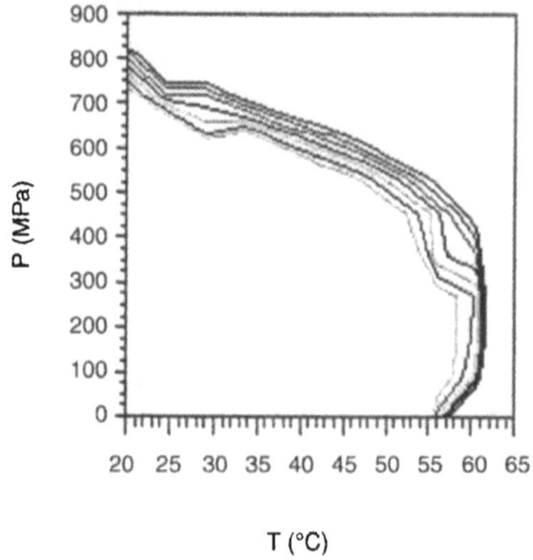

Figure 5–6 Pressure-temperature kinetic diagram for commercial orange PME in deionized water (lower line: $k_{obs} = 0.02 \text{ min}^{-1}$; upper line: $k_{obs} = 0.08 \text{ min}^{-1}$).

ity under pressure (Van den Broeck et al., 2000a). At atmospheric pressure, optimal activity was found at neutral pH and 55°C. Increasing pressure up to 300 MPa increased the enzyme activity. Maximal activity was found at about 100 to 200 MPa combined with temperature between 60°C and 65°C (Figure 5–8). Enzyme activity could be even further enhanced by addition of Ca^{++} ions (60

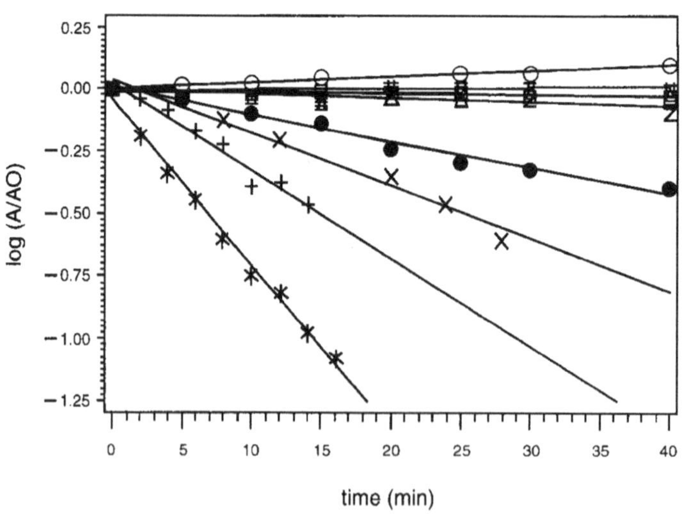

Figure 5–7 First-order inactivation of commercial tomato PME in deionized water at 60°C and pressures of 0.1 (∗), 100 (○), 200 (#), 300 (□), 500 (Δ), 700 (z), 800 (•), 850(x), and 900 (+) MPa.

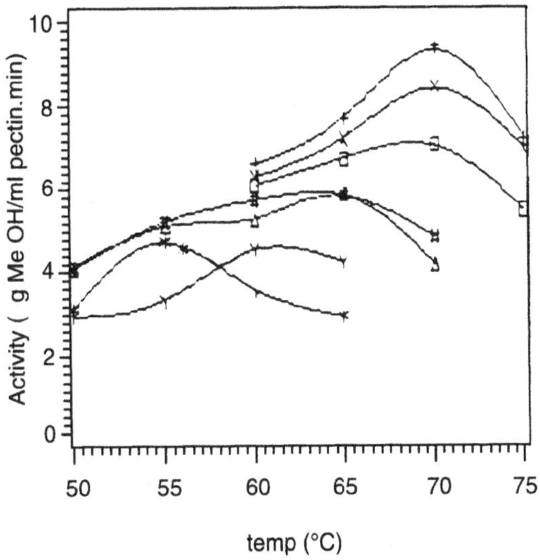

Figure 5–8 Tomato PME activity as a function of temperature in absence [0.1 (*), 100 (Δ), 200 (#), and 300 (Y) MPa] and presence of 60 mM $CaCl_2$ [100 (□), 200 (+), and 300 (x) MPa].

mM), hereby shifting the conditions for maximal activity toward higher pressure (200 to 300 MPa) and temperature (65°C to 70°C).

Lipoxygenase

Lipoxygenase (LOX) (EC 1.13.1.13) is an enzyme that catalyzes the oxygenation of polyunsaturated fatty acids into the corresponding hydroperoxides. LOX mediates, either directly or indirectly, a whole spectrum of changes in food quality including color, flavor, texture, and nutritional quality (Richardson & Hyslop, 1985; Kermasha et al., 1993). Pressure inactivation of LOX has been reported to occur between 400 and 600 MPa at room temperature (Table 5–7). Unlike for

Table 5–7 Pressure Effects at Room Temperature on LOX from Various Origins

LOX Origin	*Medium*	*Pressure Treatment*	*Activation/ Inactivation*	*Reference*
Green beans	Juice	⩾500 MPa/5 min	Inactivation	Indrawati et al., 2000a
	Whole beans	⩾500 MPa/10 min	Inactivation	Indrawati et al., 2000b
Peas	Juice	⩾400 MPa/10 min	Inactivation	Indrawati et al., 2000c
	Intact peas	⩾400 MPa/15 min	Inactivation	Indrawati et al., 2000c
Soybean	Borate buffer (pH 9, 0.2M)	>400 MPa/5 min	Inactivation	Heinisch et al., 1995
	TB (pH 9, 0.01M)	⩾525 MPa/15 min	Inactivation	Ludikhuyze et al., 1998a
	TB (pH 9, 0.01M)	⩾500 MPa/15 min	Inactivation	Indrawati et al., 1999
NM	TB (pH 7)	⩾600 MPa/2 min	Inactivation	Seyderhelm et al., 1996
	Milk	⩾600 MPa/2 min	Inactivation	Seyderhelm et al., 1996

NM = not mentioned; TB = Tris buffer

PPO and PME, activation of LOX at relatively low pressure has not yet been observed.

Seyderhelm et al. (1996) reported LOX to be markedly inactivated by a treatment at 600 MPa for 2 to 10 minutes in Tris buffer at pH 7 as well as in milk. Heinisch et al. (1995) studied the inactivation of soybean LOX in borate buffer at pH 9 at pressures up to 650 MPa and temperatures between 0°C and 75°C for 5 minutes. Based on the activity loss, these authors constructed a pressure-temperature phase diagram, revealing a slight antagonistic effect at high temperature and low pressure. Such antagonistic effect of pressure and temperature was not observed by Ludikhuyze et al. (1998a, 1998b) or Indrawati et al. (1999), who studied the inactivation of soybean LOX in Tris buffer at pressures between 0.1 and 650 MPa at temperatures between −15°C and 68°C (Figure 5–9). Maximal pressure stability of soybean LOX was observed at about 30°C, indicating that both temperature increase and temperature decrease can accelerate the inactivation at constant elevated pressure. According to both studies (Ludikhuyze et al., 1998a; Indrawati et al., 1999), pressure dependence of the first-order inactivation rate constant could be described by the Eyring equation at all constant temperatures studied. The absolute value of the activation volume was noted to decrease with increasing temperature, pointing to a reduced pressure sensitivity of the inactivation rate constant at higher temperature. Furthermore, Ludikhuyze et al. (1998a) found that pressure stability of soybean LOX is largely affected by environmental conditions. Pressure resistance increased with increasing enzyme concentration and decreased with decreasing pH (pH 9 to pH 5.4). In addition, they noted that soybean LOX was much more stable in Tris buffer than in McIllvaine

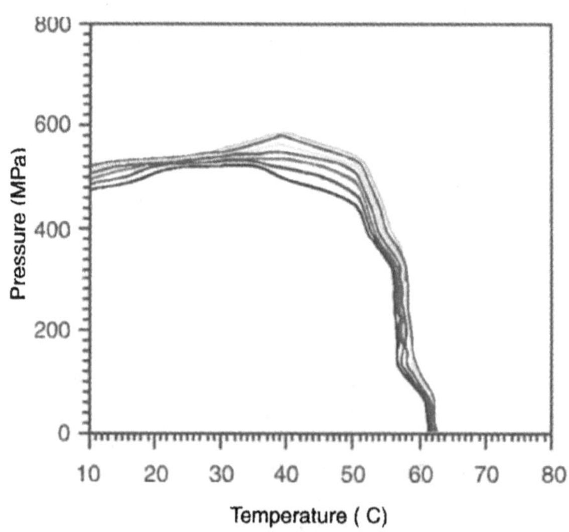

Figure 5–9 Pressure-temperature kinetic diagram for inactivation of soybean LOX in Tris buffer at pH 9 (lower line: k_{obs} = 0.03 min^{-1}; upper line: k_{obs} = 0.07 min^{-1}).

buffer. Flushing with CO_2 likewise strongly reduced pressure stability, an effect that could be completely attributed to the pH decrease caused by dissolution of CO_2 in the buffer.

LOX in crude green bean extract can be inactivated by pressure in the range of 0.1 to 700 MPa, combined with temperatures between 60°C and −10°C. Pressure dependence of the first-order inactivation rate constant could be described by the Eyring equation. The absolute value of the activation volume reached a maximum at 20°C, indicating pressure sensitivity of the inactivation rate constant to be maximal at about room temperature. To determine the temperature dependence, only the data between 20°C and 60°C were taken into account. Activation energy values decreased with increasing pressure, showing temperature sensitivity of the inactivation rate constant to be reduced at high pressure (Indrawati et al., 2000a). The level of food complexity was increased by studying inactivation of LOX in intact green beans in the same pressure-temperature domain. The threshold pressure for inactivation of LOX in intact green beans was lower than in the juice. For LOX in intact beans, the variation of pressure sensitivity with temperature was completely reversed from that of LOX in green bean juice. That is, in intact beans minimal pressure sensitivity of the inactivation rate constant was found at about 30°C (Table 5–8). As to temperature sensitivity, a decrease of the activation energy with pressure rise was noted for both LOX in green bean juice and LOX in intact green beans (Indrawati et al., 2000b). A similar study with increasing level of complexity was accomplished for LOX in green peas (Indrawati et al., in press).

Peroxidase

In vegetables, peroxidase (POD) (EC 1.11.1.7) induces negative flavor changes during storage. Moreover, POD is often used as an indicator for evaluating the efficiency of blanching processes because it is the most heat-stable

Table 5–8 Activation Volumes (cm^3/mol) for Pressure Inactivation of LOX in Green Beans at Different Constant Temperature Levels

Temperature (°C)	*Green Bean Juice*	*Intact Green Beans*
−10	−43.30 ± 5.78	ID
0	−42.59 ± 4.86	ID
10	−37.49 ± 5.03	−27.07 ± 1.99
20	−33.61 ± 4.27	ID
30	−14.75 ± 2.62	−39.30 ± 3.54
40	−16.68 ± 2.53	−28.38 ± 2.85
50	−38.62 ± 5.44	−11.31 ± 0.78

ID: Insufficient data to calculate the activation volume at the respective temperature.

enzyme in vegetables. Consequently, its inactivation indicates thermal inactivation of all other vegetable enzymes (Richardson & Hyslop, 1985; Quaglia et al., 1996).

POD is also very pressure resistant, and pressure inactivation proceeds only when POD is subjected to very high pressures. For POD from horseradish, combinations of 800 to 900 MPa with temperatures in the range of 55°C to 70°C are required to induce any significant inactivation. However, the pressure stability can be manipulated by changes in pH, with the highest stability observed at about neutral pH (Lemos, 1998). For POD extracted from carrots (pH 6 to 7), irreversible and complete loss of enzyme activity can only be achieved at 9 kbar (1 minute). The pH of the crude juice strongly affects the extent of enzyme inactivation, with a maximal effect at pH 6 (Anese et al., 1995). POD extracted from Satsuma mandarin juice retained 70% residual activity after treatment at 600 MPa and 23°C for 10 minutes (Seyderhelm et al., 1996). The maximal extent of POD inactivation in green peas was obtained at 900 MPa for 10 minutes, amounting to an activity reduction of 88%. Combination with thermal treatment enhanced the inactivation at 600 MPa, but no significant differences were noticed at 700 MPa. This may be due to the presence of isozymes, reacting in different ways to pressure as well as to heat. Moreover, further increase in treatment time seemed to have no significant effect (Quaglia et al., 1996). For POD from apples, a 5-minute treatment at a pressure exceeding 1000 MPa resulted in inactivation (Rovere et al., 1996). POD in guava puree was reported to be somewhat less pressure stable, and it could be inactivated at room temperature by a treatment of 15 minutes at 600 MPa (Yen & Lin, 1996). Tomato POD was not affected by a treatment of 500 MPa, even at 60°C. Furthermore, it was observed that the extent of POD inactivation did not vary much with the type of sample (i.e., crushed sample, whole vegetable, or liquid extract) (Crelier et al., 1998).

However, the extreme pressure resistance of POD is rebutted by the results of Cano et al. (1997). They observed POD in strawberry puree to be increasingly inactivated by treatments up to 300 MPa at 20°C for 15 minutes. Above this pressure, POD activity was slightly increased. On the other hand, above 45°C, a decrease in POD activity was found at all pressures (50 to 400 MPa). In orange juice, POD activity decreased continuously up to 400 MPa (15-minute processing time) at room temperature. At 32°C, a maximal inactivation rate was found. High pressure treatments at 32°C to 60°C adversely increased POD activity in orange juice.

In case of lactoperoxidase, the major peroxidase in milk, the barotolerance at temperatures ranging from 10°C to 30°C was strongly dependent on the surrounding medium. Pressure inactivation was much more pronounced in Tris buffer (pH 7) than in milk. In Tris buffer, initial activity was reduced by 70% at 600 MPa and 25°C for 2 minutes (Seyderhelm et al., 1996). In milk, Seyderhelm and colleagues (1996) found only 30% inactivation under the same conditions, while no inactivation was found by Ludikhuyze et al. (2001). According to Ludikhuyze, pressure up to 800 MPa combined with temperature up to 50°C did not result in any activity loss. Moreover, it has been noted that application of pressure (100 to 700 MPa) at temperatures inducing thermal inactivation at atmospheric pressure (69°C to 75°C) completely counteracts thermal inactivation. In other words, pressure exerts a protective effect against thermal inactivation.

Myrosinase

Myrosinase (MYR) (EC 3.2.3.1), especially found in *Cruciferae*, catalyzes the hydrolysis of glucosinolates, a group of sulfur-containing pseudo-glycosides. This hydrolysis results in the formation of sulfate, D-glucose, and a series of sulfur- or nitrogen-containing compounds such as isothiocyanates, thiocyanates, nitriles, and thiones. Isothiocyanates are responsible for the specific flavor and aroma of several cruciferous vegetables (Gatfield & Sand, 1983; Wilkinson et al., 1984). Recently, products formed from hydrolysis of indole glucosinolates have received special attention because of their anticarcinogenic functions (Dunford & Temelli, 1996; Wathelet et al., 1996). Hydrolysis products can also have undesirable effects owing to their pungency, bitterness, and toxic or goitrogenic activity. Hence control of MYR activity is highly desirable.

Very little information is available on pressure inactivation of MYR. Ludikhuyze et al. (1999) studied pressure inactivation of MYR from broccoli. At room temperature (20°C), inactivation occurred in the pressure range of 200 to 500 MPa and could be accurately described by a consecutive-step model. At pressures below 250 MPa, the activity remained constant, and no post-process activation was noted. Significant inactivation was observed between 300 and 500 MPa. These findings suggest that MYR is more pressure labile than the other enzymes discussed in this chapter. At 35°C, a temperature at which thermal inactivation at atmospheric pressure occurs slowly, an antagonistic effect between pressure and temperature was observed. Application of low pressure (less than 350 MPa) retarded thermal inactivation, and this effect was maximal at about 350 MPa.

Ludikhuyze et al. (2000a) investigated the effect of pressure on MYR in the presence of its substrate (sinigrin). At atmospheric pressure, MYR was optimally active at about 30°C. In the temperature range of 20°C to 40°C, application of pressure up to 100 MPa slightly enhanced the activity of MYR. The activity of MYR was completely lost at a pressure of 200 MPa (Figure 5–10), probably because of the low pressure stability of the enzyme. Indeed, it has been verified that pressure up to 400 MPa does not result in hydrolysis of the substrate in absence of the enzyme (Ludikhuyze et al., 2000a).

KINETIC MODELS TO DESCRIBE PRESSURE-TEMPERATURE INACTIVATION OF ENZYMES RELATED TO FOOD QUALITY

Pressure-temperature—induced inactivation of enzymes and microorganisms, denaturation of proteins, and gelatinization of starches often display an elliptical iso-rate contour diagram as a function of pressure and temperature (for more

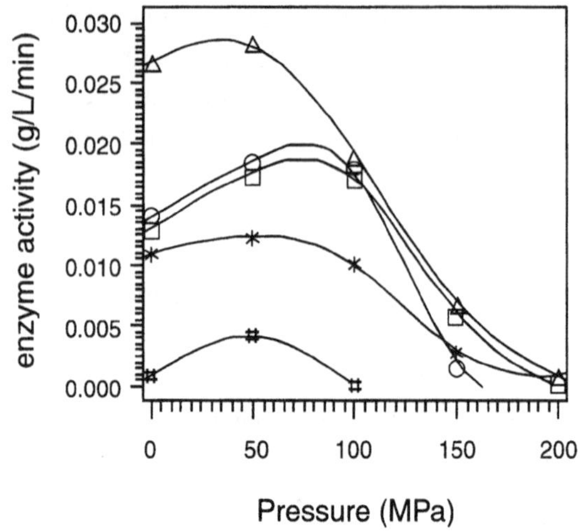

Figure 5–10 Activity of myrosinase from broccoli in phosphate buffer at pH 6.55 as a function of pressure at 20°C (∗), 25°C (□), 30°C (Δ), 40°C (○), and 50°C (#).

information, see Chapters 2 and 7). The various mathematical models that have been reported to describe such elliptical contour diagrams are discussed below. Available mathematical models and model parameter estimates to describe the pressure-temperature dependence of enzyme and microbial inactivation are summarized in Table 5–9.

Modified Thermodynamic Equation

The thermodynamic equation describing the behavior of any system undergoing a change in a pressure and temperature can be represented in the following equation (Hawley, 1971; Morild, 1981):

$$d(\Delta G) = -\Delta S dT + \Delta V dP \tag{13}$$

Because the entropy change (ΔS) and the volume change (ΔV) vary with pressure and temperature (equations 14 and 15), equation 13 can be reformulated as equation 16 (Hawley, 1971; Morild, 1981).

$$d(\Delta S) = \left(\frac{\delta \Delta S}{\delta T}\right)_P dT + \left(\frac{\delta \Delta S}{\delta P}\right)_T dP \tag{14}$$

$$d(\Delta V) = \left(\frac{\delta \Delta V}{\delta T}\right)_P dT + \left(\frac{\delta \Delta V}{\delta P}\right)_T dP \tag{15}$$

$$\Delta G = \Delta G_o + \Delta V_o(P - P_o) - \Delta S_o(T - T_o) + \frac{1}{2}\Delta\kappa(P - P_o)^2 + \Delta\zeta(P - P_o)(T - T_o) - \Delta C_p\left[T\left(\ln\frac{T}{T_o} - 1\right) + T_o\right] \tag{16}$$

Table 5–9 State-of-the-Art Kinetic Models to Describe Combined Pressure-Temperature Inactivation of Enzymes and Microorganisms under Isobaric Isothermal Conditions

Object	*Medium*	*Kinetic Model*	*Estimated Parameters*	*P/T Area Studied*	*Reference*
Enzyme inactivation					
Bacillus subtilis α amylase	Diluted in Tris HCl buffer (0.01M, pH 8.6)	$\ln(k_{obs}) = (aP^2 + bP + c) - \left[\frac{d\exp(-eP)}{R}\left(\frac{1}{T} - \frac{1}{T_{ref}}\right)\right]$	$a = -2.98*10^{-5}$; $b = 4.13*10^{-2}$; $c = -16.63$; $d = 332.1$; $e = 2.82*10^{-3}$; $T_{ref} = 40°C$	Up to 750 MPa; 25°C–73°C	Ludikhuyze et al., (1997)
Orange PME	Diluted in deionized water	$\ln(k_{obs}) = (a + bP + cP^2 + dP^3) + \left[\frac{f - eP}{R}\left(\frac{1}{T_{ref}} - \frac{1}{T}\right)\right]$	$a = -1.88$; $b = -17.55*10^{-3}$; $c = 53.27*10^{-6}$; $d = -35.95*10^{-9}$; $e = 0.348$; $f = 325.12$; $T_{ref} = 343.15K$	Up to 800 MPa; 15°C–65°C	Van den Broeck et al. (2000b)
	Diluted in citric acid buffer (5 mM, pH 3.7)	$\ln(k_{obs}) = (a + bP + cP^2 + dP^3) + \left[\frac{f - eP}{R}\left(\frac{1}{T_{ref}} - \frac{1}{T}\right)\right]$	$a = -2.39$; $b = -19.00*10^{-3}$; $c = 55.20*10^{-6}$; $d = -38.50*10^{-9}$; $e = 0.248$; $f = 193.44$; $T_{ref} = 343.15K$	Up to 800 MPa; 15°C–65°C	Van den Broeck et al. (2000b)

continues

Table 5–9 continued

Object	*Medium*	*Kinetic Model*	*Estimated Parameters*	*P/T Area Studied*	*Reference*
Avocado PPO	Diluted in phosphate buffer (0.1 M, pH 7)	$\ln(k_{obs}) = (a + bP + cP^2 + dP^3) - \left[\frac{f\exp(-eP)}{R}\left(\frac{1}{T} - \frac{1}{T_{ref}}\right)\right]$	$a = -2.42$; $b = -17.2*10^{-3}$; $c = 41.1*10^{-6}$; $d = -23.3*10^{-9}$; $e = 16.8*10^{-4}$; $f = 324.30$; $T_{ref} = 343.15K$	Up to 900 MPa; 25°C–77.5°C	Weemaes et al. (1998c)
Alkaline phosphatase	Raw milk	$\log(D_{obs}) = (aP^2 + bP + c) -((T - T_{ref})/(dP^3 + eP + f))$	$a = -1.05*10^{-5}$; $b = 3.88*10^{-3}$; $c = 3.33$; $d = -2.7*10^{-8}$; $e = 5.3*10^{-2}$; $f = 4.88$; $T_{ref} = 50°C$	Up to 725 MPa; 25°C–63°C	Ludikhuyze et al. (2000b)
Soybean LOX	Diluted in Tris HCl buffer (0.01 M, pH 9)	$\ln(k_{obs}) = (aT^2 + bT + c) - \left[\frac{(dT\exp(-eT))}{R(T+273)}(P - P_{ref})\right]$	$a = 2.66*10^{-5}$; $b = -1.39*10^{-1}$; $c = -3.12$; $d = -15.6$; $e = 7.10*10^{-2}$; $P_{ref} = 500$ MPa	Up to 650 MPa; 10°C–64°C	Ludikhuyze et al. (1998b)
		$\ln(k_{obs}) = \ln(k_0) - \frac{\Delta V_0^{\neq}}{RT}(P - P_o) + \frac{\Delta S_0^{\neq}}{RT}(T - T_o) - \frac{1}{2}\frac{\Delta\kappa^{\neq}}{RT}(P - P_o)^2 - \frac{\Delta\zeta^{\neq}}{RT}(P - P_o)(T - T_o) + \frac{\Delta C_P^{\neq}}{RT}\left[T\left(\ln\frac{T}{T_o} - 1\right) + T_o\right]$	$k_o = 1.28*10^{-2}$; $\Delta V_0^{\neq} = -34.195$; $\Delta S_0^{\neq} = -20.65$; $\Delta\kappa^{\neq} = -1.40*10^{-2}$; $\Delta\zeta^{\neq} = 0.640$; $\Delta C_p^{\neq} = 3046.55$; $T_{ref} = 298K$; $P_{ref} = 500$ MPa	Up to 650 MPa; −15°C–68°C	Indrawati et al. (1999)

Green bean LOX	In green bean juice	$k_{obs} = (aT^2 + bT + c) \exp\left[\frac{(d(T - SF)\exp(e(T - SF)))}{RT}(P - P_{ref})\right]$	$a = 3.53{*}10^{-5}$; $b = -2.03{*}10^{-2}$; $c = 2.93$; $d=2.75$; $e = -4.01{*}10^{-2}$; $SF = 259.50$; $P_{ref} = 500$ MPa; T in K	200–700 MPa; −10°C–60°C	Indrawati et al. (2000a)
	In green bean juice	$\ln(k_{obs}) = \ln(k_0) - \frac{\Delta V_0^{\neq}}{RT}(P - P_o) + \frac{\Delta S_0^{\neq}}{RT}(T - T_o) - \frac{1}{2}\frac{\Delta\kappa^{\neq}}{RT}(P - P_o)^2 - \frac{\Delta\zeta^{\neq}}{RT}(P - P_o)(T - T_o) + \frac{\Delta C_P^{\neq}}{RT}\left[T\left(\ln\frac{T}{T_o} - 1\right) + T_o\right]$	$k_o = 1.34{*}10^{-2}$; $\Delta V_0^{\neq} = -30.814$; $\Delta S_0^{\neq} = 90.63$; $\Delta\kappa^{\neq} = -0.54{*}10^{-2}$; $\Delta\zeta^{\neq} = 0.443$; $\Delta C_p^{\neq} = 2466.71$; $T_{ref} = 298$K; $P_{ref} = 500$ MPa	Up to 650 MPa; −10°C–70°C	Indrawati et al. (2000b)
	Intact green beans	$\ln(k_{obs}) = \ln(k_0) - \frac{\Delta V_0^{\neq}}{RT}(P - P_o) + \frac{\Delta S_0^{\neq}}{RT}(T - T_o) - \frac{1}{2}\frac{\Delta\kappa^{\neq}}{RT}(P - P_o)^2 - \frac{\Delta\zeta^{\neq}}{RT}(P - P_o)(T - T_o) + \frac{\Delta C_P^{\neq}}{RT}\left[T\left(\ln\frac{T}{T_o} - 1\right) + T_o\right]$	$k_0 = 2.47{*}10^{-2}$, $\Delta V_o^{\neq} = -35.672$; $\Delta S_o^{\neq} = 139.41$; $\Delta\kappa^{\neq} = -7.34{*}10^{-2}$; $\Delta\zeta^{\neq} = 0.20{*}10^{-1}$; $\Delta C_p^{\neq} = 2811.82$; $T_{ref} = 298$K; $P_{ref} = 500$ MPa	Up to 550 MPa; −10 to 70°C	Indrawati et al. (2000b)

continues

Table 5–9 continued

Object	*Medium*	*Kinetic Model*	*Estimated Parameters*	*P/T Area Studied*	*Reference*
Green pea LOX	In green pea juice	$\ln(k_{obs}) = \ln(k_0) - \frac{\Delta V_0^{\neq}}{RT}(P - P_o) + \frac{\Delta S_0^{\neq}}{RT}(T - T_o) - \frac{1}{2}\frac{\Delta\kappa^{\neq}}{RT}(P - P_o)^2 - \frac{\Delta\zeta^{\neq}}{RT}(P - P_o)(T - T_o) + \frac{\Delta C_P^{\neq}}{RT}\left[T\left(\ln\frac{T}{T_o} - 1\right) + T_o\right]$	$k_o = 2.45*10^{-2}$, $\Delta V_0^{\neq} = -38.182$; $\Delta S_0^{\neq} = 13.79$; $\Delta\kappa^{\neq} = -9.37*10^{-2}$; $\Delta\zeta^{\neq} = 1.15*10^{-1}$; $\Delta C_p^{\neq} = 1837.42$; $T_{ref} = 298K$; $P_{ref} = 500$ MPa	Up to 650 MPa; −15 to 70°C	Indrawati et al. (2000c)
	Intact green peas	$\ln(k_{obs}) = \ln(K_0) - \frac{\Delta V_0^{\neq}}{RT}(P - P_o) + \frac{\Delta S_0^{\neq}}{RT}(T - T_o) - \frac{1}{2}\frac{\Delta\kappa^{\neq}}{RT}(P - P_o)^2 - \frac{\Delta\zeta^{\neq}}{RT}(P - P_o)(T - T_o) + \frac{\Delta C_P^{\neq}}{RT}\left[T\left(\ln\frac{T}{T_o} - 1\right) + T_o\right]$	$k_0 = 5.79*10^{-2}$, $\Delta V_0^{\neq} = -50.763$; $\Delta S_0^{\neq} = -21.19$; $\Delta\kappa^{\neq} = -0.1382$; $\Delta\zeta^{\neq} = 0.23$; $\Delta C_p^{\neq} = 1058.61$; $T_{ref} = 298K$; $P_{ref} = 500$ MPa	Up to 500 MPa; −10°C–70°C	Indrawati et al. (2000c)
Microorganism inactivation *Lactobacillus casei* YIT9018	10 mM HEPES (pH 5.3)	$\log(k_{obs}) = \log(k_0) + \Delta V_o(P - P_o) - \Delta S_o(T - T_o) + \frac{1}{2}\Delta\beta(P - P_o)^2 + 2\Delta\alpha(P - P_o)(T - T_o) - \Delta Cp\left[T\left(\ln\frac{T}{T_o} - 1\right) + T_o\right]$	$\log k_o = -4.54$, $\Delta V_o = 1.14*10^{-2}$; $\Delta S_o = -1.27*10^{-2}$; $\Delta\beta = 1.41*10^{-5}$; $\Delta\alpha = -1.18*10^{-4}$; $\Delta Cp = -5.91*10^{-1}$; $T_o = 273.15K$; $P_o = 0.1$ MPa	Up to 400 MPa; 0°C–50°C	Sonoike et al. (1992)

Escherichia coli JCM1649	10 mM HEPES (pH 5.3)	$\log(k) = \log(k_0) + \Delta V_o(P - P_o) - \Delta S_o(T - T_o) + \frac{1}{2}\Delta\beta(P - P_o)^2 + 2\Delta\alpha(P - P_o)(T - T_o) - \Delta Cp\left[T\left(\ln\frac{T}{T_o} - 1\right) + T_o\right]$	$\log k_o = -4.29$, $\Delta V_o = 1.33*10^{-2}$; $\Delta S_o = -1.77*10^{-2}$; $\Delta\beta = 1.59*10^{-5}$; $\Delta\alpha = -1.03*10^{-4}$; $\Delta Cp = -5.34*10^{-1}$; $T_o = 273.15K$; $P_o = 0.1$ MPa	Up to 400 MPa; 0°C–50°C	Sonoike et al. (1992)
Saccharomyces cerevisiae (IFO0234)	0.85% NaCl	$\log(k_{obs}) = a + b(P - P_o) + c(T - T_o) + d(P - P_o)^2 + e(P - P_o)(T - T_o) + f(T - T_o)^2$	$a = -4.26$; $b = 1.25*10^{-2}$; $c = -3.37*10^{-2}$; $d = 8.55*10^{-6}$; $e = -7.55*10^{-5}$; $f = 1.42*10^{-3}$; $T_o = 273.15K$; $P_o = 0.1$ MPa	120–300 MPa; −20°C–50°C	Hashizume et al. (1995)
Zygosaccharomyces bailii	Tris HCl buffer (40 mM, pH 6.5)	$\ln(k_{obs}) = a + b(P - P_{ref}) + c(P - P_{ref})^2 + d(T - T_{ref}) + e(T - T_{ref})^2$	$a = -3.10$; $b = 37.81*10^{-3}$; $c = -5.42*10^{-5}$; $e = 55.37*10^{-3}$; $e = 3.78*10^{-3}$; $T_{ref} = 25°C$; $P_{ref} = 210$ MPa	120–320 MPa; −5°C–45°C	Reyns et al. (2000)

where ΔC_p is the heat capacity change $(T\delta\Delta S/\delta T)_P$, $\Delta\zeta$ is the thermal expansibility factor $((\delta\Delta V/\delta T)_P = -(\delta\Delta S/\delta P)_T)$, and $\Delta\kappa$ is the compressibility factor $(\delta\Delta V/\delta P)_T$. This thermodynamic model can be converted into a kinetic model (Weemaes et al., 1998c; Indrawati et al., 1999) using the transition state theory of Eyring, suggesting that enzyme inactivation is accompanied by the formation of a metastable activated state ($^{\neq}$) that exists in equilibrium with the native enzyme. This conversion is based on the substitution of equations 17 and 18 in equation 16, yielding equation 19.

$$\Delta G^{\neq} = -RT\ln(K^{\neq}) \qquad (17)$$

$$K^{\neq} = \frac{kh}{rk_B T} \qquad (18)$$

$$\ln(k_{obs}) = \ln(k_0) - \frac{\Delta V_0^{\neq}}{RT}(P - P_o) + \frac{\Delta S_0^{\neq}}{RT}(T - T_o) - \frac{1}{2}\frac{\Delta\kappa^{\neq}}{RT}(P - P_o)^2$$
$$- \frac{\Delta\zeta^{\neq}}{RT}(P - \mathrm{P}_o)(T - T_o) + \frac{\Delta C_p^{\neq}}{RT}\left[T\left(\ln\frac{T}{T_o} - 1\right) + T_o\right] \qquad (19)$$

According to Morild (1981), equation 16 should, in principle, describe most phenomena with regard to their pressure-temperature behavior. Evaluation of the unknown parameters of equation 16 (and hence of equation 19), by means of computerized curve fitting requires data from rather large pressure intervals (covering about 500 MPa) and temperature intervals (covering about 100°C). From a practical point of view, this requirement may not always be realizable and may partially explain the poor parameter estimates reported in previous studies (Hawley, 1971; Sonoike et al., 1992; Weemaes et al., 1998c).

Modified Arrhenius, Eyring, TDT, or PDT Relation

The Arrhenius, Eyring, TDT, or PDT equation can be used as a starting point to develop a mathematical model for pressure-temperature inactivation if it is valid in the entire pressure-temperature area studied. When using the Arrhenius model (equation 9) as a starting point, the pressure-dependent parameters E_a and $k_{obs,Tref}$ are replaced by mathematical expressions, reflecting their pressure dependencies (equations 20 and 21), and yielding equation (22).

$$\ln(k_{obs}) = \ln(k_{obs,Tref}) + \left(\frac{E_a}{R}\left(\frac{1}{T_{ref}} - \frac{1}{T}\right)\right) \qquad (9)$$

$$E_a = f_1(P) \qquad (20)$$

$$k_{obs,Tref} = f_2(P) \qquad (21)$$

These mathematical expressions (linear, exponential, quadratic, polynomial, or even more complex mathematical relations) can be identified by means of curve fitting.

$$\ln(k_{obs}) = \ln(f_2(P)) + \left(\frac{f_1(P)}{R}\left(\frac{1}{T_{ref}} - \frac{1}{T}\right)\right) \tag{22}$$

This approach has been successfully applied to describe the combined pressure-temperature inactivation of *Bacillus subtilis* α-amylase (BSAA) (Ludikhuyze et al., 1997), avocado PPO (Weemaes et al., 1998c), and orange PME (Van den Broeck et al., 2000b).

In an analogous way, the Eyring model (equation 10) can be used as a starting point to describe the pressure-temperature dependence of inactivation rate constants. The temperature-dependent parameters V_a and $k_{obs,Pref}$, are replaced in the Eyring equation by mathematical expressions reflecting their temperature dependence, obtaining equation 23. This method has been successfully applied to describe LOX inactivation by combined pressure and temperature treatments (Ludikhuyze et al., 1998b; Indrawati et al., 2000a).

$$\ln(k_{obs}) = \ln(f_2(T)) + \left(\frac{f_1(T)}{RT}\left(P_{ref} - P\right)\right) \tag{23}$$

For pressure-temperature inactivation of alkaline phosphatase (ALP) in raw milk, a suitable kinetic model has been successfully developed on the basis of the TDT model (Ludikhuyze et al., 2000b).

Elliptical Equation

Because the shape of a pressure-temperature kinetic diagram for microbial or enzyme inactivation is often elliptical, the mathematical equation of an ellipse (equation 24) can be used to describe the inactivation rate constant as a function of pressure and temperature. This equation is very similar to equation 19, except for the quadratic term in temperature.

$$\log(k_{obs}) = a + b(P - P_o) + c(T - T_o) + d(P - P_o)^2$$
$$+ e(P - P_o)(T - T_o) + f(T - T_o)^2 \tag{24}$$

This approach has been applied to describe inactivation of microorganisms caused by combined pressure-temperature treatments (Hashizume et al., 1995; Reyns et al., 2000).

Evaluation of the Available Inactivation Models

A disadvantage of constructing mathematical models based on the Arrhenius, Eyring, TDT, or PDT relationship by substituting the pressure- or temperature-dependent parameters with empirical equations (such as linear, exponential, quadratic, or polynomial relations) is that the resulting models are only valid within the experimental domain studied. Extrapolation to predict inactivation rates outside the experimental domain might, depending on the incorporated empirical equations, lead to an inflection of the iso-rate contour diagram, as illustrated in

Figure 5–11 for BSAA (Ludikhuyze et al., 1997), avocado PPO (Weemaes et al., 1998c), soybean and green bean LOX (Ludikhuyze et al., 1998b; Indrawati et al., 2000a), and orange PME (Van den Broeck et al., 2000b).

Because of the disadvantage of the fore-mentioned modeling approach, we endeavored to model the pressure-temperature inactivation of all enzymes related to food quality using the same mathematical expression. Both the modified thermodynamic model (equation 19) and the elliptical model (equation 24) were selected for this purpose. The estimated model parameters, the standard error, as well as some statistical information about the quality of fit and the residu-plots are summarized in Table 5–10 for all enzymes. Both for the modified thermodynamic equation and for the elliptical equation, the percentage standard error for some estimated model parameters is high. This is probably caused by the limited amount of inactivation data available and the limited range of pressure and temperature studied (Morild, 1981). In some cases, the residu-plot analysis are not randomly distributed. Notwithstanding these statistical disadvantages, the following was observed for all enzymes: (1) the different models describe a similar shape of the iso-rate contour; and (2) extrapolation of the modified thermodynamic or elliptical equation outside the experimental domain does not cause an inflection of the iso-rate contour diagram. As an example, the iso-rate contour diagrams for pressure-temperature inactivation of avocado PPO simulated using different mathematical expressions, are combined in Figure 5–12.

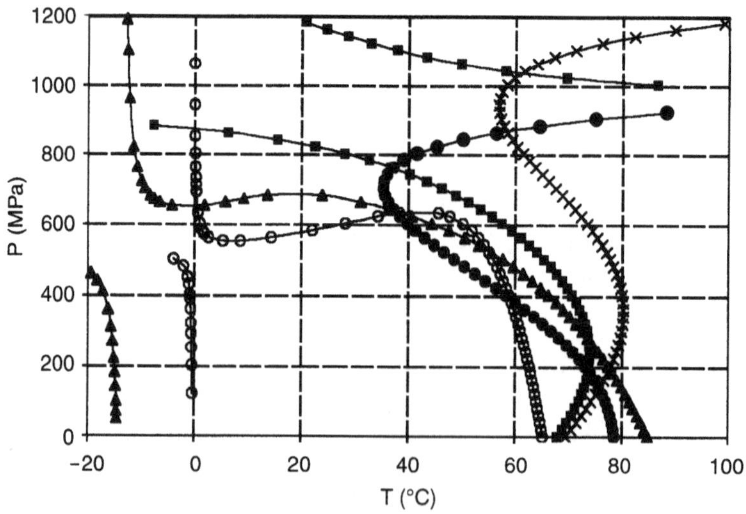

Figure 5–11 Extrapolated pressure-temperature iso-rate ($k = 7.68 \times 10^{-2}$ min^{-1}) contour diagrams using the mathematical models and model parameter estimates given in Table 5–9. BSAA (●), PME (■), PPO (×), soybean LOX (○), green bean LOX (▲).

Table 5–10 Estimated Model Parameters of Modified Thermodynamic Equation and Elliptical Equation to Describe Pressure-Temperature Inactivation of Enzymes

	Estimated Kinetic Parameters Based on Equation 19 (P_{ref} = 500 MPa and T_{ref} = 25°C)				*Estimated Kinetic Parameters Based on Equation 24 (P_o = 500 MPa and T_o = 25°C)*			
Enzyme	*Estimated Parameters*	*SE (%)*	*Fitting Quality*	*Residual Plot Analysis*	*Estimated Parameters*	*SE (%)*	*Fitting Quality*	*Residual Plot Analysis*
Bacillus subtilis α amylase (Ludikhuyze et al., 1997)	$\Delta\kappa^{\neq}=-4.55*10^{-2}$	123.91	Corrected	Temperature vs. residual* (+)	$a=-1.80$	15.10	Corrected	Temperature vs. residual* (+)
	$\Delta V_o^{\neq}=-10.62$	130.60	$r^2=0.980$,	Pressure vs. residual (+)	$b=8.93*10^{-4}$	246.57	$r^2=0.981$,	Pressure vs. residual (+)
	$\Delta S_o^{\neq}=17.00$	1238.49	SD = 0.485	Predicted ln(k) vs. residual (−)	$c=-9.63*10^{-3}$	325.68	SD = 0.204	Predicted ln(k) vs. residual (−)
	$\Delta\zeta^{\neq}=-0.52$	141.82			$d=5.75*10^{-6}$	76.79		
	$k_o=1.19*10^{-2}$	69.68			$e=1.34*10^{-4}$	86.74		
	$\Delta Cp^{\neq}=3587.54$	94.52			$f=1.25*10^{-3}$	66.61		
Commercial orange PME (Van den Broeck et al., 2000b)	$\Delta\kappa^{\neq}=-8.94*10^{-2}$	19.02	Corrected	Temperature vs. residual* (−)	$a=-2.76$	9.86	Corrected	Temperature vs. residual* (+)
	$\Delta V_o^{\neq}=-19.487$	36.88	$r^2=0.967$,	Pressure vs. residual (−)	$b=3.50*10^{-3}$	32.49	$r^2=0.97$,	Pressure vs. residual (+)
	$\Delta S_o^{\neq}=193.51$	74.14	SD = 0.511	Predicted ln(k) vs. residual (−)	$c=3.07*10^{-2}$	71.82	SD = 0.21	Predicted ln(k) vs. residual (−)
	$\Delta\zeta^{\neq}=-0.143$	217.46			$d=7.36*10^{-6}$	17.61		
	$k_o=0.17*10^{-2}$	67.65			$e=1.52*10^{-5}$	320.58		
	$\Delta Cp^{\neq}=2805.24$	68.73			$f=6.94*10^{-4}$	70.46		
Self-purified orange PME (Van den Broeck et al., 2000b)	$\Delta\kappa^{\neq}=-0.1034$	15.60	Corrected	Temperature vs. residual* (−)	$a=-2.16$	3.64	Corrected	Temperature vs. residual* (+)
	$\Delta V_o^{\neq}=-17.094$	22.67	$r^2=0.985$,	Pressure vs. residual (+)	$b=3.09*10^{-3}$	19.64	$r^2=0.987$,	Pressure vs. residual (+)
	$\Delta S_o^{\neq}=31.47$	169.68	SD = 0.401	Predicted ln(k) vs. residual (−)	$c=3.97*10^{-3}$	211.36	SD = 0.159	Predicted ln(k) vs. residual (−)
	$\Delta\zeta^{\neq}=-0.187$	122.40			$d=8.45*10^{-6}$	14.35		
	$k_o=0.67*10^{-2}$	20.19			$e=2.48*10^{-5}$	143.34		
	$\Delta Cp^{\neq}=3626.79$	32.11			$f=9.77*10^{-4}$	30.49		
Avocado PPO (Weemaes et al., 1998c)	$\Delta\kappa^{\neq}=-6.81*10^{-2}$	10.19	Corrected	Temperature vs. residual* (−)	$a=-4.47$	5.07	Corrected	Temperature vs. residual* (−)
	$\Delta V_o^{\neq}=-23.89$	18.46	$r^2=0.988$,	Pressure vs. residual (+)	$b=4.00*10^{-3}$	16.95	$r^2=0.989$,	Pressure vs. residual (+)
	$\Delta S_o^{\neq}=287.39$	30.92	SD = 0.476	Predicted ln(k) vs. residual (−)	$c=4.62*10^{-2}$	28.06	SD = 0.197	Predicted ln(k) vs. residual (−)
	$\Delta\zeta^{\neq}=0.213$	58.78			$d=5.39*10^{-6}$	9.59		
	$k_o=2.60*10^{-5}$	58.19			$e=-3.91*10^{-5}$	48.86		
	$\Delta Cp^{\neq}=2122.73$	38.90			$f=4.01*10^{-4}$	49.89		

continues

Table 5–10 continued

	Estimated Kinetic Parameters Based on Equation 19 (P_{ref} = 500 MPa and T_{ref} = 25°C)				Estimated Kinetic Parameters Based on Equation 24 (P_o = 500 MPa and T_o = 25°C)			
Enzyme	*Estimated Parameters*	*SE (%)*	*Fitting Quality*	*Residual Plot Analysis*	*Estimated Parameters*	*SE (%)*	*Fitting Quality*	*Residual Plot Analysis*
Soybean LOX	$\Delta\kappa^{\neq} = 2.17*10^{-2}$	61.89	Corrected	Temperature vs. residual* (−)	$a = -7.72*10^{-2}$	59.51	Corrected	Temperature vs. residual* (−)
(Ludikhuyze et al.,	$\Delta V_o^{\neq} = -63.321$	5.14	$r^2 = 0.990$,	Pressure vs. residual (−)	$b = 1.08*10^{-2}$	5.29	$r^2 = 0.973$,	Pressure vs. residual (−)
1998b)	$\Delta S_o^{\neq} = 5.35$	272.515	SD = 0.315	Predicted ln(k) vs. residual (−)	$c = -1.07*10^{-3}$	254.44	SD = 0.141	Predicted ln(k) vs. residual (−)
	$\Delta\zeta^{\neq} = 1.595$	8.24			$d = -1.97*10^{-6}$	56.86		
	$k_o = 7.77*10^{-3}$	8.60			$e = -2.7310^{-4}$	8.77		
	$\Delta Cp^{\neq} = 3367.33$	13.98			$f = 8.92*10^{-4}$	15.22		
Soybean LOX	$\Delta\kappa^{\neq} = -1.40*10^{-2}$	75.71	Corrected	Temperature vs. residual* (−)	$a = -1.86$	2.08	Corrected	Temperature vs. residual* (−)
(Indrawati et al.,	$\Delta V_o^{\neq} = -34.195$	6.50	$r^2 = 0.985$,	Pressure vs. residual (−)	$b = 5.75*10^{-3}$	6.83	r^2=0.983,	Pressure vs. residual (−)
1999)	$\Delta S_o^{\neq} = -20.65$	34.87	SD = 0.450	Predicted ln(k) vs. residual (−)	$c = -5.57*10^{-3}$	23.70	SD = 0.205	Predicted ln(k) vs. residual (−)
	$\Delta\zeta^{\neq} = 0.640$	8.28			$d = 1.19*10^{-6}$	7.73		
	$k_o = 1.28*10^{-2}$	8.59			$e = -1.08*10^{-4}$	9.03		
	$\Delta Cp^{\neq} = 3046.55$	6.80			$f = 8.27*10^{-4}$	7.45		
LOX in green bean	$\Delta\kappa^{\neq} = -3.24*10^{-2}$	39.81	Corrected	Temperature vs. residual* (−)	$a = -2.02$	3.09	Corrected	Temperature vs. residual* (+)
juice (Indrawati et al.,	$\Delta V_o^{\neq} = -20.267$	9.86	$r^2 = 0.977$,	Pressure vs. residual (−)	$b = 3.71*10^{-3}$	9.08	$r^2 = 0.982$,	Pressure vs. residual (+)
2000a)	$\Delta S_o^{\neq} = 80.31$	11.34	SD = 0.510	Predicted ln(k) vs. residual (−)	$c = 1.17*10^{-2}$	12.74	SD = 0.208	Predicted ln(k) vs. residual (−)
	$\Delta\zeta^{\neq} = 0.124$	72.58			$d = 2.77*10^{-6}$	35.50		
	$k_o = 1.00*10^{-2}$	16.00			$e = -2.97*10^{-5}$	49.49		
	$\Delta Cp^{\neq} = 1702.00$	15.74			$f = 4.79*10^{-4}$	15.25		
LOX in green bean	$\Delta\kappa^{\neq} = -0.54*10^{-2}$	287.04	Corrected	Temperature vs. residual* (−)	$a = -1.86$	2.62	Corrected	Temperature vs. residual* (+)
juice (Indrawati et al.,	$\Delta V_o^{\neq} = -30.814$	7.85	$r^2 = 0.988$,	Pressure vs. residual (−)	$b = 5.38*10^{-3}$	7.37	$r^2 = 0.992$,	Pressure vs. residual (−)
2000b)	$\Delta S_o^{\neq} = 90.63$	13.56	SD = 0.400	Predicted ln(k) vs. residual (−)	$c = 1.34*10^{-2}$	14.21	SD = 0.166	Predicted ln(k) vs. residual (−)
	$\Delta\zeta^{\neq} = 0.443$	23.48			$d = 3.61*10^{-7}$	346.95		
	$k_o = 1.34*10^{-2}$	11.94			$e = -8.29*10^{-5}$	21.45		
	$\Delta Cp^{\neq} = 2466.71$	13.11			$f = 6.75*10^{-4}$	13.52		

LOX in green beans in	$\Delta\kappa^{\neq} = -7.34*10^{-2}$	35.69	Corrected	Temperature vs. residual* (−)	$a = -1.59$	3.76	Corrected	Temperature vs. residual* (+)
situ (Indrawati et al.,	$\Delta V_o^{\neq} = -35.672$	12.49	$r^2 = 0.986$,	Pressure vs. residual (−)	$b = 5.95*10^{-3}$	12.79	$r^2 = 0.989$,	Pressure vs. residual (−)
2000b)	$\Delta S_o^{\neq} = 139.41$	17.18	SD = 0.44	Predicted ln(k) vs. residual (−)	$c = 2.31*10^{-2}$	17.92	SD = 0.196	Predicted ln(k) vs. residual (−)
	$\Delta\zeta^{\neq} = 0.20*10^{-1}$	800			$d = 6.31*10^{-6}$	35.57		
	$k_o = 2.47*10^{-2}$	13.36			$e = -1.95*10^{-6}$	1471.76		
	$\Delta Cp^{\neq} = 2811.82$	14.04			$f = 7.14*10^{-4}$	16.47		
LOX in green pea juice	$\Delta\kappa^{\neq} = -9.37*10^{-2}$	20.51	Corrected	Temperature vs. residual* (−)	$a = 0.41$	14.43	Corrected	Temperature vs. residual* (−)
(Indrawati et al.,	$\Delta V_o^{\neq} = -38.182$	8.82	$r^2 = 0.968$,	Pressure vs. residual (−)	$b = 6.54*10^{-3}$	8.87	$r^2 = 0.824$,	Pressure vs. residual (+)
2000c)	$\Delta S_o^{\neq} = 13.79$	85.41	SD = 0.650	Predicted ln(k) vs. residual (−)	$c = 2.71*10^{-4}$	757.44	SD = 0.196	Predicted ln(k) vs. residual (−)
	$\Delta\zeta^{\neq} = 0.115$	81.66			$d = 7.98*10^{-6}$	20.26		
	$k_o = 2.45*10^{-2}$	13.28			$e = -2.73*10^{-5}$	61.23		
	$\Delta Cp^{\neq} = 1837.42$	13.28			$f = 4.73*10^{-4}$	15.36		
LOX in green peas in	$\Delta\kappa^{\neq} = -0.1382$	29.38	Corrected	Temperature vs. residual* (−)	$a = -1.59$	3.76	Corrected	Temperature vs. residual* (+)
situ (Indrawati et al.,	$\Delta V_o^{\neq} = -50.763$	14.52	$r^2 = 0.954$,	Pressure vs. residual (−)	$b = 5.95*10^{-3}$	12.80	$r^2 = 0.986$,	Pressure vs. residual (−)
2000c)	$\Delta S_o^{\neq} = -21.19$	121.26	SD = 0.880	Predicted ln(k) vs. residual (−)	$c = 2.31*10^{-2}$	17.92	SD = 0.196	Predicted ln(k) vs. residual (−)
	$\Delta\zeta^{\neq} = 0.230$	86.64			$d = 6.31*10^{-6}$	35.57		
	$k_o = 5.79*10^{-2}$	29.37			$e = -1.95*10^{-6}$	1471.76		
	$\Delta Cp^{\neq} = 1058.61$	35.14			$f = 7.14*10^{-4}$	16.47		
Alkaline phosphatase	$\Delta\kappa^{\neq} = -0.1816$	9.23	Corrected	Temperature vs. residual* (−)	$a = -3.44$	7.65	Corrected	Temperature vs. residual* (−)
(Ludikhuyze et al.,	$\Delta V_o^{\neq} = -46.489$	17.30	$r^2 = 0.965$,	Pressure vs. residual (+)	$b = 8.16*10^{-3}$	15.63	$r^2 = 0.967$,	Pressure vs. residual (+)
2000b)	$\Delta S_o^{\neq} = 242.77$	47.09	SD = 0.669	Predicted ln(k) vs. residual (−)	$c = 3.83*10^{-2}$	45.45	SD = 0.280	Predicted ln(k) vs. residual (−)
	$\Delta\zeta^{\neq} = 0.273$	91.74			$d = 1.48*10^{-5}$	8.70		
	$k_o = 3.35*10^{-4}$	66.02			$e = -6.30*10^{-5}$	61.95		
	$\Delta Cp^{\neq} = 428.54$	314.77			$f = 4.87*10^{-5}$	700.71		

* difference between predicted ln(k) and experimentally estimated ln(k); (+) indicates a trend in the plot of residuals vs. a certain parameter; (−) indicates no trend in the plot of residuals vs. a certain parameter

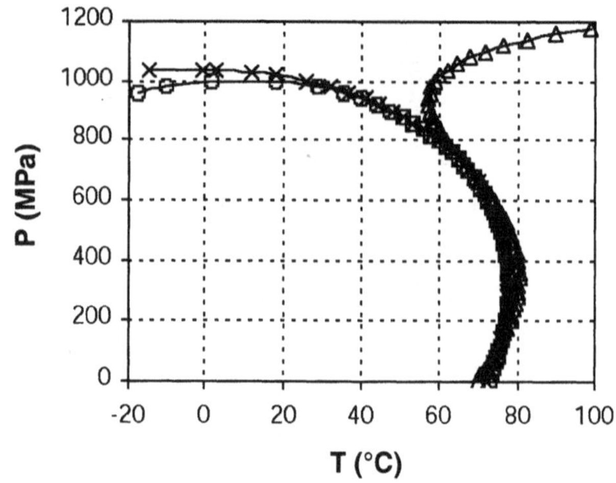

Figure 5–12 Simulated pressure-temperature iso-rate ($k = 7.68 \times 10^{-2}$ min^{-1}) contour diagrams for avocado polyphenoloxidase using a modified Arrhenius relationship (Δ), a modified thermodynamic equation (○), and an elliptical equation (×).

FROM KINETIC INFORMATION TO PROCESS ENGINEERING

Use of Pressure-Temperature Kinetic Diagrams for Process Design and Optimization

Mathematical models to describe pressure-temperature dependence of inactivation rate constants of food attributes can be used in designing and optimizing high pressure processes. In combination with the inactivation model of the food attribute under constant processing conditions, the model allows us to convert the iso-rate contour diagram in contours of an equal degree of inactivation. For several enzymatic systems, pressure-temperature contours of equal degree of inactivation were simulated by combining a first-order inactivation model with the modified thermodynamic equation 19.

For soybean LOX, the pressure-temperature levels needed to achieve different degrees of reduced enzyme activity (1, 2, and 5 log units) within a fixed total process time of 30 minutes are depicted in Figure 5–13. For example, a 30-minute treatment at 20°C at 600 MPa, 650 MPa, and 700 MPa results respectively in a 1, 2, and 5 log-unit reduction of activity. At 500 MPa, the degree of LOX inactivation can be increased from 90% (1 log unit) to 99.999% (5 log units) either by decreasing temperature from −1°C to −10°C or by increasing temperature from 59°C to 72°C. Similar examples can be elaborated for other processing times.

In the same way, the required process time at constant pressure and temperature to achieve a desired degree of inactivation after processing can be calculated and is depicted in Figure 5–14 for a 1 log-unit reduction of LOX activity. For example, a 90% reduction in soybean LOX activity can be obtained by a treat-

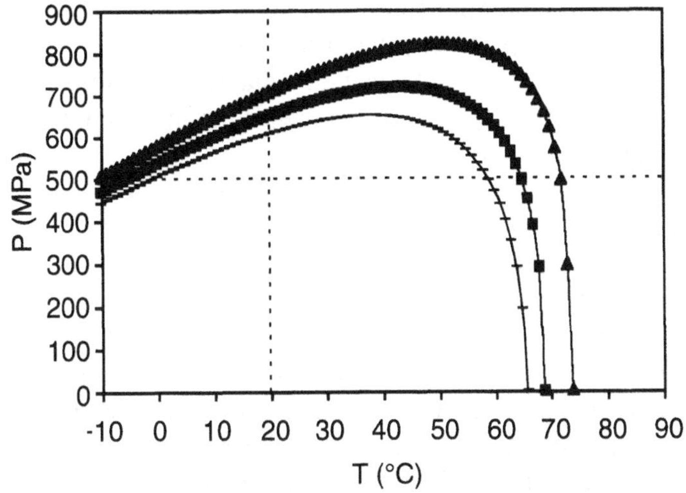

Figure 5–13 Simulated pressure-temperature diagrams (equation 19) of soybean LOX inactivation for a fixed process time of 30 minutes and different degrees of log reduction: 1 (–), 2 (■), and 5 (▲) log units.

ment of 30 minutes at 650 MPa and 40°C. Reducing the total treatment time to 5 minutes will obtain the same reduction in soybean LOX activity. However, the applicability of the principle of ultra high pressure–short time processes, in analogy with high/ultra high temperature–short time treatments in thermal process-

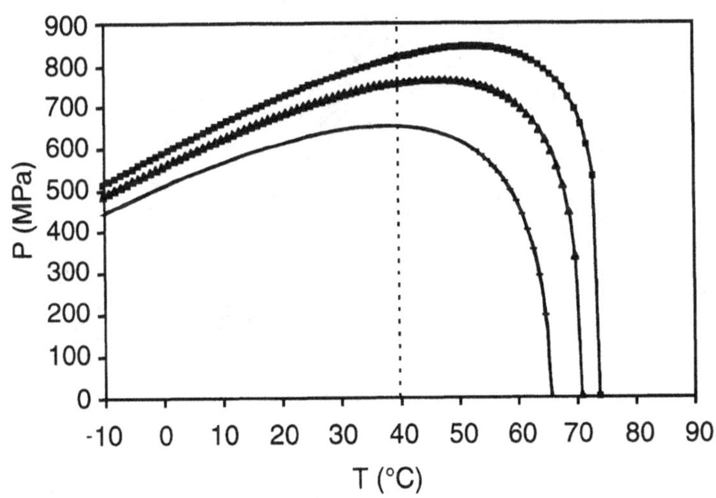

Figure 5–14 Simulated pressure-temperature diagrams (equation 19) of soybean LOX inactivation for a fixed degree of inactivation of 1 log unit and different total process times: 5 (–), 10 (■), and 30 (▲) minutes.

ing, will be limited for pressure-resistant food attributes by the technical specifications of the high pressure equipment.

In Figure 5–15, pressure-temperature diagrams resulting in a 1 log-unit reduction in enzyme activity after a treatment time of 15 minutes are combined for several food-quality–related enzymes.

The shapes of the pressure-temperature diagrams depend not only on the type of enzyme but also on the source of the enzyme (e.g., LOX from various sources). For avocado PPO, orange PME, ALP, and LOX from peas, a pronounced antagonistic effect of low pressure and high temperature can be noted. In other words, low pressure counteracts thermal inactivation of these enzymes. Such an antagonistic effect of low pressure and high temperature is not observed for pressure-temperature inactivation of BSAA or LOX from soybean and green beans. Only soybean LOX inactivation is characterized by an antagonistic effect of elevated pressure and low temperature (<40°C). The consequences of these counteracting or synergistic effects of pressure and temperature will be discussed in more detail later in this chapter.

The stability ranking for these enzymes is strongly dependent on the applied pressure and temperature. The stability ranking perceived at atmospheric pressure cannot be extrapolated toward higher pressures. At atmospheric pressure, PME from oranges has the lowest thermostability, followed by ALP and LOX

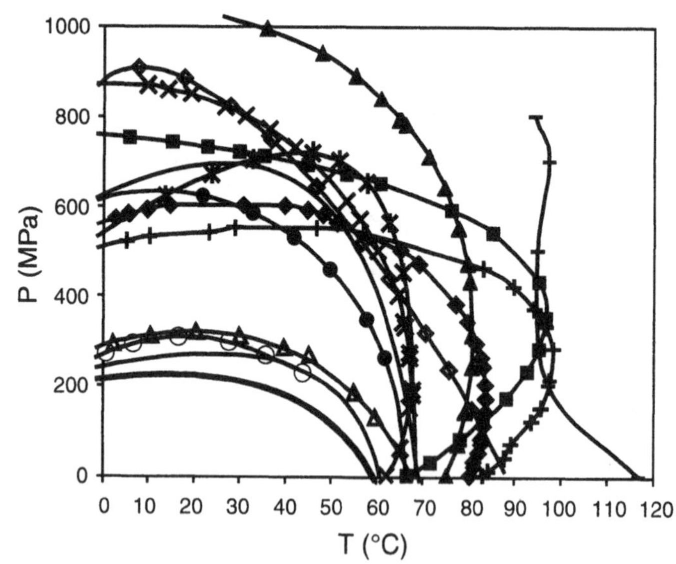

Figure 5–15 Simulated pressure-temperature combinations resulting in 6 log-unit reduction of microorganisms, 90% reduction of enzyme activity (equation 19), and 90% reduction of chlorophyll after a treatment time of 15 minutes: PPO (▲); ALP (■); BSAA (◊); pea LOX in situ (+); pea LOX in juice (◆); green bean LOX in situ (●); green bean LOX in juice (□); soybean LOX (∗); PME (×); total chlorophyl content (−); yeast (Δ); *Zygosaccharomyces bailii* (○); *Lactobacillus casei* (—); *Escherichia coli* (——).

from soybean and green beans, then avocado PPO, LOX from peas, and finally BSAA. At 400 MPa, the thermostability ranking is altered, LOX from green beans has the lowest thermostability, and ALP in raw milk is most thermostable at elevated pressure. Moreover, the difference in temperature stability among the different enzymes is higher at elevated pressure (from 55°C to 97°C at 400 MPa) than at atmospheric pressure (from 62°C to 89°C). In a similar way, pressure stability ranking at different processing temperatures (from 0°C to 90°C) can be derived from Figure 5–15. For example, at room temperature (25°C) LOX from various sources have the lowest barostability, followed by ALP, PME, and BSAA; PPO is characterized by the highest pressure stability at room temperature.

The differences in thermostability at atmospheric and elevated pressure, and the differences in pressure stability at different processing temperatures suggest that it is possible to selectively knock out certain enzymes by a scientifically based choice of pressure and temperature conditions. This could open new and challenging routes for high pressure applications in the food industry.

For process optimization, kinetic information on aspects of food safety and food quality should be combined. A theoretical case study on high pressure process optimization was elaborated by combining pressure-temperature kinetic diagrams of the afore-mentioned enzymes with pressure-temperature kinetic diagrams of microbial inactivation (Sonoike et al., 1992; Hashizume et al., 1995; Reyns et al., 2000) and of chlorophyll degradation (Van Loey et al., 1998). Pressure-temperature combinations resulting in a 6 log-unit reduction of microbial load and in 90% loss of total chlorophyll content after a processing time of 15 minutes are included in Figure 5–15. The shapes of the pressure-temperature kinetic diagrams for the various vegetative microorganisms are similar, although this is not true for the enzymes studied. Enzymes are generally more resistant than vegetative microorganisms with respect to pressure-temperature treatments. Hence, enzymes related to food quality, rather than vegetative microorganisms, can become the critical issue in defining optimal pressure processes. Moreover, at pressure-temperature combinations resulting in sufficient inactivation of food-spoiling enzymes and microorganisms, total chlorophyll content is only slightly affected. This finding supports the general statement that nutritional and sensorial quality is only minimally affected by pressure (see Chapter 6).

Implications of Dynamic Pressure-Temperature Conditions

It is generally believed that high pressure is set immediately and uniformly throughout the pressure vessel (the Pascal principle). Thus, many believe that processing time is independent of product size and geometry (Cheftel, 1991; Mertens & Deplace, 1993; Mertens, 1995), making short processing times possible. However, this statement should be interpreted with care because heat transfer through the products must also be considered. Compression of the workload results in a proportional temperature increase caused by adiabatic heating, the extent of which is dependent on thermal and physical properties of the compressed products. After compression, dissipation of heat through the metal wall induces temperature gradients inside the product. In this context, we illustrated the effect of dynamic pressure-temperature processes (characterized by an initial phase of variable pressure and temperature) on inactivation of enzymes on a theoretical basis using soybean LOX and pea LOX in situ. By incorporating correct thermal and physical properties in a numerical conductive heat transfer model (Denys et al., 2000a), we simulated temperature profiles occurring in a 15% agar gel matrix (same dimensions as the pressure vessel) during batch pressure processing in a single-vessel, high pressure apparatus (h = 300 mm; d = 50 mm; volume = 590 mL). Subsequently, we evaluated the effects of these pressure-temperature-time profiles on enzyme inactivation. First-order inactivation kinetics in combination with the modified thermodynamic model to describe pressure-temperature dependence of the inactivation rate constants were incorporated in the numerical calculations. In this way, the residual enzyme activity at different positions inside the product/vessel could be calculated for any time period by integrating the pressure-temperature-time profiles through the numerical scheme (Denys et al., 2000a; 2000b).

Figures 5–16 and 5–17 represent combinations of constant holding pressure and temperature resulting in 10% activity retention of the respective enzymes after processing for 15 minutes, both under isobaric-isothermal and dynamic conditions (using two different pressure buildup rates, 50 and 200 MPa per minute). For a pressure buildup rate of 200 MPa per minute, the difference between the geometric center and the border of the pressure vessel is shown.

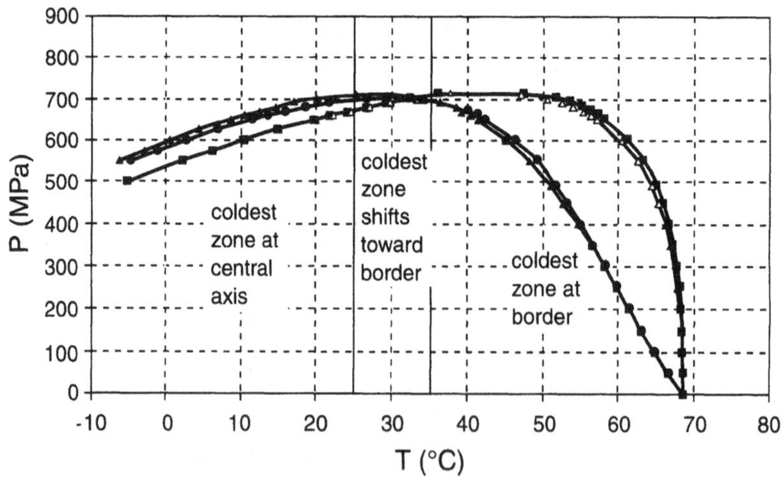

Figure 5–16 Pressure-temperature combinations resulting in 10% activity retention for soybean LOX (0.4 mg/mL in 0.01 M Tris HCl buffer at pH 9) in the sample center (closed symbols) and at the sample border (open symbols) after treatment for 15 minutes: (■) isobaric-isothermal conditions; (●) pressurization rate of 50 MPa/min; (▲/Δ) pressurization rate of 200 MPa/min.

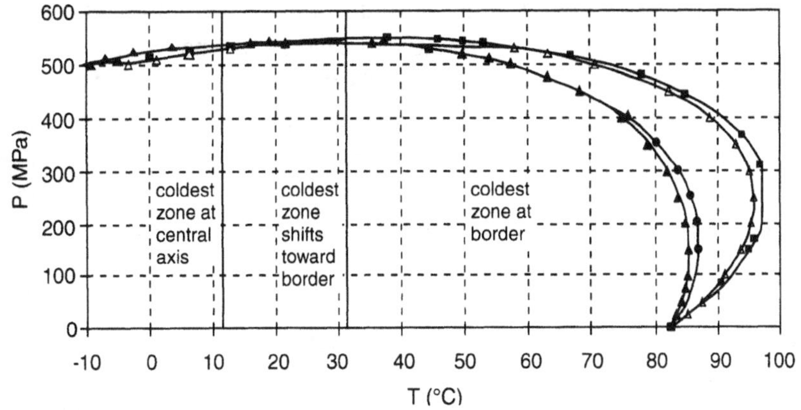

Figure 5–17 Pressure-temperature combinations resulting in 10% activity retention for pea LOX in situ in the sample center (closed symbols) and at the sample border (open symbols) after treatment for 15 minutes: (■) isobaric-isothermal conditions; (•) pressurization rate of 50 MPa/min; (▲ /Δ) pressurization rate of 200 MPa/min.

At temperatures higher than 35°C, the additional time required to reach the maximum preset pressure and the concomitant temperature increase results in increased inactivation of soybean LOX. That is, a lower pressure is sufficient to reduce enzyme activity in the center of the pressure vessel than when using isobaric-isothermal processing. Below this temperature, the reverse is observed because of the antagonistic effect of pressure and low temperature on soybean LOX inactivation. For ALP in bovine milk, where no antagonistic effect of pressure and low temperature is observed, the extent of inactivation induced by dynamic pressure processes was consistently higher than that induced by isobaric-isothermal processes (results not shown). For pea LOX in situ, similar conclusions as for soybean LOX can be drawn, with the following differences: (1) the breakpoint temperature is 20°C, and (2) the differences between isobaric-isothermic and dynamic processing conditions (reduced inactivation under dynamic conditions below the breakpoint temperature) are not as dramatic. At the border of the pressure vessel, no differences between isobaric-isothermal and dynamic processing conditions are observed. This can be explained by the rapid dissipation of heat throughout the metal vessel wall. In this case, the time required to reach the maximum pressure does not significantly affect the activity retention of the enzymes considered.

Evaluation of Process Uniformity on the Basis of Enzyme Inactivation Kinetics

As illustrated previously, dissipation of heat through the metal wall of the high pressure vessel results in temperature gradients, both in the processed product and the entire pressure vessel. The different pressure-temperature-time profiles perceived at different positions in the vessel can result in nonuniform enzyme/

microbial inactivation and uneven nutritional/sensorial quality degradation. For this reason, design of safe pressure processes must be based on the pressure-temperature history at the "coldest point" (zone where the integrated impact of combined pressure and temperature is lowest) in the pressure vessel. As is illustrated in Figures 5–16 and 5–17, one can expect the coldest point in the pressure vessel to vary, depending on the processing conditions. To verify this, preset pressure-temperature conditions along the center 10% activity retention contour were selected for both enzyme systems, and activity retentions at different positions in the pressure vessel (a grid matrix of 15 nodes in horizontal direction starting from the geometric center and 19 nodes in vertical direction, Figure 5–18) were calculated. Based on the residual activity distribution profiles, the coldest point/zone for each pressure-temperature combination tested was determined. An initial homogeneous temperature distribution was used as a starting point in order to account for heat transfer effects caused by increases or decreases in pressure.

When considering soybean LOX inactivation at elevated pressure combined with temperatures between −5°C and 25°C, the coldest zone is situated along the central axis, from the geometric center to about half the height of the vessel. Between 25°C and 35°C, the coldest zone consistently shifts toward the border of the vessel. For instance, at 30°C, the coldest zone is situated along vertical axis h, while at 35°C it is along vertical axis m. At temperatures higher than 35°C, the outline of the vessel always shows the lowest process impact. For pea LOX inactivation in situ, a similar conclusion can be drawn, except that the coldest zone starts to shift from the central axis toward the border of the vessel at a somewhat lower temperature (15°C). These findings are likewise indicated on Figures 5–16 and 5–17.

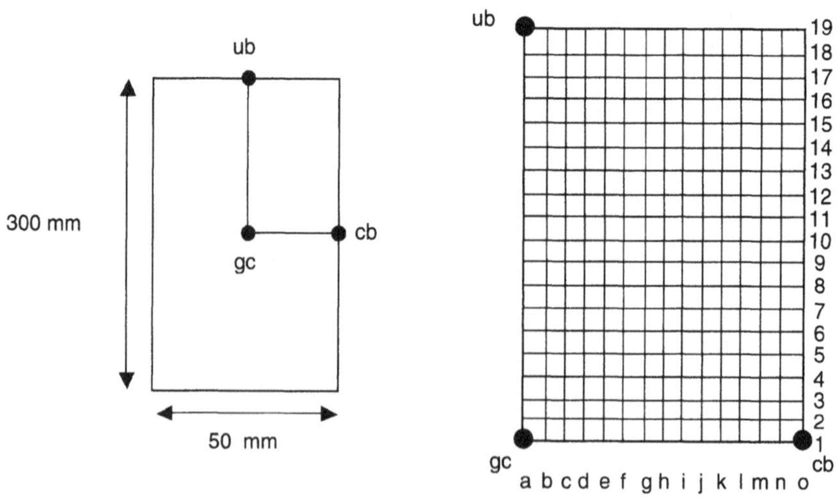

Figure 5–18 Schematic overview of the pressure vessel and the grid used for cold point determination.

In addition to nonuniformity within a single process, one can observe nonuniformity between different processes, resulting in the same impact on enzyme activity (Denys et al., 2000a, 2000b). This nonuniformity is more or less pronounced, depending on the pressure-temperature inactivation/degradation kinetics of the attribute being investigated. For two enzyme systems, BSAA and soybean LOX, nonuniformity of batch high pressure processing was examined on both a theoretical and an experimental basis. To compare uniformity of different pressure processes, trials with the same processing time were selected because longer holding times result in a more uniform inactivation/degradation of the attribute considered. Different pressure-temperature combinations resulting in an average residual activity of 50% after 25 minutes of processing are summarized in Table 5–11. In addition to the average activity retention, the residual activity retention at different positions in the vessel (Figure 5–18) after the respective processes was calculated. Box plots, representing the spatial variation in residual activity (the degree of uniformity or nonuniformity) of these processes are shown in Figure 5–19.

Because the difference between the minimal and maximal residual enzyme activity is a measure for process uniformity, all processes considered for BSAA inactivation are characterized by a pronounced nonuniformity. In the case of LOX inactivation, a clear variation in process uniformity can be observed. Processes 1 to 6 are quite uniform, whereas processes 7 to 12 are characterized by an extreme nonuniformity. The influence of the inactivation kinetics of the target attribute on the uniformity of the pressure-temperature process can be explained by the shape of the iso-rate contours for inactivation of BSAA and LOX. In the case of LOX, there seems to be an antagonistic effect of pressure and temperature in the high pressure–low temperature range (antagonistic domain).

Table 5–11 Simulated Pressure-Temperature Combinations Resulting in an Average Residual Activity of 50% for BSAA and Soybean LOX after 25 Minutes of Processing

	Bacillus subtilis α-*amylase*				*Soybean LOX*			
Process	*T (°C)*	*P (MPa)*	A_{min}	A_{max}	*T (°C)*	*P (MPa)*	A_{min}	A_{max}
1	25	568.8	29.39	60.62	5	518.4	44.49	59.85
2	30	526.6	28.18	61.25	10	528.8	44.42	59.64
3	35	485.0	26.98	61.87	15	539.0	44.45	58.87
4	40	444.0	26.02	62.41	20	549.2	45.22	56.50
5	45	402.3	25.26	62.89	25	557.9	47.15	52.24
6	50	358.5	24.72	63.27	30	561.7	43.01	51.27
7	55	311.5	24.52	63.52	35	553.3	30.82	56.16
8	60	259.8	25.13	63.36	40	521.3	18.47	63.07
9	65	200.1	26.89	62.60	45	455.3	11.56	68.93
10	70	126.9	31.86	60.06	50	354.9	11.28	70.92
11					55	227.2	18.41	67.39
12					60	76.3	37.63	56.85

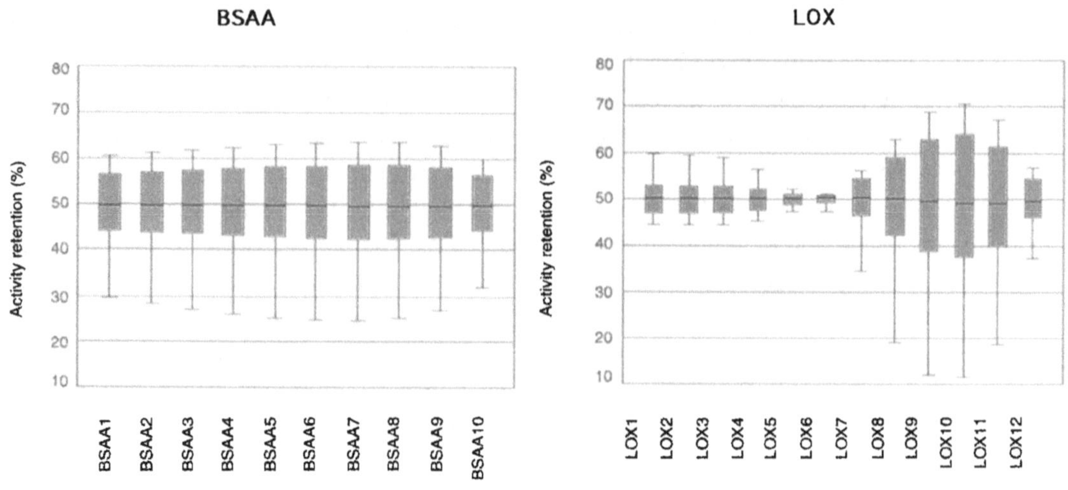

Figure 5–19 Box plots representing the uniformity/nonuniformity of BSAA and LOX inactivation for the pressure-temperature processes outlined in Table 5–11.

In this range, the inactivation rate constant is nearly temperature independent (horizontal contour line in Figure 5–16). Processes with holding temperature in the antagonistic domain (e.g., process 3) will be much more uniform with regard to LOX inactivation than processes at higher holding temperature (e.g., process 10). In the case of BSAA, no antagonistic effect of pressure and temperature was observed. This observation explains the similarity in process uniformity at all pressure-temperature conditions considered.

Following the theoretical evaluation, Denys et al. (2000b) carried out some validation experiments by submitting the respective enzyme systems, placed at different positions in the vessel, to some of the afore-mentioned processes (Table 5–11). For BSAA, they carried out processes 1, 4, and 8. For LOX, they carried out processes 3, 8, and the most uniform process (27°C/560.3 MPa). After these processes, we determined residual enzyme activity values experimentally and compared them to the calculated ones. The processing conditions (pressure buildup rate, maximal pressure reached, average pressure during holding time, and temperature of pressure vessel) obtained through data acquisition were incorporated in the model, and residual enzyme activities at the sample positions were calculated by computer. We obtained a good agreement between experimentally and calculated residual BSAA activities for process 1 (see Figure 5–20) and for the other two processes. For LOX, the agreement between simulated and experimentally determined LOX activity retention was much lower (Figure 5–21). A possible explanation for this deviation between calculated and experimental residual activities is the occurrence of cold denaturation during decompression, which is not accounted for in the inactivation model derived from experiments in the pressure range 50 to 650 MPa at temperatures between 10°C and 64°C.

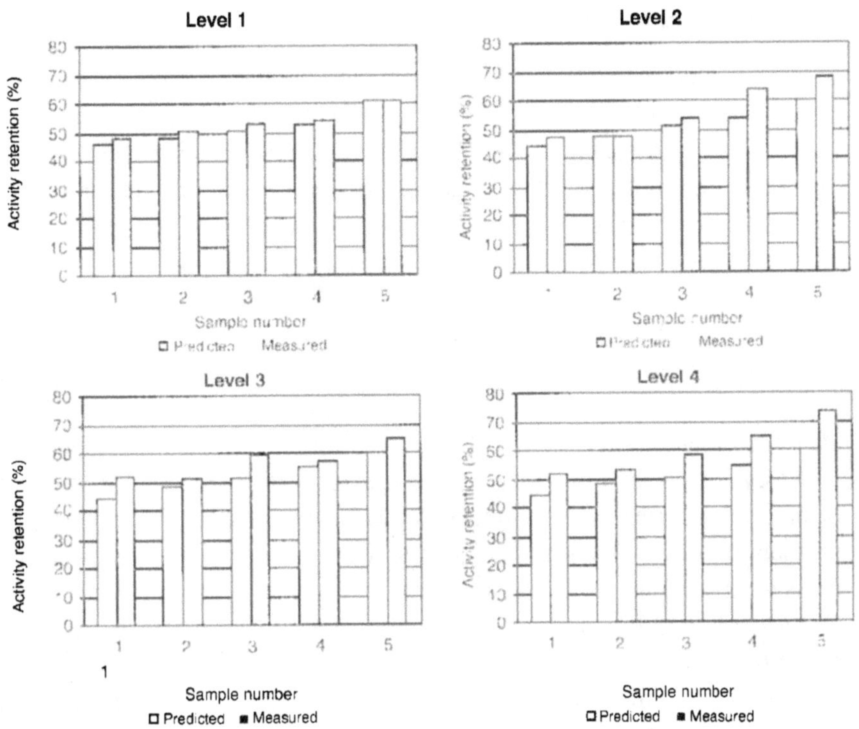

Figure 5–20 Comparison of simulated and experimentally determined BSAA activity retentions for process 1.

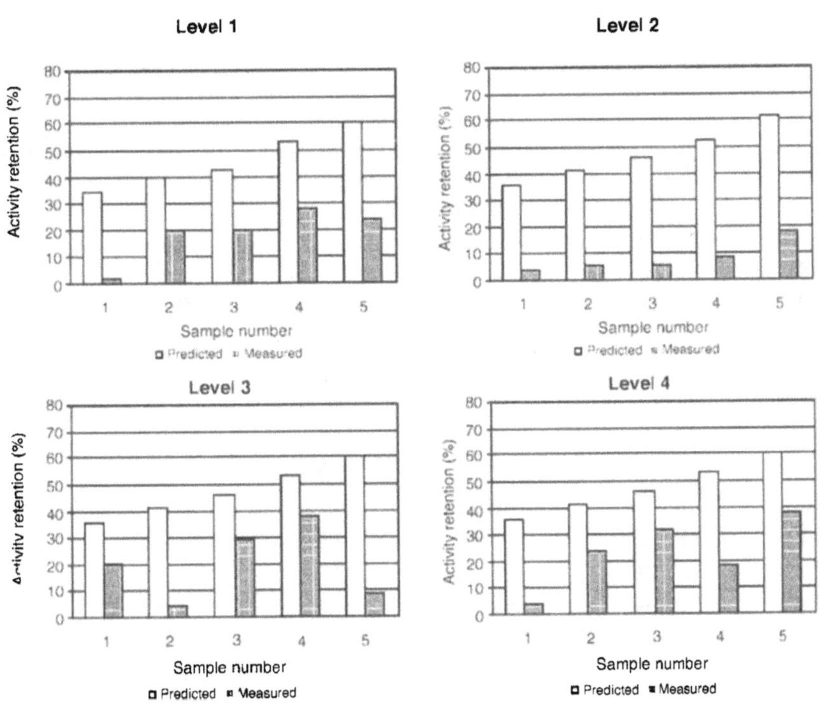

Figure 5–21 Comparison of simulated and experimentally determined LOX activity retentions for process 3.

Effect of Pressure Cycling on Enzyme Inactivation Kinetics

Finally, it is worthwhile to investigate the effect of pressure cycling, a dynamic process involving violent pressure and temperature changes in function of process time, on enzymes. To date, it is unclear whether multicycle or pulsated pressure treatments will be useful in food preservation and processing. Contradictory results regarding the effect of successive application of pressure have been reported. With respect to microbial inactivation, most authors reported none or only slight effects of multicycling (Timson & Short, 1965; Sale et al., 1970; Butz et al., 1990). However, Aleman et al. (1996) reported otherwise. Curl and Jansen (1950a, 1950b) observed that successive application of pressure resulted in an increase in the percentage of enzyme inactivation, which could be attributed merely to pressure cycling after elimination of any temperature effect.

We evaluated the effect of pulsated pressure treatments on inactivation of soybean LOX and pea LOX in situ on a theoretical basis. As for one-step, dynamic pressure processes, we simulated temperature evolution occurring during pulsated pressure treatments on the basis of the heat transfer model. By incorporating the afore-mentioned kinetic models, we calculated enzyme activity retention at different positions in the vessel. The extra inactivation effect of a three-step cycle as compared to a single-step process at the same holding pressure and for a total holding time of 15 minutes is presented in Tables 5–12 and 5–13.

Table 5–12 Activity Retention of Soybean LOX (0.4 mg/mL in 0.01M Tris HCl buffer at pH 9) after Single- and Three-Cycle Processes

	Center of the Vessel			*Border of the Vessel*		
T (°C)	*1 Cycle**	*3 Cycles***	*Extra Inactivation (%)*	*1 Cycle**	*3 Cycles***	*Extra Inactivation (%)*
−5	0.77	0.37	51.95	—	—	—
0	8.49	6.31	25.68	—	—	—
5	23.34	20.25	13.24	1.07	0.92	14.02
10	37.04	34.22	7.61	8.82	8.1	8.16
15	45.92	43.62	5.00	22.64	21.39	5.52
20	49.92	47.99	3.87	35.69	34.17	4.26
25	49.55	47.81	3.51	44.75	43.1	3.69
30	45.02	43.32	3.78	49.63	47.84	3.61
35	36.27	34.54	4.77	50.81	48.80	3.96
40	23.86	22.19	6.99	48.61	46.27	4.81
45	10.66	9.42	11.63	43.04	40.25	6.48
50	2.13	1.68	21.13	34.09	30.81	9.62
55	0.07	0.04	42.86	22.39	18.89	15.63
60	—	—	—	10.37	7.57	27.00
65	—	—	—	2.35	1.23	47.66

* 600 MPa for 15 minutes
** 600 MPa for 5 minutes for each cycle

Table 5–13 Activity Retention of Pea LOX In Situ after Single- and Three-Cycle Processes

	Center of the Vessel			*Border of the Vessel*		
T (°C)	*1 Cycle**	*3 Cycles***	*Extra Inactivation (%)*	*1 Cycle**	*3 Cycles***	*Extra Inactivation (%)*
−10	49.69	47.53	4.34	36.09	34.88	3.36
0	63.21	61.65	2.47	55.13	53.91	2.21
10	68.96	67.68	1.86	65.24	64.06	1.81
20	69.60	68.33	1.82	69.30	68.08	1.76
30	65.96	64.46	2.27	69.17	67.75	2.05
40	57.57	55.61	3.40	65.28	63.50	2.73
50	43.52	40.93	5.95	57.17	54.82	4.11
60	24.59	21.76	11.51	44.00	40.98	6.86
70	7.34	5.61	23.57	26.32	23.04	12.46
75	2.51	1.66	33.86	17.15	14.21	17.14
80	0.49	0.26	46.94	9.24	7.04	23.81
85	0.04	0.01	75	3.74	2.51	32.88

*450 MPa for 15 minutes
**450 MPa for 5 minutes for each cycle

For both soybean LOX and pea LOX, the extra effect induced by pressure cycling is most pronounced at subzero or high temperatures. In addition, for a preset combination of pressure and temperature, the effect of pressure cycling is higher in the center than at the border of the vessel. That is, pressure cycling especially induces extra inactivation in areas where the inactivation rate constant is sensitive to changes in temperature. These findings support the conclusions previously drawn by Ludikhuyze (1998) that the additional inactivation effect observed during pressure cycling is mainly due to more extensive temperature variations occurring during multicycle processes.

CONCLUSION

The influence of pressure, temperature, and medium conditions on enzyme activation/inactivation is clear. Because pressure resistance is not unambiguously related to thermal resistance, knowledge previously gathered about thermal treatment cannot directly be transposed to pressure or pressure-temperature treatment. Moreover, enzyme activation at low pressure can lead to problems related to food quality (e.g., color or flavor changes). On the other hand, enzyme activation may be used to create products with new, improved, or unique functional properties that cannot be produced by thermal technologies.

Knowledge about the kinetics of pressure-induced changes (or pressure-temperature-induced changes) in enzymes related to food quality is indispensable from an engineering point of view. By means of some limited theoretical and experimental case studies, the potential of using kinetic models in process design and optimization and in evaluation of process uniformity has been clearly shown.

GLOSSARY

Abbreviation

ALP—alkaline phosphatase.
BSAA—*Bacillus subtilis* α-amylase.
HACCP—hazard analysis of critical control points.
LOX—lipoxygenase.
MYR—myrosinase.
PDT—pressure death time.
POD—peroxidase.
PME—pectinmethylesterase.
PPO—polyphenoloxidase.
TDT—thermal death time.

Symbols

$\Delta\kappa$—compressibility factor $(\delta/\Delta V/\delta P)_T$.
$\Delta\zeta$—expansibility factor $(\delta\Delta V/\delta T)_P = -(\delta\Delta S/\delta P)_T$.
ΔC_p—heat capacity change $(T\delta\Delta S/\delta T)_P$.
ΔG—change in Gibbs free energy.
A—enzyme activity.
ΔS—entropy change.
ΔV—volume change.
a,b,c,d,e,f—model parameters.
D—D value.
E_a—activation energy (kJ/mol, K).
f_1, f_2—mathematical function.
h—Planck constant.
I—irreversibly inactivated enzyme form.
k—inactivation rate constant.
$K^{\#}$—equilibrium constant.
n—reaction order.
N—native enzyme form.
P—pressure (MPa or N/cm^2).
rk_B—Boltzman constant.
R—gas constant (8.314 J/mol, K).
SF—scaling factor.
t—time.
T—temperature (°K or °C).
U—reversibly unfolded enzyme form.
V_a—activation volume (cm^3/mol).
z_P—z_P value (MPa).
z_T—z_T value (°C).

Subscripts/Superscripts

0—at initial conditions.
1—first reaction step/enzyme form.
2—second reaction step/enzyme form.
obs—observed value.
l—labile fraction.
refP—at reference pressure.
refT—at reference temperature.
rf—residual fraction.
s—stable fraction.

REFERENCES

Ahern, J.T., & Klibanov, A.M. (1985). The mechanism of irreversible enzyme inactivation at 100°C. *Science, 228,* 1280–1284.

Aleman, G.D., Ting, E.Y., Mordre, S.C., Hawes, A.C., Walker, M., Farkas, D.F., & Torres, D.A. (1996). Pulsed high pressure treatments for pasteurisation of pineapple juice. *J. Food Sci., 61,* 388–390.

Anese, M., Nicola, M.C., Dall'Aglio, G., & Lerici, C.R. (1995). Effect of high pressure treatments on peroxidase and polyphenoloxidase activities. *J. Food Biochem., 18,* 285–293.

Aoyama, Y., Asaka, M., & Nakanishi, R. (1994). Effect of high pressure on activity of some oxidizing enzymes. In T. Yano, R. Matsuno, & K. Nakamura (Eds.), *Developments in Food Engineering* (pp. 861–863). Glasgow: Blackie.

Arrhenius, S. (1889). Über die Reactionsgeschwindigkeit bei der Inversion von Rohrzucker durch Säuren. *Z. Phys. Chem., 4,* 226–248.

Asaka, M., & Hayashi, R. (1991). Activation of polyphenoloxidase in pear fruits by high pressure treatment. *Agric. Biol. Chem., 55,* 2439–2440.

Asaka, M., Aoyama, Y., Nakanishi, R., & Hayashi, R. (1994). Purification of a latent form of polyphenoloxidase from La France pear fruit and its pressure-activation. *Biosci. Biotechnol. Biochem., 58,* 1486–1489.

Balny, C., & Masson, P. (1993). Effects of high pressure on proteins. *Food Rev. Int., 9,* 611–628.

Barbosa-Canovas, G.V., Pothakamury, U.R., Palou, E., & Swanson, B.G. (Eds.). (1997). *Nonthermal Preservation of Foods.* New York: Marcel Dekker.

Basak, S., & Ramaswamy, H.S. (1996). Ultra high pressure treatment of orange juice: A kinetic study on inactivation of pectin methyl esterase. *Food Res. Int., 29,* 601–607.

Berg, B. (1996). High pressure treatment of enzymes. In T. Ohlsson (Ed.), *High Pressure Processing of Food and Food Components—A Literature Survey and Bibliography* (pp. 1–7). Göteborg: Kompendiet.

Bigelow, W.D. (1921). The logarithmic nature of thermal death time curves. *J. Inf. Dis., 29*(5), 528–536.

Butz, P.V., Ries, J., Traugott, U., Weber, H., & Ludwig, H. (1990). Hochdruckinactivierung von Bakterien und Bakteriensporen. *Pharm. Ind., 52*(4), 487–491.

Butz, P., Koller, W.D., Tauscher, B., & Wolf, S. (1994). Ultra-high pressure processing of onions: chemical and sensory changes. *Lebensm. Wiss. Technol., 27,* 463–467.

Cano, M.P., Hernandez, A., & De Ancos, B. (1997). High pressure and temperature effects on enzyme inactivation in strawberry and orange products. *J. Food Sci., 62,* 85–88.

Castellari, M., Matricardi, L., Arfelli, G., Rovere, P., & Amati, A. (1997). Effects of high pressure processing on polyphenoloxidase enzyme activity of grape musts. *Food Chem., 60,* 647–649.

Cheftel, J.-C. (1991). Applications des hautes pressions en technologie alimentaire. *Ind. Aliment. Agric., 108*(3), 141–153.

Cheftel, J.-C. (1992). Effects of high hydrostatic pressure on food constituents: An overview. In C. Balny, R. Hayashi, K. Heremans, & P. Masson (Eds.), *High Pressure and Biotechnology,* Colloque INSERM (Vol. 224, pp. 195–209). Montrouge: John Libbey Eurotext.

Crelier, S., Tâche, M.-C., Raemy, A., Renken, A., & Raetz, E. (1995). High pressure for the inactivation of enzymes in food products: Thermal and HHP treatment of tomato pectin methyl esterase. Poster presentation at *International Symposium on High Pressure Effects on Foods*, July 30–Aug. 4, Budapest, Hungary.

Crelier, S, Robert, M.-C., & Juillerat, M.-A. (1998). Effect of a high pressure treatment on the texture and enzyme activities of selected vegetables. Poster presentation at *International Conference on High Pressure Bioscience and Biotechnology*, Aug. 30–Sep. 3, Heidelberg, Germany.

Curl, L., & Jansen, E.F. (1950a). Effect of high pressures on trypsin and chymotrypsin. *J. Biol. Chem., 184,* 45–54.

Curl, L., & Jansen, E.F. (1950b). Effect of high pressures on pepsin and chymotrypsinogen. *J. Biol. Chem., 185,* 713–723.

Damodaran, S. (1988). Amino acids, peptides and proteins. In O.R. Fennema (Ed.), *Food Chemistry* (3rd ed., pp. 321–430). New York: Marcel Dekker.

Darnell, J. Loddish, H., & Baltimore, D. (Eds.). (1992). *Molecular Cell Biology.* New York: W.H. Freeman.

Denys, S., Ludikhuyze, L., Van Loey, A., & Hendrickx, M. (2000a). Modelling conductive heat transfer and process uniformity during batch high-pressure processing of foods. *Biotechnol. Prog., 16*(1), 92–101.

Denys, S., Van Loey, A., & Hendrickx, M.E. (2000b). A modeling approach for evaluating process uniformity during batch high hydrostatic pressure processing: Combination of a numerical heat transfer model and enzyme inactivation kinetics. *Innov. Food Sci. Emerg. Technol, 1*(1), 5–19.

Dörnenburg, H., & Knorr, D. (1997). Evaluation of elicitor- and high-pressure-induced enzymatic browning utilizing potato (*Solanum tuberosum*) suspension cultures as a model system for plant tissues. *J. Agric. Food Chem., 45,* 4173–4177.

Dunford, N.T., & Temelli, F. (1996). Effect of supercritical CO_2 on myrosinase activity and glucosinolate degradation of canola. *J. Agric. Food Chem., 44,* 2372–2376.

Eshtiaghi, M.N., & Knorr, D. (1993). Potato cubes response to water blanching and high hydrostatic pressure. *J. Food Sci., 58,* 1371–1374.

Ferrar, P.H., & Walker, J.R.L. (1996). Inhibition of diphenol oxidases: A comparative study. *J. Food Biochem., 20,* 15–30.

Fersht, A. (1985). The basic equations of enzyme kinetics. In A. Fersht (Ed.), *Enzyme Structure and Mechanism* (2nd ed., pp. 98–120). New York: W.H. Freeman.

Funtenberger, S., Dumay, E., & Cheftel, J.-C. (1997). High pressure promotes β-lactoglobulin aggregation through SH/S interchange reactions. *J. Agric. Food Chem., 45,* 912–921.

Gatfield, I.L., & Sand, T.A. (1983). Coupled enzymatic procedure for the determination of myrosinase activity. *Lebensm. Wiss. Technol., 16,* 73–75.

Giovane, A., Quagliuolo, L., Castoldo, D., Servillo, L., & Balestrieri, C. (1990). Pectin methyl esterase from atinidia chinensis fruits. *Phytochemistry, 29,* 2821–2823.

Gomes, M.R.A., & Ledward, D.A. (1996). Effect of high-pressure treatments on the activity of some polyphenoloxidases. *Food Chem., 56,* 1–5.

Goodner, J.K., Braddock, R.J., & Parish, M.E. (1998). Inactivation of pectinesterase in orange and grapefruit juices by high pressure. *J. Agric. Food Chem., 46,* 1997–2000.

Gould, G.W., & Sale, J.H. (1970). Initiation of germination of bacterial spores by hydrostatic pressure. *J. Gen. Microbiol., 60,* 335–346.

Gross, M., & Jaenicke, R. (1994). Proteins under pressure: The influence of high hydrostatic pressure on structure, function and assembly of proteins and protein complexes. *Eur. J. Biochem., 221,* 617–630.

Hashizume, C., Kimura, K., & Hayashi, R. (1995). Kinetic analysis of yeast inactivation by high pressure treatment at low temperatures. *Biosci. Biotech. Biochem., 59,* 1455–1458.

Hawley, S.A. (1971). Reversible pressure-temperature denaturation of chymotrypsinogen. *Biochemistry, 10,* 2436–2442.

Heinisch, O., Kowalski, E., Goossens, K., Frank, J., Heremans, K., Ludwig, H., & Tauscher, B. (1995). Pressure effects on the stability of lipoxygenase: Fourier Transform Infrared spectroscopy and enzyme activity study. *Z. Lebensm. Unters. Forsch., 201,* 562–565.

Hendrickx, M., Ludikhuyze, L., Van den Broeck, I., & Weemaes, C. (1998). Effects of high pressure on enzymes related to food quality. *Trends Food Sci. Technol., 9,* 197–203.

Heremans, K. (1982). High pressure effects on proteins and other biomolecules. *Ann. Rev. Biophys. Bioeng., 11,* 1–21.

Heremans, K. (1993). The behaviour of proteins under pressure. In R. Winter & J. Jonas (Eds.), *High Pressure Chemistry, Biochemistry and Material Science* (pp. 443–469). Dordrecht: Kluwer Academic.

Hernández, A., & Cano, M.P. (1998). High-pressure and temperature effects on enzyme inactivation in tomato puree. *J. Agric. Food Chem., 46,* 266–270.

Indrawati, I., Van Loey, A., Ludikhuyze, L., & Hendrickx, M. (1999). Soybean lipoxygenase inactivation by pressure at subzero and elevated temperature. *J. Agric. Food. Chem., 47,* 2468–2474.

Indrawati, I., Van Loey, A., Ludikhuyze, L., & Hendrickx, M. (2000a). Kinetics of pressure inactivation at subzero and elevated temperature of lipoxygenase in crude green beans (*Phaseolus vulgaris* L.) extract. *Biotechnol. Prog., 16*(1), 109–115.

Indrawati, I., Ludikhuyze, L.R., Van Loey, A.M., & Hendrickx, M.E. (2000b). Lipoxygenase inactivation in green beans (*Phaseolus vulgaris* L.) due to high pressure treatment at subzero and elevated temperatures. *J. Agric. Food Chem., 48*(5), 1850–1859.

Indrawati, I., Van Loey, A., Ludikhuyze, L., & Hendrickx, M. Lipoxygenase inactivation in green peas (*Pisum sativum*) due to high pressure treatment at subzero and elevated temperatures. *J. Food Sci.* In press.

Irwe, S., & Olsson, I. (1994). Reduction of pectinesterase activity in orange juice by high pressure treatment. In R.P. Singh & F.A.R Oliveira (Eds.), *Minimal Processing of Foods and Process Optimization—An Interface* (pp. 35–42). Boca Raton: CRC Press.

Jolibert, F., Tonello, C., Sagegh, P., & Raymond, J. (1994). Les effets des hautes pressions sur la polyphénol oxydase des fruits. *Bios Boissons, 251,* 27–35.

Kermasha, S., Bisakowski, B., Ramaswamy, H., & Van de Voort, F.R. (1993). Thermal and microwave inactivation of soybean lipoxygenase. *Lebensm. Wiss. Technol., 26,* 215–219.

Knorr, D. (1993). Effects of high hydrostatic pressure on food safety and quality. *Food Technol., 47,* 156–161.

Lambrecht, H.S. (1995). Sulfite substitutes for the prevention of enzymatic browning in foods. In C.Y. Lee & J.R. Whitaker (Eds.), *Enzymatic Browning and Its Prevention* (pp. 313–323). Washington, DC: American Chemical Society.

Lemos, M.A.C.L. (1998). Kinetic studies on the thermal inactivation of horseradish peroxidase under extreme conditions: Application to the development of measuring systems of processing impact on food quality and safety. Ph.D. Dissertation, Escola Superior De Biotechnologica, Departemento de Estudos Graduados, Porto, Portugal.

Lencki, R.W., Arul, J., & Neufeld, R.J. (1992). Effect of subunit dissociation, denaturation, aggregation, coagulation and decomposition on enzyme inactivation kinetics: I. First-order behavior. *Biotechnol. Bioeng., 40,* 1421–1426.

Ludikhuyze, L.R., Van den Broeck, I., Weemaes, C.A., Herremans, C.H., Van Impe, J.F., Hendrickx, M.E., & Tobback, P.P. (1997). Kinetics for isobaric-isothermal inactivation of *Bacillus subtilis* α-amylase. *Biotechnol. Prog., 13,* 532–538.

Ludikhuyze, L. (1998). High pressure technology in food processing and preservation: A kinetic case study on the combined effect of pressure and temperature on enzymes. Ph.D. Dissertation No. 362, Faculty of Agricultural and Applied Biological Sciences, Katholieke Universiteit Leuven.

Ludikhuyze, L., & Hendrickx, M. (2000). [Effects of pressure & temperature on activity of lactoperoxidase in milk and whey.] Unpublished data.

Ludikhuyze, L., Indrawati, I., Van den Broeck, I., Weemaes, C., & Hendrickx, M. (1998a). Effect of combined pressure and temperature on soybean lipoxygenase: I. Influence of extrinsic and intrinsic factors on isobaric-isothermal inactivation kinetics. *J. Agric. Food Chem., 46,* 4074–4080.

Ludikhuyze, L., Indrawati, I., Van den Broeck, I., Weemaes, C., & Hendrickx, M. (1998b). Effect of combined pressure and temperature on soybean lipoxygenase: II. Modeling inactivation kinetics under static and dynamic conditions. *J. Agric. Food Chem., 46,* 4081–4086.

Ludikhuyze, L., Ooms, V., Weemaes, C., & Hendrickx, M. (1999). Kinetic study of the irreversible thermal and pressure inactivation of myrosinase from broccoli (*Brassica oleracea* L. cv. Italica). *J. Agric. Food Chem., 47,* 1794–1800.

Ludikhuyze, L., Rodrigo, L., & Hendrickx, M. (2000a). The activity of myrosinase from broccoli (*Brassica oleracea* L. cv. Italica): Influence of intrinsic and extrinsic factors. *J. Food Prot., 63*(3), 400–403.

Ludikhuyze, L.R., Claeys, W., & Hendrickx, M. (2000b). Combined pressure-temperature inactivation of alkaline phosphatase in bovine milk: A kinetic study. *J. Food Sci., 65,* 155–160.

Ludikhuyze, L.R., Claeys, W., & Hendrickx, M. (2001). Effect of temperature and pressure on lactoperoxidase activity in bovine milk and acid whey. *J. Dairy Res.*, in press.

Matser, A.M., Knorr, E.R., & Bartels, P.V. (1998). Effect of high isostatic pressure on polyphenoloxidase and colour of mushrooms. Poster presentation at *International Conference on High Pressure Bioscience and Biotechnology*, Aug. 30–Sep. 3, Heidelberg, Germany.

Mertens, B., & Knorr, D. (1992). Developments of nonthermal processes for food preservation. *Food Technol., 46,* 124–133.

Mertens, B., & Deplace, G. (1993). Engineering aspects of high pressure technology in the food industry. *Food Technol., 47,* 164–169.

Mertens, B. (1995). Hydrostatic pressure treatment of food: Equipment and processing. In G.W. Gould (Ed.), *New Methods of Food Preservation* (pp. 135–158). Glasgow: Blackie Academic & Professional.

Miyagawa, K., Sannoe, K., & Suzuki, K. (1964). Studies of Taka amylase A under high pressure: II. Recovery of enzymic activity of pressure inactivated Taka amylase A and its enhancement by retreatment at moderate pressure. *Arch. Biochem. Biophys., 106,* 467–474.

Morild, E. (1981). The theory of pressure effects on enzymes. *Adv. Prot. Chem., 34,* 93–166.

Mozhaev, V.V., Heremans, K., Frank, J., Masson, P., & Balny, C. (1994). Exploiting the effects of high hydrostatic pressure in biotechnological applications. *Trends Biotechnol., 12,* 493–501.

Mussa, D.M., & Ramaswamy, H.S. (1997). Ultra high pressure pasteurisation of milk: Kinetics of microbial destruction and changes in physico-chemical characteristics. *Lebensm. Wiss.Technol., 30,* 551–557.

Ogawa, H., Fukuhisa, K., Kubo, Y., & Fukumoto, H. (1990). Pressure inactivation of yeasts, molds, and pectinesterase in Satsuma mandarin juice: Effects of juice concentration, pH, and organic acids and comparison with heat sanitation. *Agric. Biol. Chem., 54,* 1219–1225.

Pace, C.N. (1990). Conformational stability of proteins. *Trends Biochem. Sci., 15,* 14–17.

Parish, M., & Olsson, I. (1994). Isostatic high pressure processing of orange juice. In R. Ahvenainen, T. Mattila-Sandholm, & T. Ohlsson (Eds.), *Minimal Processing of Foods (VTT Symposium 142),* (pp. 93–102). Espoo, Finland: VTT Offsetpaino.

Parish, M.E. (1998). High pressure inactivation of *Saccharomyces cerevisiae,* endogenous microflora and pectin methylesterase in orange juice. *J. Food Saf.,* 18, 57–65.

Penniston, J.T. (1971). High hydrostatic pressure and enzyme activity: Inhibition of multimeric enzymes by dissociation. *Arch. Biochem. Biophys., 142,* 322–332.

Quaglia, G.B., Gravina, R., Paperi, R., & Paoletti, F. (1996). Effect of high pressure treatments on peroxidase activity, ascorbic acid content and texture in green peas. *Lebensm. Wiss. Technol., 29,* 552–555.

Reyns, K.M., Soontjes, C.C., Cornelis, K., Weemaes, C.A., Hendrickx, M.E., & Michiels, C.W. (2000). Kinetic analysis and modeling of combined pressure-temperature inactivation of the yeast *Zygosaccharomyces bailii. J. Food Microbiol.,* 56, 199–210.

Richardson, T., & Hyslop, D.B. (1985). Enzymes. In O.R. Fennema (Ed.), *Food Chemistry* (2nd ed., pp. 373–476). New York and Basel: Marcel Dekker.

Rizvi, A.F., & Tong, C.H. (1997). Fractional conversion for determining texture degradation kinetics of vegetables. *J. Food Sci., 62,* 1–7.

Robert, C.M., Cadet, F.R., Rouch, C.C., Pabion, M., & Richard-Forget, F. (1995). Kinetic study of the irreversible thermal deactivation of palmito (*Acantophoenix rubra*) polyphenol oxidase and effect of pH. *J. Agric. Food Chem., 43,* 1143–1150.

Rovere, P., Tosoratti, D., & Maggi, A. (1996). Prove di sterilizzazione a 15000 bar per ottenere la stabilità microbiologica ed enzimatica (Sterilising trials to 15000 bar to obtain microbiologic and enzymatic stability). *Ind. Aliment., 35,* 1062–1065.

Sale, A.J.H., Gould, G.W., & Hamilton, W.A. (1970). Inactivation of bacterial spores by hydrostatic pressure. *J. Gen. Microbiol., 60,* 323–334.

Seyderhelm, I., Bogulawski, S., Michaelis, G., & Knorr, D. (1996). Pressure-induced inactivation of selected food enzymes. *J. Food Sci., 60,* 164–168.

Seymour, T.A., Preston, J.F., Wicker, L., Lindsay, J.A., & Marschall, M.R. (1991). Purification and properties of pectinesterases from marsh white grapefruit pulp. *J. Agric. Food Chem., 39,* 1080–1085.

Silva, J.L., & Weber, G. (1993). Pressure stability of proteins. *Ann. Rev. Phys. Chem., 44,* 89–113.

Sonoike, K., Setoyama, T., Kuma, Y., & Kobayashi, S. (1992). Effect of pressure and temperature on the death rates of *Lactobacillus casei* and *Escherichia coli.* In C. Balny, R. Hayashi, K. Heremans, & P. Masson (Eds.), *High Pressure and Biotechnology,* Colloque INSERM (Vol. 224, pp. 297–301). Montrouge: John Libbey Eurotext.

Stryer, L. (Ed.). (1988). *Biochemistry.* New York: Freeman.

Takahashi, Y., Ohta, H., Yonei, H., & Ifuku, K. (1993). Microbicical effect of hydrostatic pressure on Satsuma Mandarin juice. *Int. J. Food Sci. Technol., 29,* 95–102.

Timson, W.J., & Short, A.J. (1965). Resistance of microorganisms to hydrostatic pressure. *Biotechnol. Bioeng., 7,* 139–159.

Van den Broeck, I. (2000). Kinetics of temperature and pressure inactivation of pectinesterase from oranges and tomatoes. Ph.D. Dissertation No. 415, Faculty of Agricultural and Applied Biological Sciences, Katholieke Universiteit Leuven, Leuven, Belgium.

Van den Broeck, I., Ludikhuyze, L.R., Weemaes, C.A., Van Loey, A.M., & Hendickx, M.E. (2000a). Effect of temperature and pressure on tomato pectinesterase activity. *J. Agric. Food Chem., 48,* 551–558.

Van den Broeck I., Ludikhuyze, L.R., Van Loey, A.M., and Hendrickx, M.E. (2000b). Inactivation of orange pectinesterase by combined high pressure and temperature treatments: A kinetic study. *J. Agric. Food Chem., 48*(5), 1960–1970.

Van Loey, A., Ooms, V., Weemaes, C., Van den Broeck, I., Ludikhuyze, L., Indrawati, I., Denys, S., & Hendrickx, M. (1998). Thermal and pressure-temperature degradation of chlorophyll in broccoli (*Brassica oleracea* L. *italica*) juice: A kinetic study. *J. Agric. Food Chem., 46,* 5289–5294.

Wathelet, J.-P., Mabon, N., Foucart, M., & Marlier, M. (1996). Influence du blanchiment sur la qualité du chou de Bruxelles (*Brassica oleracea* L. *cv. gemmifera*). *Sci. Aliment., 116,* 393–402.

Weale, K.E. (1967). Chemical equilibria in liquids at high pressure. In M.B. Donald (Ed.), *Chemical Reactions at High Pressure* (pp. 131–174). Birkenhead: Willmer Brothers.

Weemaes, C., Rubens, P., De Cordt, S., Ludikhuyze, L., Van den Broeck, I., Hendrickx, M., Heremans, K., & Tobback, P. (1997). Temperature sensitivity and pressure resistance of mushroom polyphenoloxidase. *J. Food Sci., 62,* 261–266.

Weemaes, C., Ludikhuyze, L., Van den Broeck, I., & Hendrickx, M. (1998a). High pressure inactivation of polyphenoloxidases. *J. Food Sci., 63,* 873–877.

Weemaes, C., Ludikhuyze L., Van den Broeck, I., & Hendrickx, M. (1998b). Effect of pH on pressure and thermal inactivation of avocado polyphenoloxidase: A kinetic study. *J. Agric. Food Chem., 46,* 2785–2792.

Weemaes, C., Ludikhuyze, L., Van den Broeck, I., & Hendrickx, M. (1998c). Kinetics of combined pressure-temperature inactivation of avocado polyphenoloxidase. *Biotechnol. Bioeng., 60,* 292–300.

Weemaes, C., Ludikhuyze, L., Van den Broeck, I., & Hendrickx, M. (1999a). Influence of pH, benzoic acid, glutathione, EDTA, 4-hexylresorcinol, and sodium chloride on the pressure inactivation kinetics of mushroom polyphenoloxidase. *J. Agric. Food Chem., 47,* 3526–3530.

Weemaes, C., Ludikhuyze, L., Van den Broeck, I., & Hendrickx, M. (1999b). Kinetic study of antibrowning agents and pressure inactivation of avocado PPO. *J. Food Sci., 64,* 826–827.

Whitaker, J.R. (Ed.). (1972). *Principles of Enzymology for the Food Sciences.* New York: Marcel Dekker.

Wilkinson, A.P., Rhodes, M.J.C., & Fenwick, G.R. (1984). Myrosinase activity of cruciferous vegetables. *J. Sci. Food Agric., 35,* 543–552.

Yen, G.-C., & Lin, H.-T. (1996). Comparison of high pressure treatment and thermal pasteurisation effects on the quality and shelf life of guava puree. *Int. J. Food Sci. Technol., 31,* 205–213.

Chapter 6

Effects of High Pressure on Chemical Reactions Related to Food Quality

Linda Ludikhuyze and Marc E. G. Hendrickx

INTRODUCTION

In response to the growing consumer demand for high-quality, nutritious food products, the food industry is interested in high pressure technology as a potential food processing/preservation method that minimally affects sensory and nutritional quality attributes. High pressure allows inactivation of pathogenic/spoilage microorganisms and food-spoiling enzymes while leaving most attributes of food quality intact (Hayashi, 1989; Cheftel, 1991; Mertens, 1992; Knorr, 1993; Galazka & Ledward, 1995; Thakur & Nelson, 1998). This advantage is attributed to the fact that high pressure keeps covalent bonds intact and affects only noncovalent bonds (such as hydrogen, ionic, and hydrophobic bonds). Important characteristics of high-quality foods are texture, flavor, color, and nutritive value. The first three properties are correlated with purchase and consumption quality and determine consumer's acceptance of the product. Nutritive value (i.e., vitamins, minerals, and other nutrients) is a hidden quality. Chemical or biochemical reactions occurring in food products can bring about undesirable changes in or deterioration of these attributes of food quality during preservation/processing treatments and subsequent storage (Table 6–1).

Reactions leading to such undesirable changes include nonenzymatic and enzymatic browning, lipid hydrolysis or oxidation, protein denaturation, hydrolysis or cross-linking, polysaccharide hydrolysis or synthesis, and degradation of natural pigments. In the context of conventional preservation/processing methods, a great deal of research has been carried out to investigate the effects of various processing variables (e.g., temperature, time, heating rate, pH, water activity) on reactions leading to deterioration of food products in order to design processes so that they induce the required microbial and enzymatic inactivation while retarding or inhibiting specific deteriorative reactions. Knowledge about the effects of pressure is indispensable for implementation of high pressure tech-

The authors would like to thank the National Fund for Scientific Research and the European Commission (project FAIR-CT96-1175) for their financial support.

Table 6–1 Overview of Important Quality Attributes of Foods and Undesirable Changes They Can Undergo during Preservation, Processing, or Storage

Quality Attribute	*Undesirable Change*
Color	Development of off-colors (e.g., darkening, bleaching, browning)
Flavor	Development of off-flavors (e.g., hydrolytic or oxidative rancidity, cooked or caramel flavors)
Texture	Loss of solubility or water-holding capacity; toughening or softening
Nutritive value	Loss or degradation of vitamins, nutrients, minerals, proteins, or lipids

nology as an alternative preservation and processing method in food industry. The most likely route that the food industry will take is to combine high pressure processing with other processing methods, especially moderately elevated temperature, because bacterial spores and some enzymes are very pressure stable (Sale et al., 1970; Hendrickx et al., 1998; Smelt, 1998). This chapter reviews information in the literature on the effects of pressure used alone and in combination with temperature on the various aspects of food quality and concomitant deteriorative reactions. It covers changes in quality caused by the pressure treatment and changes in quality during storage time.

THE EFFECT OF HIGH PRESSURE ON THE COLOR OF FOOD PRODUCTS

Fruit and Vegetable Products

For many fruit and vegetable products, high pressure treatment largely preserves fresh color. Chemical and spectrophotometrical analysis of the color of pressure-processed orange juice showed that color values were not significantly altered by pressure treatment and during subsequent storage time (Donsi et al., 1996). Color parameters indicated pressure-treated fruit jam to be superior to conventionally treated jam in brightness (L-value) and redness (a-value) (Watanabe et al., 1991). Pressure treatment of strawberries resulted in a high retention of red color (Matser & Bartels, 1999). The color of tomato juice (in terms of a-value/b-value ratio) increased by pressure treatment in comparison to conventional hot break processes; the highest a/b ratio was observed at pH 4.5, irrespective of the pressure applied. This effect was purely physical in nature (compacting and homogenizing effects) and not caused by differences in lycopene content of pressure-treated and heat-treated samples (Poretta et al., 1995).

Pressurized guava puree (600 MPa, 25°C, 15 minutes) retained the original color of freshly extracted puree. During storage at 4°C for 60 days, lightness and greenness fell continuously in all samples, especially in the untreated puree. The color of pressurized puree samples was much more stable (Yen & Lin, 1996). Avocado puree treated at pressures between 345 and 689 MPa (10 to 30 minutes) had a color equivalent to the freshly prepared puree (i.e., a green color with colorimetric a-values between −7.9 and −7.3) (Mermelstein, 1997; Lopez-Malo et al., 1998). However, during storage several changes occurred. Less than 1% variation in lightness (L-value) was observed with samples at pH 4.1 treated at 689 MPa for 20 minutes and stored at 5°C. Higher decrease in lightness was found when purees were treated at lower pressure or higher pH or stored at higher temperature. For hue angle, net color difference, and a-value, more pronounced changes were observed. During storage, green contribution to color gradually decreased (i.e., the a-value became positive because of severe browning as a result of residual polyphenoloxidase (PPO) activity. The longest acceptability storage time (time to reach the limits of acceptance) could be reached with small initial pH, high pressure resulting in larger PPO inactivation, and lower storage temperature (Lopez-Malo et al., 1998). Similar results were obtained for banana puree. High pressure treatment (689 MPa for 10 minutes) preserved the initial color of banana puree; however, during storage several color changes occurred because of residual PPO activity that induced enzymatic browning of the puree (Palou et al., 1999).

For broccoli juice, investigators have performed detailed quantitative kinetic studies regarding the effect of combined high pressure and temperature on chlorophyll content and green color (Van Loey et al., 1998; Weemaes et al., 1999). With regard to green color, a common change in processed green vegetables is the degradation of chlorophyll (green) to pheophytin (gray-brown), a process highly associated with quality loss and thus undesirable to consumers. For this reason, minimizing chlorophyll degradation and greenness loss is important to food processors. When green vegetables were pressure processed at low temperature, an extreme stability of chlorophyll was observed. This same phenomenon has been noted for other pigments such as annatto, carotene, anthocyanins, and hibiscus extract (Kimura et al., 1994; Donsi et al., 1996; Salanski et al., 1997). Even after treatment of 4 hours at 800 MPa and 40°C, no reduction in chlorophyll a and b content was found. This observation can be attributed to the stability of the covalent structure of chlorophyll to high pressure arising from the negligible compressibility of covalent bonds. Only when high pressure is combined with temperatures higher than 50°C, does significant reduction of chlorophyll content become noticeable, following a first-order model. From calculation of degradation rate constants and activation energies, it was concluded that chlorophyll a is less pressure-stable and more temperature-sensitive than chlorophyll b. In addition, for both chlorophylls individually, as well as for total chlorophyll content, a synergistic effect of pressure and temperature was noted (Figure 6–1). However, where a temperature increase of 10°C markedly increased the chlorophyll degradation rate, a pressure increase of 100 MPa only had a minor effect (Van Loey et al.). Subsequently, a similar study was performed measuring spectrophotometrically the a-color value of broccoli juice after treatment. At low temperature (less than 40°C), no significant reduction of green color was noted, even after treatment at 800 MPa for 180 minutes. When combining high pressure with temperatures of 50°C to 60°C, some greenness loss was observed. At temperatures between 70°C and 80°C, greenness loss was noticed at all pressures studied (0.1 to 850 MPa).

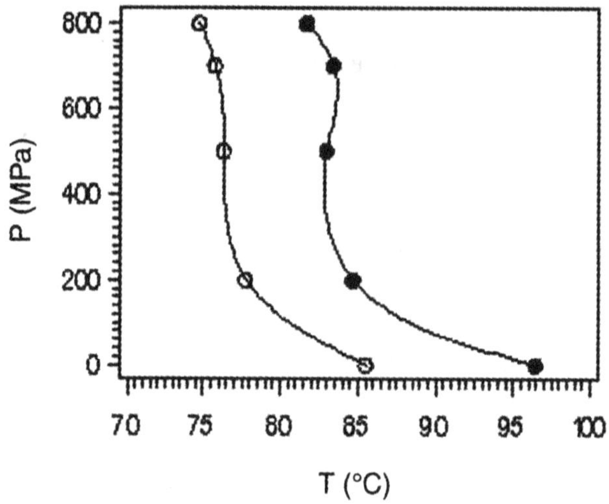

Figure 6–1 Pressure-temperature combinations resulting in 90% loss of total chlorophyll content after process time of 50 minutes (•) and 100 minutes (○) at the selected pressure-temperature combination.

This loss could be described by two consecutive first-order degradation steps, indicating that both pheophytin and pyropheophytin formation occurred. Both degradation steps were influenced by pressure in different ways (Figures 6–2 and 6–3). At pressure below 150 MPa, pressure and temperature acted synergistically with respect to the first degradation step and antagonistically with respect to the second (Weemaes et al., 1999). Hence, to minimize chlorophyll degradation and consequent loss of greenness in broccoli juice during processing, it is recommended to perform high pressure treatment at temperatures below 50°C.

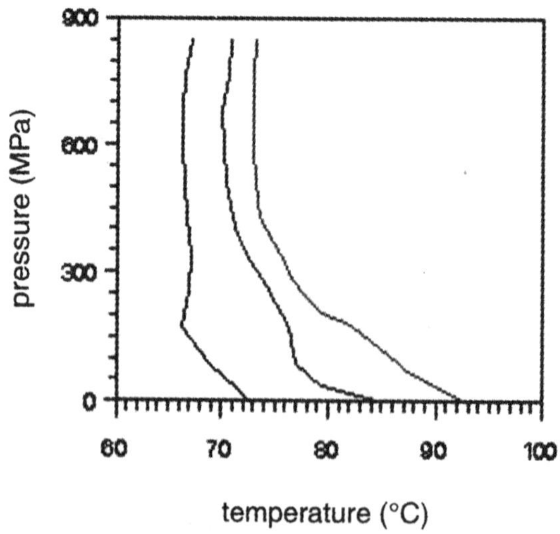

Figure 6–2 Pressure-temperature kinetic diagram for chlorophyll to pheophytin conversion in broccoli juice. From left to right: $k = 0.01$, 0.02, and 0.03 min^{-1}.

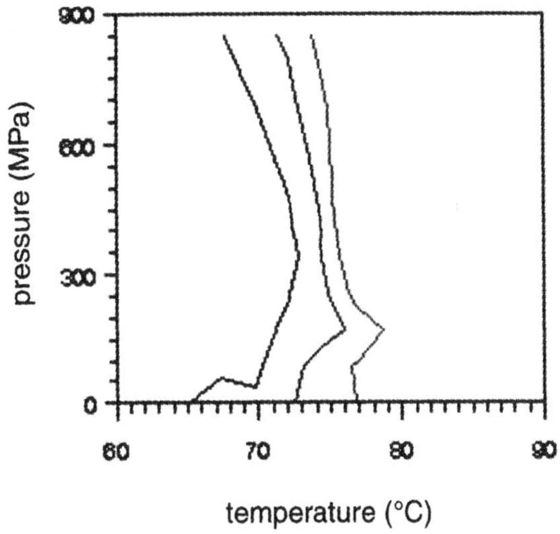

Figure 6–3 Pressure-temperature kinetic diagram for the degradation of pheophytin in broccoli juice. From left to right: k = 0.00025, 0.0005, and 0.00075 min^{-1} .

On the contrary, high pressure has been found to induce discoloration in mushrooms and onions, because of the activity of PPO, an enzyme responsible for browning (Butz et al., 1994; Matser & Bartels, 1999). PPO is a soluble enzyme predominantly localized in the cytosol of plant cells or bound to particulate cell fractions. In intact cells, phenolic compounds are confined to vacuoles and spatially separated from the enzyme. When the cell and tonoplast are disrupted, enzyme and substrate are brought together and phenolic oxidation products are formed, resulting in the formation of brown complexes. Microscopic studies revealed severe damage of the vacuoles of onion epidermis cells treated at 300 MPa at 25°C. At this low pressure, PPO, which is very pressure-resistant (see Chapter 5), is not inactivated and can catalyze the oxidation of the released phenolic compounds, resulting in enzymatic browning (Butz et al., 1994).

Milk and Dairy Products

Milk color is not affected by pressure to a large extent. Investigators have reported that L-values showed consistent changes, indicating some progressive change in brightness, whereas a- and b-values remained unchanged. Overall, color degradation could be described by a first-order model, indicating that higher pressure and longer treatment time resulted in more pronounced color changes. However, at pressures sufficient to reduce the microbial load to an acceptable level, only minor color changes were observed (Mussa & Ramaswamy, 1997). Similar findings were reported by Johnston et al. (1992). After pressurization for 1 hour at 300 MPa, the L-value (lightness) of milk decreased from 85.6 to 67.0, while higher pressure caused little further decrease. This increased translucence suggests that the micelle structures are altered and disintegrated into smaller fragments as a result of pressure treatment. Nevertheless, increased translucence is less pronounced than that which occurs when micelle disintegration is induced by calcium-chelating agents (Johnston et al., 1992; Mussa & Ramaswamy, 1997). Adapa et al. (1997) compared the effect of pressurization on color of concentrated and unconcentrated skim milks. In both cases, mean values for lightness were lower in pressurized samples than in unpressurized ones, because of the reduced light-scattering ability. In unconcentrated milk, pressurization led to lower a-values (greener) and b-values (bluer). In concentrated milk, pressure treatment imparted greener color only.

Meat

The effect of high pressure on meat and meat products has been studied by many authors because high pressure processing may offer potential to preserve and restructure meat, provided that the red color can be maintained. The intensity of meat visual color is due to its total myoglobin content, and its stability is determined by the occurrence of myoglobin in its ferrous oxygenated form. Different colors observed in meat are related to the relative proportions of oxymyoglobin (bright red), myoglobin (dark red), and metmyoglobin (gray-brown). Factors that increase the rate of myoglobin oxidation and hence meat discoloration include partial oxygen pressure, oxygen consumption by the tissues, concentration of multivalent ions, temperature, light, pH, microflora, and pressure. Indeed, application of pressure even at refrigerated temperatures (5°C to 10°C) induces drastic changes in the color of red meat (Cheftel & Culioli, 1997). A detailed quantitative study on pressure-induced changes in color and myoglobin in minced beef at 10°C was performed by Carlez et al. (1995). The color of pressure-processed meat samples became paler above 200 MPa (increase in L-value) and gray-brown above 400 MPa (decrease in a-value). Similar results were obtained at 20°C. Shigehisa et al. (1991) reported that L-values of pork slurries started to increase between 100 and 200 MPa, reached a maximum between 300 and 400 MPa, and did not change with further increase in pressure up to 600 MPa. On the other hand, the decrease in a-value was moderate between 100 and 200 MPa and progressed until 600 MPa. Similar findings were noted with regard to redness of beef meat (Nose et al., 1994).

Thus, meat discoloration through pressure processing appears to result from (1) a whitening effect (increase in L-value) in the pressure range of 200 to 350 MPa, possibly as a result of globin denaturation or heme displacement or release, and (2) a loss of red color (decrease in a-value) caused by oxidation of ferrous myoglobin into ferric myoglobin at or above 400 MPa (Cheftel & Culioli, 1997). Both degradation steps appear to depend more on critical pressure thresholds than on processing time.

During storage, untreated control samples turned gray-brown as a result of partial oxidation of oxymyoglobin into metmyoglobin. For samples pressurized at 500 MPa, L-, a-, and b-values did not change during storage, corresponding to a stable gray-brown color. On the other hand, meat samples processed at 350

MPa were pink and progressively lost their reddish color during storage; they turned a gray-brown color, resembling that of samples treated at 500 MPa. In absence of oxygen, the loss of red color (i.e., the oxidation of myoglobin) did not take place, but the whitening of the meat could not be prevented (Carlez et al., 1995). In addition, whitening and color changes were observed in low- and high-fat meat emulsions. In both cases, pressurization caused increased lightness, which became more pronounced with higher pressure and longer treatment time. In low-fat emulsions, pressure treatment at 300 MPa (10 to 20 minutes) reduced a- and b-values. In high-fat emulsions treated for 10 minutes between 100 and 300 MPa, similar results were obtained; opposite effects were found when pressurization time was prolonged to 20 minutes. In minced meat samples, pressurization up to 150 MPa did not affect L-, a-, and b-color values. However, higher pressure again resulted in increased lightness and reduced redness (Colmenero et al., 1997).

Thus, it can be concluded that pressure processing of fresh red meat is a viable alternative only when subsequent cooking is carried out before presenting the products for sale or consumption (as is the case with ready-to-eat meals). In contrast, high pressure processing of cured or white meats is unlikely to cause serious color problems.

Maillard Browning in Model Systems

The Maillard reaction, which is manifested by the development of brown color in processed food products, is known to be strongly dependent on temperature and pH. However, only a few studies have been carried out with regard to the effect of pressure. Tamoaka et al. (1991) reported development of brown color to be suppressed when pressures of 200 to 400 MPa were applied to a xylose-lysine system at 50°C in sodium carbonate buffer at pH 8.2. Using a glucose-lysine system at pH 8 and 10.1, Hill et al. (1996) observed that pressure treatment at 600 MPa enhanced Maillard browning; however, at pH 5.1 and 6.5, application of pressure depressed the rate of browning. Both for pressurized and unpressurized solutions, browning rate constants clearly increased with increasing pH. At atmospheric pressure, the rate increased about 40 times for each unit of pH; at 600 MPa, the rate increased about 1000 times. It is postulated that pressure enhances the rate of Maillard browning but, at pH lower than 7, this enhancement is overridden by the pressure-induced pH decrease. Under pressure, solutions will be at lower pH than at normal pressure, with concomitant reduction of the browning reaction (Hill et al., 1996).

THE EFFECTS OF HIGH PRESSURE ON THE FLAVOR OF FOOD PRODUCTS

Fruit and Vegetable Products

The potential of high pressure as a new processing technology for the production of fruit juices partly arises from the fact that fresh flavor can be maintained during pressure treatment. Ogawa et al. (1990) and Takahashi et al. (1993) reported that Satsuma mandarin juice maintained freshness and original flavor during high pressure processing. Results of extensive sensory testing of orange, grapefruit, and apple juices allowed Mermelstein (1999) to conclude that tasters could not differentiate between the flavor of pressure-processed and unprocessed juice made from the same raw material.

Researchers have shown a particular interest in the effects of high pressure on orange juice. Ogawa et al. (1992) stated that high pressure treatment of orange juice does not affect sensory properties. According to Parish (1994), an experienced sensory panel could not perceive differences between orange juice that was pressure treated for 30 seconds at 500 MPa and unpressurized juice. In addition, while the control juice showed signs of fermentation after 2 weeks at 6°C and became organoleptically unacceptable, pressure-treated juice remained free from fermentation for 5 weeks. In a more detailed study, Parish (1998) evaluated the effects of pressure in the range of 500 to 700 MPa on Valencia orange juice. Sensory panels generally judged the flavor of pressure-treated juices to be significantly closer to that of fresh/frozen control juices than thermally treated ones, although some significant differences in flavor between pressurized and control juices were observed. Also during storage, pressure-treated juices maintained their flavor advantages over heat-treated ones. Addition of even minimal heating (50°C to 60°C) became detrimental to overall sensory quality. According to Bignon (1996), sensory qualities of pressure-processed orange juice resemble those of freshly squeezed juice and remain so for more than 30 days. Thus, the retention of fresh flavor is a major advantage of cold, high pressure processing.

For fruit jams (e.g., strawberry jam) high pressure was found to retain fresh flavor much more than traditional thermal processing (Watanabe et al., 1991; Kimura et al., 1994).

Unlike for orange juice, the flavor of tomato juice and onions is largely affected by pressure. Sensory panels judged pressure-treated onions to smell less intensely than fresh ones and rather like cooked or fried onions (Butz et al., 1994). For tomato juice, samples treated at different pressure-temperature-time combinations proved to be absolutely inedible owing to a strong rancid taste (Poretta et al., 1995).

In addition to reports on the effect of pressure on overall flavor of fruit and vegetable products, there are reports in the literature containing detailed information about individual aroma compounds. Donsi et al. (1996) performed a gas chromatographic study on individual flavor components of orange juice. In pressure-treated orange juice, the content of limonene—the least stable aroma component, which is drastically changed by heat—is similar to that of fresh juice. In addition, pressure treatment did not cause significant modification of other aroma components such as terpinene, cymene, linalool, and myrcene. Head space gas chromatographic analysis of pressure-treated fruit jam indicated that most original flavor compounds (e.g., trans-2-hexenol, linalool, ethyl butyrate, methyl butyric acid) were retained to a much greater degree than with traditional heat treatment. This excellent quality was retained unchangeably during storage for 2 to 3 months at 5°C. On the other hand, storage at 25°C resulted in rapid quality deterioration because of dissolved oxygen and remaining enzyme activity (Watanabe et al., 1991; Kimura et al., 1994).

For onions, pressure treatment was reported to diminish dipropylsulfide concentration (a compound responsible for pungency and the characteristic odor of fresh onions) and to increase transpropenyldisulfide and 3,4-dimethylthiophene concentrations, leading to a change in flavor that seemed like braised or fried onions. Simultaneously, 2-methyl-pent-2-enal, the main aldol condensation product of propanal resulting from alliinase action, was increased because of pressure-induced cell injury and decompartmentalization (Butz et al., 1994).

Pressure treatment of tomato juice resulted in a decrease in hexanol and a marked increase in trans-2-hexenal and n-hexanal, the latter being responsible for fresh tomato flavor when present in a concentration of about 1 to 2 mg/kg. Higher concentrations impart a rancid flavor (Poretta et al., 1995).

In pressurized peach, various alcohols and aldehydes not present in fresh peach were identified, probably produced by enzyme-induced oxidation of unsaturated fatty acids. The concentration of benzaldehyde, in particular, was markedly increased. It is believed that benzaldehyde is enzymatically released from prunasin by disruption of fruit tissue during pressurization. Indeed, the enzyme activity of emulsin was found to remain after pressure treatment, and the adiabatic temperature increase may accelerate the enzymatic reaction. This benzaldehyde largely contributes to the flavor quality of pressure-treated peaches. Further, it was found that the concentration of γ-decalactone, likewise contributing to characteristic peach flavor, was not significantly changed by pressure treatment (Sumitami et al., 1994).

Yen and Lin (1999) studied the changes in volatile components of guava juice during pressure processing (25°C, 600 MPa, 15 minutes) and subsequent storage at 4°C and 25°C, using gas chromatography/mass spectrometry. Esters were the most volatile fraction in guava juice and alcohols the second most volatile, with ethyl-1-hexanoate being the most important flavor component. When compared to fresh juice, volatile compounds were not significantly changed by pressure; however, during storage at 4°C and 25°C, a decrease was noted. Total alcohols and esters as well as the contents of the individual components were relatively close to that of fresh juice after storage at 4°C for 30 days. Pressure-treated juice stored at 4°C maintained flavor stability better than untreated juice, probably because enzymes are partially inhibited. After storage for 60 days, alcohol concentration was markedly increased, caused not to post-processing contamination but by enzymatic activity. Significant changes were observed when the juice was stored at 25°C for 30 days. Methanol, ethanol, ethyl acetate, methyl-1-propionate, and 2-ethylfurfuran concentration increased while that of other compounds decreased.

Milk and Dairy Products

The effect of pressure on the flavor of milk and dairy products has only been studied to a very limited extent. High pressure processes at 400 to 500 MPa for 3 to 15 minutes followed by refrigeration were found to result in a shelf-life comparable to that of thermal pasteurization; however, this report made no mention of sensory properties of the pressure-processed milk (Rademacher & Kessler, 1996). Drake et al. (1997) reported that flavor scores from trained dairy panels were similar for pasteurized and pressurized milk cheeses; both types of cheeses were criticized for bitter and high-acid flavor defects, attributed to consistently low salt percentages. Raw milk cheese, on the other hand, scored even lower with respect to flavor because of fermented/fruity flavor defects caused by the presence of undesirable bacteria in raw milk.

Meat

Very little information is available on the effect of high pressure on meat flavor, although it has been reported that pressurized beef meat has a sweeter taste than control meat (Cheftel & Culioli, 1997). Suzuki et al. (1994) studied the effect of pressure on water-soluble compounds responsible for taste and meaty flavor. The amount of peptides and amino acids estimated as phenol-reagent-positive materials apparently increased with increasing pressure up to 300 MPa (5 minutes at 2°C). When muscle samples were stored at 2°C for 7 days, peptides and amino acids increased in the extracts of both untreated and pressurized muscles. The content of inosinic acid, a flavor enhancer, was reduced by pressurization. High-performance liquid chromatography (HPLC) analysis of soluble peptides revealed no significant changes in any fraction from pressurized muscles up to 200 MPa and a significant decrease in the peptide fraction from muscles pressurized at 300 to 400 MPa. It can be concluded that pressurization (up to 300 MPa) causes similar changes in the content of taste-related amino acids and peptides as meat conditioning. A combination of pressure-induced denaturation of muscle proteins and release of proteolytic enzymes from lysosomes could explain these observed changes (Cheftel & Culioli, 1997).

THE EFFECTS OF HIGH PRESSURE ON TEXTURE OF FOOD PRODUCTS

This section focuses on the effects of high pressure on the texture of fruits and vegetables, milk and dairy products, and meats. Changes in textural properties of starch- and protein/polysaccharide-based systems as well as the underlying mechanisms are discussed in detail in Chapter 7.

Fruit and Vegetable Products

High pressure treatment is often reported to render fruits and vegetables more pliable and to increase their softness. However, pressures up to 350 MPa can be applied to plant systems without any major effect on texture and structure (Knorr, 1995). Kasai et al. (1995, cited by Basak & Ramaswamy, 1998) found the relative hardness (ratio of hardness of pressure-treated vegetables to untreated vegetables) to increase substantially with the magnitude and retention time of the applied pressure. Relative hardness of vegetables treated at 400 MPa for 120 minutes was in the range of 2 to 3 after 30 minutes of cooking. For potato cubes, tissue softening induced by pressure was comparable to that induced by hot water blanching (Eshtiagi et al., 1994).

An extensive study on the effect of pressure on the texture of fruits and vegetables has been carried out by Basak and Ramaswamy (1998). They observed the change in firmness of treated samples to be dependent on both pressure level and pressurization time. In general, the softening curves revealed that texture changes caused by pressure occurred in two phases: (1) a sudden loss as a result of the pulse action of pressure, followed by (2) further loss or gradual recovery during pressure-holding phase. The instantaneous pressure softening was clearly dependent on the pressure level and on the type of product, and the sensitivity ranking was dependent on pressure intensity (Figures 6–4 and 6–5).

Pear was most pressure sensitive at 100 MPa; celery, at pressure higher than 200 MPa. At low pressure (100 MPa), slight loss of firmness was caused by compaction of cellular structures without disruption. At higher pressure (greater than 200 MPa), severe texture loss occurred because of rupture of cellular membranes and consequent loss of turgor pressure. Pressure holding time clearly affected overall texture of processed products. Many fruits and vegetables gradually recovered most of the instantaneous loss in texture after a holding time of 30 to 60 minutes at 100 to 200 MPa, and some became more firm than their fresh counterparts. However, for carrots and green pepper, further texture loss was observed at 200 MPa.

In many cases, pressure-treated vegetables did not soften during subsequent cooking, which is attributed to the action of pectinmethylesterase (PME). Pressure treatment leads to partial inactivation of PME (see Chapter 5) and disruption of structures, causing enzyme and substrate to come into contact with each other. Firming effects are thus mainly the result of PME-initiated deesterification of cell wall pectin, leading to cross-linking with divalent ions. Increased compactness of cellular structure following degassing of the tissue may also be involved. At higher pressure, products may be severely damaged, leading to permanent loss of texture.

Consistency of tomato products refers to their viscosity and ability of their solid portion to remain in suspension throughout the shelf life of the product. A high-consistency product has almost no syneresis (no separation of juice pulp and serum) (Poretta, 1996). Poretta et al. (1995) reported that high pressure treatment clearly affected viscosity of tomato juice and gave rise to a jelly-like, translucent structure caused by protein-tissue coagulation and compacting. For pressures above 700 MPa, viscosity increased with increasing pH, indicating that certain pH ranges enhance enzyme activity and result in viscosity changes

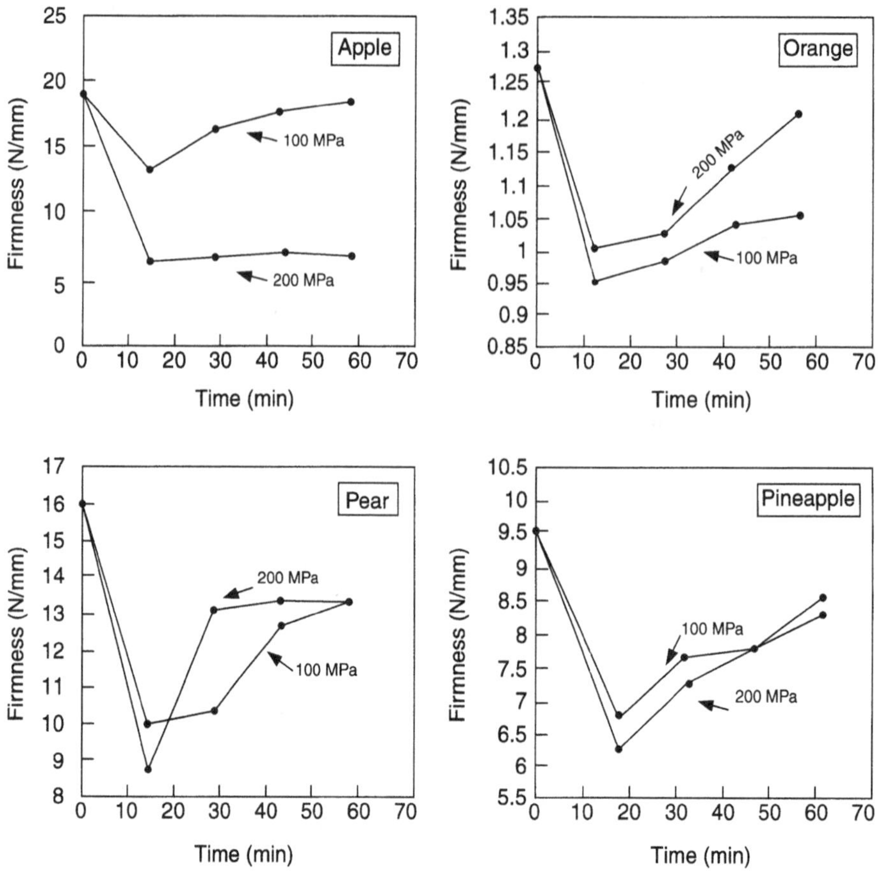

Figure 6–4 Effect of pressure level and treatment time on the firmness of fruits.

(Poretta et al., 1995). For guava puree, turbidity, viscosity, and cloud content were not significantly affected by pressure treatment (400 to 600 MPa). However, some decrease in turbidity, viscosity, and cloud content was observed during storage for 20 days (Yen & Lin, 1996).

Milk and Dairy Products

High pressure treatment only slightly influenced milk viscosity. The building up of milk viscosity can be described by a first-order reaction. The small negative activation volume indicates that the rate of buildup of milk viscosity increases at higher pressure levels. Increase in viscosity has been explained to be the result of casein micelle disintegration. However, viscosity increase may not be a serious drawback to pressure processing of milk, because the viscosity of milk is much less sensitive to pressure changes than are the microorganisms present in milk (Mussa & Ramaswamy, 1997).

Pressurized milk cheeses scored significantly lower texture scores than either raw or pasteurized milk cheeses. They were judged to have a pasty and weak tex-

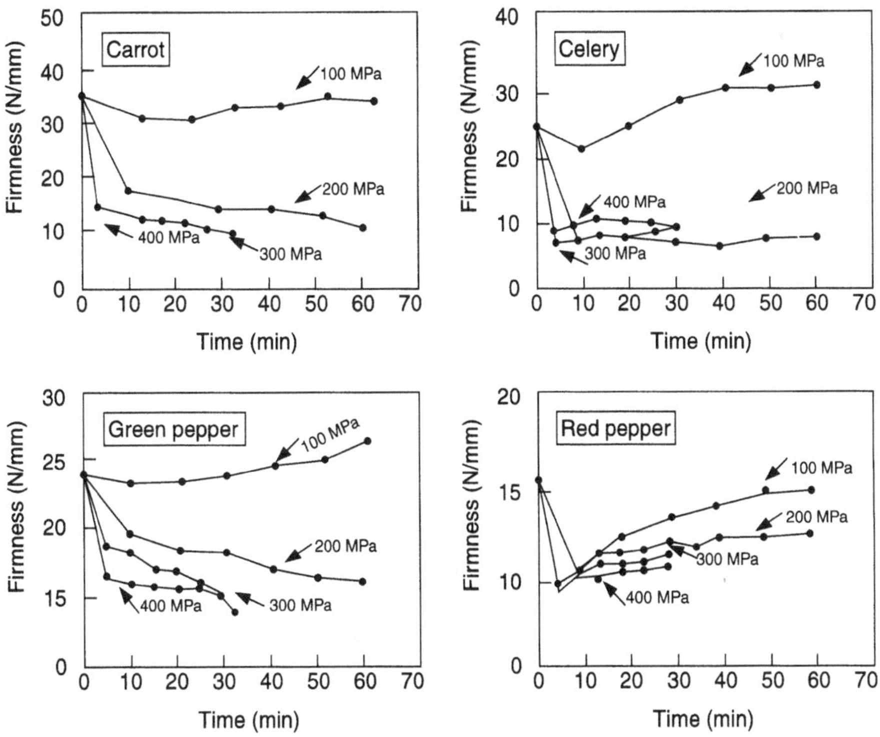

Figure 6–5 Effect of pressure level and treatment time on the firmness of vegetables.

ture, along with a whey-tainted flavor. These defects can be attributed to the larger moisture content and wet weight yield of pressurized cheeses (Drake et al., 1997). On the other hand, this higher moisture content may be useful in the manufacture of reduced-fat cheese (Datta & Deeth, 1999).

Meat

When various muscles are subjected to pressure, very firm and contracted raw meat is obtained. However, after cooking, pressurized meat is more tender and has higher moisture content and lower shear values than nonpressurized meat, mainly as a result of lower contraction and less drip loss. Sensory analysis shows pressurized meat to be less juicy, but more tender than control meat. The fact that tenderization instead of toughening occurs indicates damage of the sarcomere structure.

Many authors have reported brief exposure of pre-rigor meat to pressures in the range of 100 to 200 MPa to be effective for tenderization (Elgasim & Kennick, 1980; Ohmori et al., 1991). Post-rigor application of pressure, which is far more important commercially, has no beneficial effect on beef tenderness when carried out at temperatures lower than 30°C; in addition, tenderness, juiciness, and shear strength do not improve upon cooking meat afterwards. On the other hand, pressure (150 MPa) combined with heat (55°C to 60°C) is a very effective method to counteract toughening induced by cold-shortening (Bouton et al., 1977; Ohmori et al., 1991). This tenderization effect was attributed to modification of myofibrillar structure but not of connective tissue (Cheftel & Culioli, 1997). Application of higher pressure up to 500 MPa allowed meat tenderization without any additional heating (Suzuki et al., 1990), but it modified meat color and appearance. It can be concluded that pressure clearly offers some potentials with respect to meat tenderization, especially for cured and white meats or for ready-to-eat meals where color defects induced by pressure are negligible.

THE EFFECTS OF HIGH PRESSURE ON NUTRITIVE VALUE AND HEALTH COMPONENTS OF FOOD PRODUCTS

Fruit and Vegetable Products

Many authors have observed that the vitamin content of fruit and vegetable products is not significantly affected by high pressure processing. According to Bignon (1996), the contents of vitamin C, A, B_1, B_2, E, and folic acid are not changed by pressure treatment. About 95% of the vitamin C content of fresh strawberries was preserved in pressure-processed strawberry jam, which is about 1.7 times more than in heat-processed jam (Byrne, 1993; Kimura et al., 1994). In strawberry coulis/nectar, ascorbic acid content was maintained after pressure treatment of 400 MPa for 30 minutes at 20°C. In addition, the profiles of decreased nutritive value during storage of pressurized and untreated strawberry coulis were similar, indicating that pressure neither accelerated nor slowed down the degradation kinetics during storage (Sancho et al., 1999). Initial content of ascorbic acid of guava puree was not altered by pressure (400 to 600 MPa for 15 minutes) or heat treatment (88°C to 90°C for 24 seconds). In untreated and 400-MPa–treated puree, the ascorbic acid content began to decline after 10 and 20 days, respectively, while in heated and 600-MPa–treated puree, it remained constant during 30 and 40 days of storage (Yen & Lin, 1996). For freshly squeezed orange juice and mandarin orange juice, high pressure (up to 600 MPa) did not change the initial vitamin C content (Ogawa et al., 1990; Takahashi et al., 1993; Matser & Bartels, 1999). Donsi et al. (1996) studied the effect of pressure (200 to 500 MPa) on several vitamins (vitamin C, B_6, B_2, B_1, and niacin), sugars (sucrose, fructose, and glucose), and organic acids (malic acid, citric acid, and isocitric acid) in orange juices and found no substantial modifications at any of the pressure levels tested.

A more detailed kinetic study on pressure-temperature stability of ascorbic acid in buffer systems and natural environments (orange and tomato juice) was performed by Van den Broeck et al. (1998). Pressure-temperature degradation of ascorbic acid at temperatures from 65°C to 80°C in combination with pressure of 850 MPa followed first-order kinetics and proceeded more rapidly with increasing temperature, indicating that pressure and temperature act synergistically.

Ascorbic acid was clearly more sensitive to pressure in orange juice than in tomato juice. In addition, these authors noted that thermal sensitivity of ascorbic acid degradation was not dependent on the pressure level (Figure 6–6). These experiments demonstrated that ascorbic acid is only unstable at a combination of very high pressure with elevated temperature. Therefore, the authors concluded that mild treatments, which are of importance to the food industry, do not affect ascorbic acid.

Currently, relationships between human cancer incidence and dietary habits are being studied with a great deal of interest. Diets rich in fruits and vegetables are found to be associated with low incidence of many types of human cancer. However, the antimutagenic activity of most fruits and vegetables is considerably reduced upon heat treatment. Therefore, Butz et al. (1997a) performed a detailed study on the effect of pressure on antimutagenic activity of fruits and vegetables. Strawberry and grapefruit exerted moderate antimutagenic activity, even after pressure or heat treatment. Carrots, cauliflower, kohlrabi, leeks, and spinach were characterized by strong antimutagenic potencies that were sensitive to heat but not to pressure. The antimutagenic activities of beets and tomatoes were affected by pressure, but only at extreme conditions, (600 MPa at 50°C or 800 MPa at 35°C). In general, pressure processing of vegetable juices has less effect on their antimutagenicity than thermal processing, which destroys much of their protective potential.

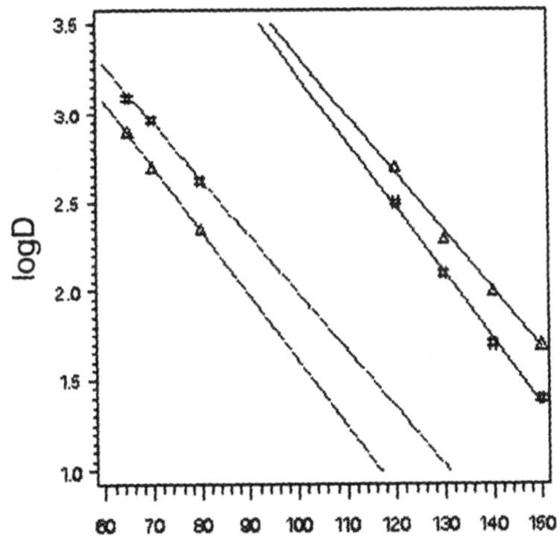

Figure 6–6 Temperature dependence of the D-value for thermal degradation of L-ascorbic acid in squeezed oranges at 0.1 (—Δ—) and 850 (- - Δ - -) MPa and in squeezed tomatoes at 0.1 (—#—) and 850 (- - # - -) MPa.

Milk and Dairy Products

β-lactoglobulin is a food allergen that is present in milk. When milk concentrates are added to modified milk for infants, the goal is to eliminate β-lactoglobulin without lowering the level of α-lactalbumin. Selective elimination of β-lactoglobulin from whey concentrates is possible by hydrolysis with thermolysin at elevated pressure. α-Lactalbumin is resistant to hydrolysis due to the presence of four disulfide bonds. When thermolysin digestion is performed under pressure, β-lactoglobulin is hydrolyzed much faster and more completely than at atmospheric pressure, while α-lactalbumin is not affected (Hayashi et al., 1987).

High pressure (400 to 1000 MPa) did not cause any variation in vitamin C content in egg yolk. When egg yolk was pressurized at 20°C and 200, 400, or 600 MPa for 30 minutes, ascorbate concentrations remained largely unchanged (92.6%, 101.3%, and 102.7%, respectively) (Sancho et al., 1999). Similar findings were reported for thiamin content of egg yolk (Hayashi et al., 1989).

Butz et al. (1997b) studied the effect of high pressure on aspartame in milk, using aspartame solutions of about 0.5 g/L, corresponding to the concentration in commercial diet chocolate milk. A treatment of 3 minutes at 60°C and 600 MPa induced 50% loss of active substance, while the non-sweet aspartylphenylalanine and diketopiperazine formed; the molar sum of these compounds is the symmetrical reflection of the aspartame degradation curve (Figure 6–7).

In buffer solution, smaller effects were observed indicating an additional effect of the milk matrix. The effects in aqueous solution were still smaller, because of

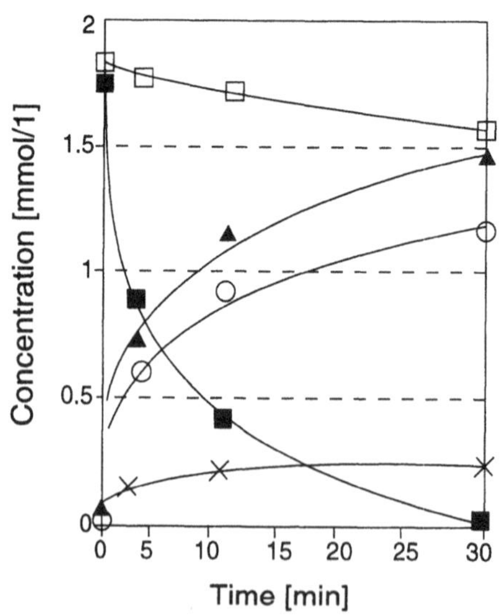

Figure 6–7 Pressure treatment of aspartame solutions of 0.5 g/L in full-cream milk at 600 MPa and 60°C: (■) aspartame; (□) aspartame control at 60°C; (▲) molar sum of diketopiperazine and aspartylphenylalanine; (○) diketopiperazine; (×) aspartylphenylalanine.

self-ionization of water resulting in lower pH and stabilization of aspartame. Hence, especially in dairy products of neutral pH containing aspartame (such as chocolate milk and ice cream), high pressure treatment may present problems. During pressure treatment of 600 MPa at 60°C and pH 7, 1.15 mmol diketopiperazine could be present after 5 minutes. Thus, a human individual of 40 kg drinking 1 L of pressure-treated chocolate milk would ingest the upper limit of diketopiperazine. Compensation of the pressure-related aspartame loss by higher initial concentration would result in even higher diketopiperazine concentrations in the end product.

Meat

Information on the effect of high pressure on the nutritional value and health components in meat is very scarce. Sancho et al. (1999) found that thiamin in pork meat was not degraded by pressure treatment at 125 to 600 MPa for 10 minutes to 18 hours at 20°C. Although it is unlikely that partial denaturation of muscle proteins and structures could reduce protein digestibility or biological value, this should be confirmed experimentally. It would also be useful to evaluate whether the nutritional availability of vitamins and iron is modified after pressure treatment (Cheftel & Culioli, 1997).

THE EFFECT OF HIGH PRESSURE ON LIPID OXIDATION IN FOOD PRODUCTS

Lipid oxidation is a major cause of quality deterioration of many food products, impairing both flavor and nutritional value. In addition, an increase in peroxidizing lipids may pose health risks, since they have been implicated in the development of coronary heart disease and cancer. Thus, it is important to consider the effect of high pressure on modification of the lipidic fraction of food products.

Vegetable Oils

Severini et al. (1997) studied pressure-induced lipid oxidation in seed oils and olive oil. The peroxide value of olive oil was not markedly changed because of pressure treatment; on the other hand, for seed, sunflower, and grape-stone oil, a more evident increase was found. The latter two samples had the highest level of unsaturated fatty acids, which affect lipid oxidation; for this reason, they were least resistant against oxidation. The para-anisidine value, an indication of secondary oxidation products with a carbonylic functional group, generally increased after pressure treatment. It can be concluded that olive oil samples are generally more resistant to oxidation than seed oil samples. Hence, to prolong the shelf life of complex foods containing a lipid fraction, it is advisable to use olive oil instead of seed oil when using high pressure processing.

Fish and Meat

Pressure treatment (800 MPa for 20 minutes) of rendered pork fat ($a_w = 0.44$) reduced the induction period (length of the initial stage of very slow oxidation) from that found for the control samples. In addition, during storage, pressure-treated samples showed higher peroxide values, and the effect became more pronounced with higher pressure. At this low water activity, pressure is catalytic to oxidation. Pork fat contains about 1.5 ppm iron and 0.4 ppm copper. Under pressure, these transition metals can be freed from complexes and act as powerful pro-oxidants. However, at water activity levels higher than 0.55, the reverse was noted; high pressure inhibited lipid oxidation due to the destruction of peroxides. In addition, the free metal ions become hydrated and lost their catalytic power. However, at high temperature the inhibiting/protective effect of pressure on lipid oxidation diminished. When the water activity level is between 0.4 and 0.55, the catalytic effect of the freed metal ions probably overrides the inhibiting effect of peroxide destruction (Cheah & Ledward, 1995).

In a subsequent study, Cheah & Ledward (1997) found that addition of citric acid prior to pressure treatment of rendered pork fat, washed fibers, and minced pork meat reduced the extent of pressure-induced catalysis, suggesting that citric acid may chelate metal ions released during pressure treatment. However, citric acid could not completely inhibit the increased rate of lipid oxidation in washed fibers and minced pork meat. Addition of citric acid shifted the pH from 6.5 to 4.9. In this case, the effectiveness of citric acid as a chelating agent might be counteracted by the increase in lipid oxidation due to the decrease in pH. On the other hand, in the presence of di-sodium ethylene diamine tetra-acetate (Na_2EDTA), pressure catalyzed oxidation in washed fibers and minced pork was totally inhibited, although the pH decreased. This indicates that transition metal ion catalysis is the major factor in increasing lipid oxidation in pressurized meat samples (Cheah & Ledward, 1997).

Dissing et al. (1997) have reported a more quantitative description of oxidation kinetics in meat subjected to high pressure. Turkey meat, which is rather susceptible to oxidation due to its relatively large content of membrane-associated phospholipids in combination with a low endogenous level of tocopherols, was used for this purpose. Pressure treatment (100 to 500 MPa at 10 to 30 minutes) enhanced lipid oxidation in turkey thigh muscle during subsequent chilled storage as compared to nontreated meat. The extent of enhancement was mainly dependent on the pressure level applied, at least above a certain threshold pressure. Indeed, at 100 MPa, pressurization time had a significant effect on the extent of lipid oxidation. However, only the most intense pressure treatment resulted in lipid oxidation comparable to that induced by thermal treatment.

For cod muscle exposed to 200 to 600 MPa for 15 to 30 minutes, the peroxide value of the extracted lipids increased with increased pressure and processing time. Even more pronounced effects of pressure were observed for mackerel muscle lipids (Ohshima et al., 1993). Likewise a mixture of sardine oil and defatted sardine meat showed increased oxidation when treated at 100 MPa for 30 to 60 minutes. On the other hand, when extracted sardine oils were treated with pressure up to 500 MPa, only minimal oxidation occurred (Tanaka et al., 1991). Thus, it can be concluded that isolated extracted marine lipids are generally quite stable against oxidation under high pressure up to 600 MPa. By contrast, the oxidation of lipids in fish muscle is accelerated by pressure treatment because of the presence of metal ions in the fish meat, which play an important role in catalyzing lipid autoxidation (Ohshima et al.).

In conclusion, pressure-induced lipid oxidation may limit the usefulness of high pressure technologies for meat-based products, unless suitable packaging or antioxidants are used. Metal chelators that can effectively remove the initiators of catalysis have been proposed as the most appropriate antioxidants to prevent lipid oxidation in pressure-processed meats.

CONCLUSION

Most review articles that point out the potential of high pressure as a novel food preservation and processing method state that pressure-treated foods keep their original flavor, color, and freshness. However, although detailed, systematic data on the effects of pressure on microbial and enzymatic inactivation are reported in the literature (see Chapters 3, 4, and 5), the effects of pressure on individual aspects of food quality are much less documented. Information in the literature is mainly fragmentary, and systematic, quantitative data are scarce. Such detailed information is indispensable to establish industrial high pressure processes for preservation and production of food products that comply with the Novel Food Regulation and hazard analysis of critical control points (HACCP) principles. It is important to ensure that these new technologies increase overall food quality or induce new functional properties (and thus address consumers' concerns) within the constraints of microbial safety (and thus address legislators' concerns). In addition, it is important to address concerns about possible occurrence of toxic or allergenic compounds in pressure-treated products. For all of these reasons, there is a need for more systematic studies of chemical effects involved in high pressure processing of food products.

REFERENCES

Adapa, S., Schmidt, K.A., & Toledo, R. (1997). Functional properties of skim milk processed with continuous high pressure throttling. *J. Dairy Sci., 80,* 1941–1948.

Basak, S., & Ramaswamy, H.S. (1998). Effect of high pressure processing on texture of selected fruit and vegetables. *J. Text. Stud., 29,* 587–601.

Bignon, J. (1996). Cold pasteurizers Hyperbar for the stabilization of fresh fruit juices. *Fruit Processing, 2,* 46–48.

Bouton, P.E., Ford, A.L., Harris, P.V., Macfarlane, J.J., & O'Shea, J.M. (1977). Pressure-heat treatment of post-rigor muscle: Effects on tenderness. *J. Food Sci., 42,* 132–135.

Butz, P., Edenharder, R., Fister, H., & Tauscher, B. (1997a). The influence of high pressure processing on antimutagenic activities of fruit and vegetable juices. *Food Res. Int., 30*(3/4), 287–291.

Butz, P., Fernandez, A., Fister, H., & Tauscher, B. (1997b). Influence of high hydrostatic pressure on aspartame: Instability at neutral pH. *J. Agric. Food Chem., 45,* 302–303.

Butz, P., Koller, D., & Tauscher, B. (1994). Ultra-high pressure processing of onions: Chemical and sensory changes. *Lebensm. Wiss. Technol., 27,* 463–467.

Byrne, M. (1993). The heat is off. *Food Eng. Int.,* 35–37.

Carlez, A., Veciana-Nogues, T., & Cheftel, J.-C. (1995). Changes in colour and myoglobin of minced beef meat due to high pressure processing. *Lebensm. Wiss.Technol., 28,* 528–538.

Cheah, P.B., & Ledward, D.A. (1995). High-pressure effects on lipid oxidation. *JAOCS, 72*(9), 1059–1063.

Cheah, P.B., & Ledward, D.A. (1997). Catalytic mechanism of lipid oxidation following high pressure treatment in pork fat and meat. *J. Food Sci., 62,* 1135–1138.

Cheftel, J.-C. (1991). Applications des hautes pressions en technologie alimentaire. *Ind. Aliment. Agric., 108*(3), 141–153.

Cheftel, J.-C., & Culioli, J. (1997). Effects of high pressure on meat: A review. *Meat Sci., 46*(3), 211–236.

Colmenero, F.J., Carballo, J., Fernandez, P., Baretto, G., & Solas, M.T. (1997). High-pressure-induced changes in the characteristics of low-fat and high-fat sausages. *J. Sci. Food Agric., 75,* 61–66.

Datta, N., & Deeth, H.C. (1999). High pressure processing of milk and dairy products. *Austr. J. Dairy Technol., 54,* 41–48.

Dissing, J., Bruun-Jensen, L., & Skibsted, L.H. (1997). Effect of high pressure treatment on lipid oxidation in turkey thigh muscle during chill storage. *Z. Lebensm. Unters. Forsch., 205,* 11–13.

Donsi, G., Ferrari, G., & di Matteo, M. (1996). High pressure stabilization of orange juice: Evaluation of the effects of process conditions. *Ital. J. Food Sci., 2,* 99–106.

Drake, M.A., Harrison, S.L., Asplund, M., Barbosa-Canovas, G., & Swanson, B.G. (1997). High pressure treatment of milk and effects on microbiological and sensory quality of cheddar cheese. *J. Food Sci., 62*(4), 843–845.

Elgasim, E.A., & Kennick, W.H. (1980). Effect of pressurization of pre-rigor beef muscles on protein quality. *J. Food Sci., 45,* 1122–1124.

Eshtiagi, M.N., Stute, R., & Knorr, D. (1994). High pressure and freezing pretreatment effects on drying, rehydration, texture and color of green beans, carrots and potatoes. *J. Food Sci., 59,* 1168–1170.

Fennema, O.R., & Tannenbaum, S.R. (1985). Introduction to food chemistry. In O.R. Fennema (Ed.), *Food Chemistry* (2nd ed., pp. 1–21). New York and Basel: Marcel Dekker.

Galazka, V.B., & Ledward, D.A. (1995). Developments in high pressure food processing. *Food Technol. Int. Europe,* 123–125.

Hayashi, R. (1989). Application of high pressure to food processing and preservation: Philosophy and development. In W.E.L. Spiess & H. Schubert (Eds.), *Engineering and Food* (pp. 815–826). London: Elsevier Applied Science.

Hayashi, R., Kawamura, Y., & Kunugi, S. (1987). Introduction of high pressure to food processing: Preferential proteolysis of β-lactoglobulin in milk. *J. Food Sci., 52,* 1107–1108.

Hayashi, R., Kawamura, Y., Nasaka, T., & Okinaka, O. (1989). Applications of high pressure to food processing: Pressurization of egg white and yolk, and properties of gels formed. *Agr. Biol. Chem., 53,* 2935–2939.

Hendrickx, M., Ludikhuyze, L., Van den Broeck, I., & Weemaes, C. (1998). Effects of high pressure on enzymes related to food quality. *Trends Food Sci. Technol., 9,* 197–203.

Hill, V.M., Ledward, D.A., & Ames, J.M. (1996). Influence of high hydrostatic pressure and pH on the rate of Maillard browning in a glucose-lysine system. *J. Agric. Food Chem., 44,* 594–598.

Johnston, D.E., Austin, B.A., & Murphy, R.J. (1992). Effects of high hydrostatic pressure on milk. *Milchwissenschaft, 44*(12), 760–763.

Kimura, K., Ida, M., Yoshida, Y., Ohki, K., Fukumoto, T., & Sakui, N. (1994). Comparison of keeping quality between pressure-processed and heat-processed jam: Changes in flavor components, hue and nutritional elements during storage. *Biosci. Biotechnol. Biochem., 58,* 1386–1391.

Knorr, D. (1993). Effects of high-hydrostatic-pressure processes on food safety and quality. *Food Technol., 47,* 156–161.

Knorr, D. (1995). High pressure effects on plant derived foods. In D.A. Ledward, D.E. Johnston, R.G. Earnshaw, & A.P.M. Hasting (Eds.), *High Pressure Processing of Foods* (pp. 123–135). Loughborough: Nottingham University Press.

Lopez-Malo, A., Palou, E., Barbosa-Canovas, G.V., Welti-Chanes, J., & Swanson, B.G. (1998). Polyphenoloxidase activity and color changes during storage of high hydrostatic pressure treated avocado puree. *Food Res. Int., 31*(8), 549–556.

Matser, A.M., & Bartels, P.V. (1999). Nieuwe eigenschappen met bekende ingrediënten. *VMT, 13,* 30–32.

Mermelstein, N.H. (1997). High-pressure processing reaches the US market. *Food Technol., 51,* 95–96.

Mermelstein, N.H. (1999). High pressure pasteurization of juice. *Food Technol., 53,* 86–90.

Mertens, B. (1992). Under pressure. *Food Manufacture, 11,* 23–24.

Mussa, D.M., & Ramaswamy, H.S. (1997). Ultra high pressure pasteurization of milk: Kinetics of microbial destruction and changes in physico-chemical characteristics. *Lebensm. Wiss. Technol., 30,* 551–557.

Nose, M., Iwaki, A., & Hattori, M. (1994). Development of processing system of meat and meat products by introduction of high pressure treatment. In R. Hayashi, S. Kunugi, S. Shimada, & A. Suzuki (Eds.), *High Pressure Bioscience* (pp. 270–272). Kyoto: San-Ei Suppan.

Ogawa, H., Fukuhisa, K., & Fukumoto, H. (1992). Effect of hydrostatic pressure on sterilisation and preservation of citrus juice. In C. Balny, R. Hayashi, K. Heremans, & P. Masson (Eds.), *High Presssure and Biotechnology,* Colloque INSERM (Vol. 224, pp. 269–278). Montrouge: John Libbey Eurotext.

Ogawa, H., Fukuhisa, K., Kubo, Y., & Fukumoto, H. (1990). Pressure inactivation of yeasts, moulds and pectinesterase in Satsuma mandarin juice: Effects of juice concentration, pH and organic acids and comparison with heat sanitation. *Agric. Biol. Chem., 54,* 1219–1225.

Ohmori, T., Shigehisa, T., Taji, S., & Hayashi, R. (1991). Effect of high pressure on the protease activities in meat. *Biol. Chem., 55,* 357–361.

Ohshima, T., Ushio, H., & Koizumi, C. (1993). High pressure processing of fish and fish products. *Trends Food Sci. Technol., 4,* 370–375.

Palou, E., Lopez-Malo, A., Barbosa-Canovas, G.V., Welti-Chanes, J., & Swanson, B.G. (1999). Polyphenoloxidase activity and color of blanched and high hydrostatic pressure treated banana puree. *J. Food Sci., 64,* 42–45.

Parish, M.E. (1994). Isostatic high pressure processing of orange juice. In R. Ahvenainen, T. Matilla-Sandholm, & T. Ohlsson (Eds.), *Minimal Processing of Foods (VTT symposium 142)* (pp. 93–102). Espoo, Finland: VTT Offsetpaino.

Parish, M.E. (1998). Orange juice quality after treatment by thermal pasteurization and isostatic high pressure. *Lebensm. Wiss. Technol., 31,* 439–442.

Poretta, S. (1996). New finding on tomato products. *Fruit Processing, 2,* 58–65.

Poretta, S., Birzi, A., Ghizzoni, C., & Vicini, E. (1995). Effects of ultra-high hydrostatic pressure treatments on the quality of tomato juice. *Food Chem., 52,* 35–41.

Rademacher, B., & Kessler, H.G. (1996). High pressure inactivation of micro-organisms and enzymes in milk and milk products. In Abstracts of Poster presentations, *High Pressure Processing of Foods Symposium,* (p. 10). Cologne: Flair Flow Europe.

Salanski, P., Jurczak, J., Fonberg-Broczek, M., & Szczepek, J. (1997). Influence of ultra high pressure on chosen natural pigments. In K. Heremans (Ed.), *High Pressure Research in the Biosciences and Biotechnology,* (pp. 343–345). Leuven: Leuven University Press.

Sale, A.H.J., Gould, G.W., & Hamilton, W.A. (1970). Inactivation of bacterial spores by hydrostatic pressure. *J. Gen. Microbiol., 60,* 323–334.

Sancho, F., Lambert, Y., Demazeau, G., Largeteau, A., Bouvier, J.-M., & Narbonne, J.-F. (1999). Effects of ultra-high hydrostatic pressure on hydrosoluble enzymes. *J. Food Eng., 39,* 247–253.

Severini, C., Romani, S., Dall'Aglio, G., Rovere, P., Conte, L., & Lerici, C.R. (1997). High pressure effects on oxidation of extra virgin olive oils. *Int. J. Food Sci., 3*(9), 183–191.

Shigehisa, T., Ohmori, T., Saito, A., Taji, S., & Hayashi, R. (1991). Effects of high hydrostatic pressure on characteristics of pork slurries and inactivation of microorganisms associated with meat and meat products. *Int. J. Food Microbiol., 12,* 207–216.

Smelt, J.P.P.M. (1998). Recent advances in the microbiology of high pressure processing. *Trends Food Sci. Technol., 9,* 152–158.

Sumitami, H., Suekane, S., Nakatani, A., & Tatsuka, K. (1994). Changes in the composition of volatile compounds in high pressure treated peach. *J. Agric. Food Chem., 42,* 785–790.

Suzuki, A., Homma, N., Fukuda, A., Hirao, K., & Uryu, T. (1994). Effects of high pressure treatment on the flavor-related components in meat. *Meat Sci., 37,* 369–379.

Suzuki, A., Watanabe, M., Iwamura, K., Ikeuchi, Y., & Saito, M. (1990). Effect of high pressure on the ultrastructure and myofibrillar protein of beef skeletal muscle. *Agric. Biol. Chem., 54,* 3085–3091.

Takahashi, Y., Ohta, H., Yonei, H., & Ifuku, Y. (1993). Microbicidal effect of hydrostatic pressure on Satsuma mandarin juice. *Int. J. Food Sci. Technol., 28,* 95–102.

Tamoaka, T., Itoh, N., & Hayashi, R. (1991). High pressure effect on the Maillard reaction. *Agric. Biol. Chem., 55,* 2071–2074.

Tanaka, M., Zhuo, X.Y., Nagashima, Y., & Taguchi, T. (1991). Effect of high pressure on the lipid oxidation in sardine meat. *Bull. Jap. Soc. Sci. Fish., 57,* 957–963.

Thakur, B.R., & Nelson, P.E. (1998). High pressure processing and preservation of food. *Food Rev. Int., 14,* 427–447.

Van den Broeck, I., Ludikhuyze, L., Weemaes, C., Van Loey, A., & Hendrickx, M. (1998). Kinetics of isobaric-isothermal degradation of L-ascorbic acid. *J. Agric. Food Chem., 46,* 2001–2006.

Van Loey, A., Ooms, V., Weemaes, C., Van den Broeck, I., Ludikhuyze, L., Indrawati, I., Denys, S., & Hendrickx, M. (1998). Thermal and pressure-temperature degradation of chlorophyll in broccoli (*Brassica oleracea* L. *italica*) juice: A kinetic study. *J. Agric. Food Chem., 46,* 5289–5294.

Watanabe, M., Arai, E., Kumeno, K., & Homma, K. (1991). A new method for producing non-heated jam sample: The use of freeze concentration and high pressure sterilisation. *Biol. Chem., 55,* 2175–2176.

Weemaes, C., Ooms, V., Ludikhuyze, L., Van den Broeck, I., Van Loey, A., & Hendrickx, M. (1999). Pressure-temperature degradation of green color in broccoli juice. *J. Food Sci., 64,* 504–508.

Yen, G.-C., & Lin, H.-T. (1996). Comparison of high pressure treatment and thermal pasteurization effects on the quality and shelf life of guava puree. *Int. J. Food Sci. Technol., 31,* 205–213.

Yen, G.-C., & Lin, H.-T. (1999). Changes in volatile flavor components of guava juice with high pressure treatment and heat processing and during storage. *J. Agric. Food Chem., 47,* 2082–2087.

CHAPTER 7

Effects of High Pressure on Protein- and Polysaccharide-Based Structures

Martin Michel and Karin Autio

INTRODUCTION

Since the early 1980s, high hydrostatic pressure has been evaluated as an alternative to classical heat treatment food-processing technologies for energy-saving reasons. The main focus of high pressure applications has been food preservation at moderate temperatures, and Japan has been the main country to apply high pressure technologies (Dumoulin et al., 1998). However, the baroresistance of spores (for review see Smelt, 1998) and the high investment costs have until recently prevented food manufacturers from using this technology widely. In the past five years, high pressure treatment has been identified as a potential area for food texture engineering because of its influence on food ingredient properties. Unlike thermal pasteurization or sterilization, pressure treatment is generally considered to be less destructive to the food and food ingredients.

Pressure affects molecules (e.g., inactivation/activation of enzymes, protein denaturation); can change the state of molecular organization (e.g., crystallization of fat, phase changes of phospholipids, denaturation-induced aggregation and gelation, gelatinization of crystalline starch, depolymerization); and can influence the architecture of food tissues (e.g., cell wall breakage caused by the presence of compressible intercellular air spaces). With respect to the final food properties, the induced changes can either be positive (e.g., gel formation or inactivation of undesired enzymes) or negative (e.g., activation of undesired enzymes or loss of texture). This chapter discusses the effects of pressure on food raw materials, starch dispersions, and proteins and pectins (single and mixed) and discusses its potential use in food structure engineering.

This work was performed in the frame of the European Project FAIR-CT-1175, "Combined High Pressure Thermal Treatments of Foods: A Kinetic Approach to Safety and Quality Evaluation." The work carried out at Nestec was financially supported by the Swiss Federal Institute for Education and Science, contract no. 95.0895. The authors thank Dr. Elizabeth Prior for English corrections and Professor N. Garti, Professor V. Tolstoguzov, and Dr. M. Leser for helpful scientific input.

PRESSURE-RELATED ALTERATIONS IN FOOD RAW MATERIALS

Pressure acts simultaneously on the tissue's morphology, cell organelles, membranes, and at the molecular level. Because fruits and vegetables vary with respect to their tissue anatomy and composition, pressure effects differ significantly from one plant species to another. A major hurdle for high pressure treatment of plant tissues is the presence of intercellular air spaces. Because of the high compressibility of air, the tissue is severely compressed, resulting in cell wall breakage (if the cell wall is not flexible enough), membrane disruption, loss of compartmentalization, and liberation of cellular compounds. To avoid such negative effects, it is generally recommended to fill intercellular spaces with an incompressible liquid such as water (which is only compressible by 15% at 600 MPa) or a buffer before high pressure treatment.

Several studies have shown that pressure treatment of vegetables can cause both firming (Fuchigami et al., 1995; Kasai et al., 1995a, 1995b; Stute et al., 1996) and softening of texture (Basak & Ramaswamy, 1998). Table 7–1 shows the relative loss of firmness of fruits and vegetables upon pressure treatment. In celery and apple, great changes occur at pressures of 200 MPa and higher. The initial texture loss is followed either by a gradual recovery or further loss during the pressure hold period. At pressures of 100 and 200 MPa, fruits such as oranges, pears, and pineapples recover most of the initial texture loss during a pressure holding time of 30 to 60 minutes, whereas apples do not show any recovery at 200 MPa. With vegetables such as carrots, celery, and green and red peppers, samples treated at 100 MPa were, despite the initial texture loss, firmer after the treatment than the corresponding fresh samples. Celery and red peppers showed recovery at all pressures, whereas carrots and green pepper showed texture loss at pressures of 200 MPa and higher. Potatoes have been reported to become softer at pressures higher than 125 MPa (Crelier et al., 1998).

Currently it is not clear what mechanisms are responsible for the textural changes. In intact cells, polygalacturonase or pectinmethylesterase (PME) are

Table 7–1 Instantaneous Loss of Texture at Various Pressure Levels

	Texture loss (%)			
Sample	*100 MPa*	*200 MPa*	*300 MPa*	*400 MPa*
Carrots	13.5	47.0	57.5	59.0
Celery	16.5	69.0	71.0	67.0
Green peppers	6.0	15.5	19.0	30.0
Red peppers	32.5	34.0	37.0	35.0
Apples	36.5	68.5	—	—
Pears	43.0	44.0	—	—
Pineapples	33.5	40.5	—	—
Oranges	29.0	29.0	—	—

bound to the cell walls. Upon high pressure treatment, the enzymes might be liberated. It has been suggested that the hardening of some plants in high pressure treatment is the result of insufficient inactivation of liberated PME (Basak & Ramaswamy, 1996). Because of the complexity of food tissues Knorr proposed to use tissue or cell cultures as plant model systems to study the impact of pressure (Knorr, 1994). Using *Chenopodium rubrum* and *Morinda citrifolia*, Dornenburg and Knorr (1993) found that viability was completely lost at pressures of 110 MPa and 75 MPa, respectively.

Ultrastructural changes in cauliflower and spinach treated with high hydrostatic pressure were investigated by cryo-fracture scanning electron microscopy (Prestamo & Arroyo, 1998). High pressure changed cell permeability and allowed the movement of water and metabolites from inside to outside of the cell. Cauliflower was more suitable than spinach for high hydrostatic pressure treatment. Spinach leaves lost nutrients after treatment as a consequence of damage to the molecular structure of the lipid-peptide complex and disruption of the phosphatidic acid bilayer membrane. Because the potential of high pressure processing for meat was extensively reviewed by Cheftel and Culioli (1997), it will not be discussed here.

BEHAVIOR OF STARCH DISPERSIONS UNDER PRESSURE

Starch granules are semicrystalline particles that are insoluble in water at room temperature. The sizes and forms are specific to each plant species, and the diameters of granules vary from 2 to 100 μm. Starch granules are composed of two polysaccharides, amylose and amylopectin. Amylose is mainly linear, consisting of about 100 to 10,000 glucose monomers linked by 1–4 α-bonds. Amylopectin is highly branched with 1–6 α-links. The ratio of these two components varies according to the plant source. In normal cereal starches, amylose content is in the range of 25% to 31% (Shannon & Garwood, 1984). Corn, sorghum, rice, barley, and potato have mutants whose starches consist mainly of amylopectin. These waxy mutants contain from 0% to 8.3% amylose. Amylose has not been shown to contribute to the crystallinity of the native granule. With sufficient heat, a starch water system or a starch-containing raw material undergoes a series of dramatic changes, referred to as gelatinization and pasting. At a certain temperature, the starch granule loses its crystallinity; at another temperature the anisotropic order disappears, which can be seen as a loss of birefringence. Some starch leaches out of the swollen granules and dissolves in the surrounding aqueous phase. Gel formation during cooling and increase of rigidity during storage are two separate reactions with different kinetics. The short-term changes are related to the formation of a solubilized amylose gel in the continuous phase, and long-term hardening is attributed to crystallization of short amylopectin chains within the granule (Miles et al., 1985; Orford et al., 1987; Ring et al., 1987).

Starches in aqueous environments can be gelatinized under pressure, but the conditions (pressure level, treatment time, temperature, and water content) required for gelatinization depend considerably on the origin of the starch (Autio & Stolt, 1998). Starches with high amylose/amylopectin ratios are very resistant to high pressure. Pressure-induced physicochemical changes are very similar to those induced during heat treatment, but rheological properties differ greatly.

Pressure-Induced Rheological and Microstructural Changes

According to Stolt et al. (1999), the viscosity of a waxy cornstarch suspension started to increase slightly during a 110-minute pressure treatment at 450 MPa at 30°C, and the higher the applied pressure the faster the increase (Figure 7–1). After the consistency index (*K*-value) had reached a value of 20 to 25 Pas, no more changes were observed during further pressurization. A sigmoidal-type model was found to adequately correlate the *K* values obtained at different holding times and processing pressures with the *F* value of the process. The *F* value represents an equivalent processing time at specified reference pressure and temperature conditions, defined analogously to traditional thermal processing terminology. The storage modulus (G') of the pressure-treated starch dispersions behaved almost similarly to the *K* value as a function of holding time. Within the pressure range 450 to 500 MPa at 50°C, the rate of viscosity increase as a function of holding time was higher than at 30°C. However, the maximum viscosity was higher at 30°C than at 40°C or 50°C with longer treatment time. Stute et al. (1996) reported that a 5% waxy cornstarch dispersion was completely gelatinized at 600 MPa and 20°C after a holding time of 15 minutes. Waxy cornstarch was the only starch that lost its granular structure under pressure (Stute et al., 1996; Autio et al., in press).

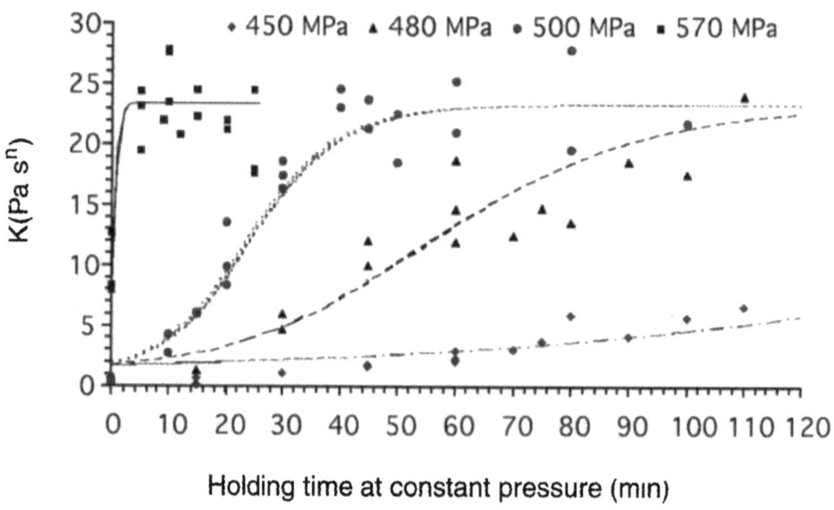

Figure 7–1 Predicted and experimental *K* values as a function of holding time and processing pressure at 30°C for 10% waxy cornstarch.

High-amylose cornstarch with 56% amylose was most resistant to high pressure. No viscosity increase was observed at a pressure of 900 MPa, even after a 50-minute holding time. Microstructural examination showed that no detectable swelling of starch granules or loss of birefringence was observed under these severe conditions (Autio et al., in press). High-amylose cornstarch has been shown to be B-type on X-ray diffractogram. B-starches have been shown to be more resistant to pressure treatment than A-starches (such as normal cereal starches) (Ezaki & Hayashi, 1992).

Only very minor rheological changes were observed in pressure-treated barley starches (Stolt et al., 2000). The *K* value increased slightly with increasing pressure and holding time. At 400 MPa, the increase in viscosity was very slight, even after a holding time of 75 minutes. At 600 MPa, maximum viscosity was reached almost directly after the pressure buildup.

Microstructural examination revealed that barley starch lost its birefringence at a lower pressure than did waxy cornstarch. The granular structure of barley starch seems to be very resistant under pressure.

Researchers have reported that the microstructural and rheological properties of 10% heat-treated barley starch are completely different from those of the pressure-treated samples. During heat treatment, amylose leached out and formed a rather rigid gel during cooling (Autio, 1990; Stolt et al., 2000). At 25%, barley starch formed a rigid paste under pressure. In Figure 7–2, the storage moduli (G') of the pressure-treated samples are presented as a function of holding time and pressure. The rate of structure formation increased with increasing pressure, as expected. At 550 MPa, the maximum G' value was about 23 kPa. It was reached immediately after the initial pressure buildup (0 minutes) and remained constant

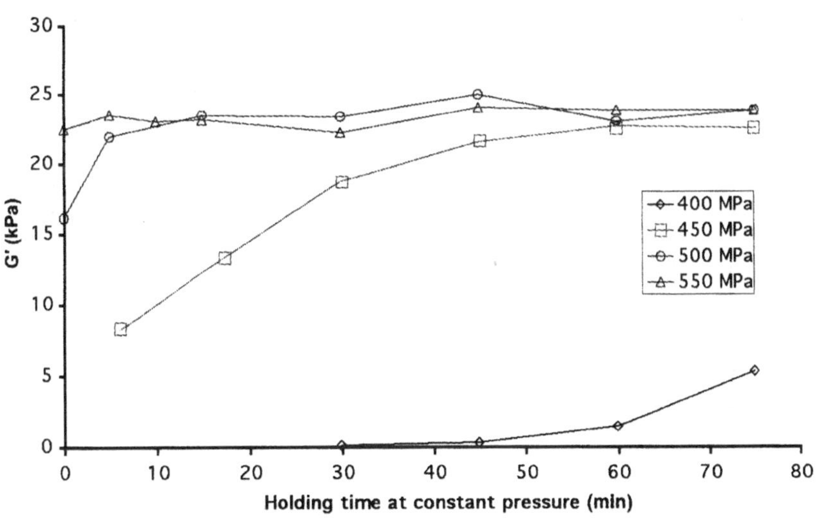

Figure 7–2 Storage moduli of 25% barley starch dispersions as a function of pressure holding time at different pressure levels.

during prolonged pressurization. The phase angle decreased with increasing G' value and was about 3 degrees when G' was at maximum. The corresponding heat-treated samples were more gel-like, with a G' value of about 11 kPa and a phase angle of about 5 degrees. Microstructural examination showed that the pressure-induced gel was composed of closely packed, only slightly swollen starch granules. In the heat-induced gel, the starch granules were much more swollen, amylose and amylopectin were phase separated, and some amylose had leached out of the granule (Autio et al., in press).

Potato starch gelatinizes at pressures higher than 650 MPa at 30°C. In contrast to heated potato starch paste, pressure-treated 10% potato starch formed a very rigid (3.5 kPa) and quite elastic (phase angle less than 10 degrees) gel at 30°C at pressures higher than 650 MPa. The higher the applied pressure, the faster the G' value of the pressure-treated samples increased (Figure 7–3).

Microscopic examination in situ and after pressure release has shown that, in contrast to heated samples, potato starch granules swell enormously and evenly under pressure (Autio et al., in press). The high-pressure treatment probably destroys the dense outer layer of the potato starch granule, allowing improved swelling.

Autio et al. (in press) reported that a 10% potato starch dispersion was completely gelatinized at 900 MPa at 30°C after a 30-minute holding time, whereas after a similar treatment many Maltese crosses were still observed from a 25% potato starch dispersion. At 10% concentration, pressure-treated potato starch granules were more evenly hydrated than at 25% concentration. The rheological properties of pressure-treated 10% and 25% barley starch dispersions varied greatly (Stolt et al., 2000). Pressure-treated 10% barley starch dispersions resem-

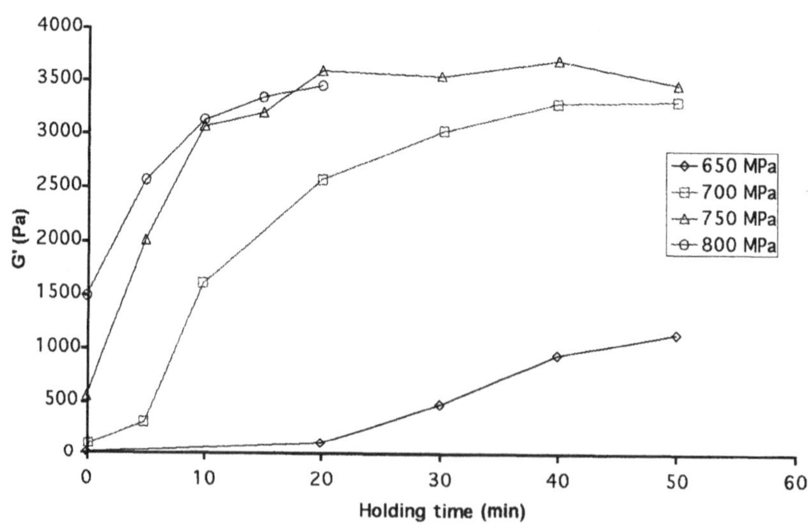

Figure 7–3 Storage moduli of 10% potato starch as a function of treatment time at different pressures.

bled the untreated sample; the pressure-treated 25% dispersion was very hard, in contrast to the untreated sample.

In conclusion, high-pressure–induced physicochemical changes of starches—such as loss of crystallinity, loss of anisotropic order, hydration, and rise in viscosity—are very similar to those induced by heat. The major difference, in the case of cereal starches, is the lack of amylose leaching during high pressure treatment, which explains the great difference in rheological properties. In the case of potato starch, swelling of granules increases under pressure, which is the reason for the high rigidity of pressure-treated starch gels.

Retrogradation of Pressure-Treated Starches

Retrogradation is the general term used to describe the structural changes that occur when the gelatinized starch is cooled to room temperature and stored. Miles et al. (1985) suggested that the increase in gel strength during cooling and storage of starch dispersions consists of two separate mechanisms with different kinetics. Short-term changes have been attributed to formation of amylose gels, and long-term changes to crystallization of short amylopectin chains within the granule (Orford et al., 1987; Ring et al., 1987). Heat-treated starches from different botanical sources differ in the extent of recrystallization of amylopectin. This long-term increase followed the order: pea, potato, corn, wheat.

In the work of Stolt et al. (2000), the retrogradation behavior of 25% barley starch dispersions treated at 550 MPa at 30°C for 10 or 60 minutes was followed by differential scanning calorimetry (DSC). After 1 day of storage at 4°C, a peak resulting from recrystallization of amylopectin was observed. The area of the peak increased with increasing storage time. Similar results were obtained at different holding times under pressure. When the extent of recrystallization of amylopectin in pressure-treated gels was compared to that of a heat-treated sample, no differences were observed. Morris (1990) has suggested that the stiffening effect of recrystallization of amylopectin within the granules results in reinforcement of the amylose gel matrix. In pressure-treated barley starch gels, the starch granules were only slightly swollen and amylose was not separated from amylopectin. Further investigation is required on the effect of pressure treatment and subsequent storage on the rigidity of starch gels.

INFLUENCE OF PRESSURE ON PECTIN

Hydrocolloids are important ingredients in most modern foods. Hydrocolloids are known to self-associate in aqueous solutions, to impart viscosity, and to gel. Water-soluble (aggregated) polysaccharides are used as texture builders to give body to liquid formulations and to produce gels. Recently, hydrocolloids have been used as emulsifiers rather than stabilizers in oil-in-water emulsions (Garti & Reichmann, 1993). The effect of pressure on hydrocolloids has been studied only to a limited extent, and Yen and Lin (1998) report that hydrocolloid solutions (colloids) are not affected by hydrostatic pressure. Pectin is defined as a mixture of heteropolysaccharides consisting predominantly of partially methoxylated galacturonic acid residues (Pilnik, 1990). Pectins are present in relatively large portions in plant tissue, where they play a role in cell wall architecture. The degree of methyl esterification is defined as moles of methoxyl groups per 100 moles of galacturonosyl residues. Pectins are referred to as high-methoxyl pectins if the degree of methyl esterification is higher than 50%.

Figure 7–4 shows that the degree of methyl esterification of the high-methoxyl pectin (extracted from citrus fruits) did not change when the solutions were pressure treated at pH 5.0; when the pH was 7.0, high pressure treatment caused the degree of methyl esterification to fall from 65% to below 60%. Pectin has optimal stability at pH 3.5 to 4.0. However, at high pH and elevated temperatures a demethylation by β-elimination occurs (Albersheim et al., 1960). These findings suggest that pectin is not affected chemically by pressure at pH 5.0, and its solubility does not change within the concentration range used. Keeping these results in mind, the observed reduction in degree of methyl esterification at pH 7.0 seems to be a purely pH-driven demethylation. These results also explain the unchanged macroscopic appearance of pectin solutions after high pressure treatment.

Viscosity measurements (Figure 7–5) showed that pressure influences the apparent viscosity of a 1.5% high-methoxyl JMJ-pectin solution (containing one-third sucrose). Pressure treatment at 400 MPa and 25°C led to an almost tenfold viscosity increase of a 1.5% JMJ-pectin solution at low shear rates and doubled the viscosity at high shear rates. The effect on viscosity was lower at 800 MPa. Both pressure treatments (400 MPa and 800 MPa) at room temperature resulted in shear thinning that was not observed in the untreated sample, indicating a structure buildup destroyed upon shear impact. Thus, a pressure treatment of 400 MPa seems to have a greater impact on viscosity than one of 800 MPa. The reason for this phenomenon is currently unknown, but might be related to changes in the interactions at the molecular level. No influence of pressure on viscosity was observed if pressure was applied at 40°C (Figure 7–5).

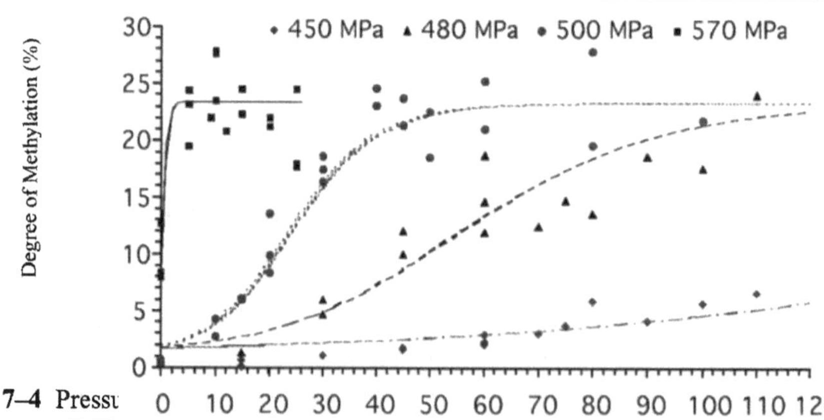

Figure 7–4 Pressu ectin solution (Kopenhagen JMJ).

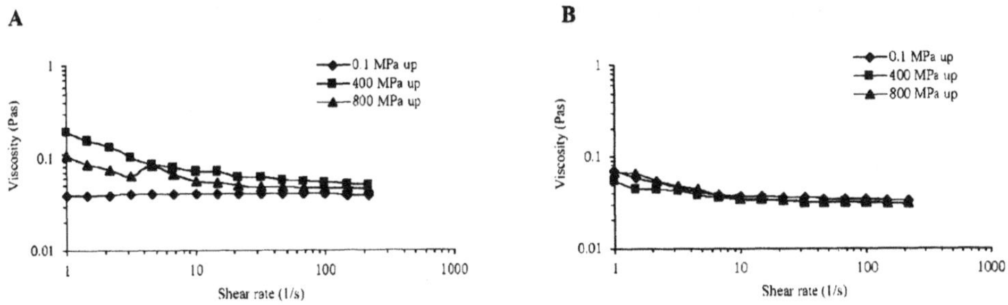

Figure 7–5 Apparent viscosity of a 1.5% JMJ-pectin solution pressure treated at 25°C (**A**) and 40°C (**B**). Flow curves were performed at 22°C using 50 mm parallel plate geometry with a 1000 μm gap size.

An analysis of the elastic (G′) and viscous moduli (G″) showed that G″ is much greater than G′ (Table 7–2), indicating that the pectin solution shows a predominantly "solution-like" behavior at all pressures. Pressure increases both the elastic and the viscous moduli. Michel et al. (1998) observed a much more pronounced viscosity increase upon pressure treatment with 1.5% sugar-free high methoxyl apple pectin (degree of methyl esterification 70% to 75%). Thus, the pressure influence might be dependent on the pectin source (e.g., apple or citrus) (Rolin et al., 1998) and concentration, or the sucrose present in JMJ-pectin may have a baro-protective influence on pectin, as known for β-lactoglobulin (Dumay et al., 1994). The underlying mechanisms for the viscosity changes that result from pressure treatment are currently unclear and remain hypothetical. A possible explanation is that pressure influences the intermolecular interactions of pectin. In the high-methoxyl pectin, the binding forces between the chains are hydrogen bridges and hydrophobic forces between the methoxyl groups (Oakenfull & Scott, 1984, 1985). Pectin molecules are also known to have highly branched regions called "hairy" or "ramified" regions, which are built by galacturonosyl residues that contain almost all neutral sugars present in pectin (Schols et al., 1997). De Vries et al. (1982) found that the hairy regions consist of rhamnose and galacturonic acid in a rather different ratio (0.1 to 0.9). The exact composition of these hairy regions is still unknown.

Chou and Kokini (1987) reported that pectin solutions with viscosities higher than 10 mPas are already above the so-called critical entanglement concentration. Pectin solutions with viscosities higher than 30 mPas, even at the highest shear rates, are in the concentration range where entanglement occurs. Pressure might have an influence on the intermolecular entanglement, especially in the hairy regions. If pressure treatment increases the number of entanglements, the viscosity would consequently be higher. Furthermore, pressure might influence the hydrophobic interactions between the pectin molecules. The lower viscosity after pressure treatments at 40°C could be explained by the observed pH-dependent pectin demethylation, leading to fewer molecule entanglements.

Table 7–2 Elastic (G′) and Viscous (G″) Moduli of a 1.5% JMJ-Pectin Solution as a Function of Pressure and Temperature

Processing Conditions	*G′ (Pa/1.0 Hz)*	*G′ (Pa/10 Hz)*	*G″ (Pa/1.0 Hz)*	*G″ (Pa/10 Hz)*
0.1 MPa/25°C/10 min	0.053	0.844	0.3142	2.327
0.1 MPa/40°C/10 min	0.029	0.510	0.2708	2.076
400 MPa/25°C/10 min	0.022	0.561	0.2671	0.258
400 MPa/40°C/10 min	0.058	0.505	0.2722	2.124
800 MPa/25°C/10 min	0.194	1.502	0.6810	2.929
800 MPa/40°C/10 min	0.028	0.519	0.2708	2.076

PRESSURE EFFECTS ON PROTEIN FUNCTIONALITY

In recent years, the use of high pressure to modify protein functionality or specific structure formation (denaturation, aggregation, gel formation) has been recognized as a potential new application for food texture engineering. Most of the published articles deal with the formation of gels made from proteins in binary solutions (Van Camp & Huyghebaert, 1995; Pittia et al., 1996). Messens et al. (1997) reviewed the use of high pressure to modify the functionality of food proteins. The effects of high pressure treatment on protein solutions are mainly related to the rupture of protein noncovalent interactions and the subsequent reformation of intra- and intermolecular bonds within or between protein molecules.

Food proteins are known to interact with water (influence on hydration properties, water adsorption and retention, wettability, swelling, adhesion, dispersability, solubility affecting precipitation, gelation), to interact with other macromolecules (interaction or complexion with proteins and polysaccharides), and to adsorb onto surfaces (surface properties and emulsion stability).

The native structure of globular proteins is stabilized by noncovalent (ionic, hydrogen, hydrophobic) and covalent (disulphide) bonds. Temperature, pH, ionic strength, and high pressure affect intra- and interprotein interactions, and change conformation leading to protein denaturation, aggregation, or gelation. For example, pressure induces significant β-lactoglobulin aggregation into soluble oligomers and polymers, linked by hydrophobic interactions and disulfide bonds (Kanno et al., 1998). These interactions depend on the pH, type of buffer, holding time, and temperature, as well as pressure (Balny & Masson, 1993; Cheftel & Dumay, 1996; Iametti et al., 1997; Galazka et al., 1999b). Felipe et al. (1997) demonstrated that β-lactoglobulin is rather susceptible to high pressure, whereas α-lactalbumin and bovine serum albumin are less sensitive to high pressure treatment. Various studies have demonstrated significant differences between heat-induced and pressure-induced protein denaturation and aggregation in a number of food proteins (for review see Funtenberger et al., 1995; Heremans, 1997; Heremans et al., 1997, 1999). For example, the structure and properties of heat- and pressure-induced gels of β-lactoglobulin are very different (Van Camp & Huyghebaert, 1995). The properties of protein gels produced thermally or by pressure depend strongly on the conditions of formation (heating rate, pH, ionic strength, holding time, concentration), which influence the degree of unfolding of the proteins. In whey protein mixtures, the changes in conformation of β-lactoglobulin seem to be the rate limiting for heat-induced aggregation (Law, 1996; Relkin et al., 1997). Generally pressure-induced gels are weaker, less elastic, and more exudative than heat-induced gels (Cheftel & Dumay, 1996). Galazka et al. (1999a) showed that pressure increased the surface activity of both β-lactoglobulin and bovine serum albumin (BSA), and these authors attributed this phenomenon to a higher protein surface hydrophobicity.

Another pressure effect on proteins is the depolymerization of casein micelles (Gaucheron et al., 1997; Schrader et al., 1997) and calcium-caseinate systems (Lee et al., 1996). Dissolution of colloidal calcium-phosphate-casein complexes upon high pressure treatment is considered to be the origin of the disintegration of the micelles. Figure 7–6 shows the effect of pressure treatment on casein micelles. Pressure drastically reduces the average micelle size, and disintegration of the micelles is greater at higher pressures. In addition, the dispersion becomes less turbid, resulting from reduced light scattering caused by the smaller micelles (Anema et al., 1997).

Protein Denaturation

The degree of protein denaturation is pressure and/or temperature dependent (Figures 7–7, 7–8 and 7–9). Tedford et al. (1998) reported that all four components of the whey protein (BSA, α-lactalbumin, β-lactoglobulin A, and β-lactoglobulin B) denatured gradually as the pressure increased. However, the denaturation rates were different for each of the four whey protein isolate (WPI)

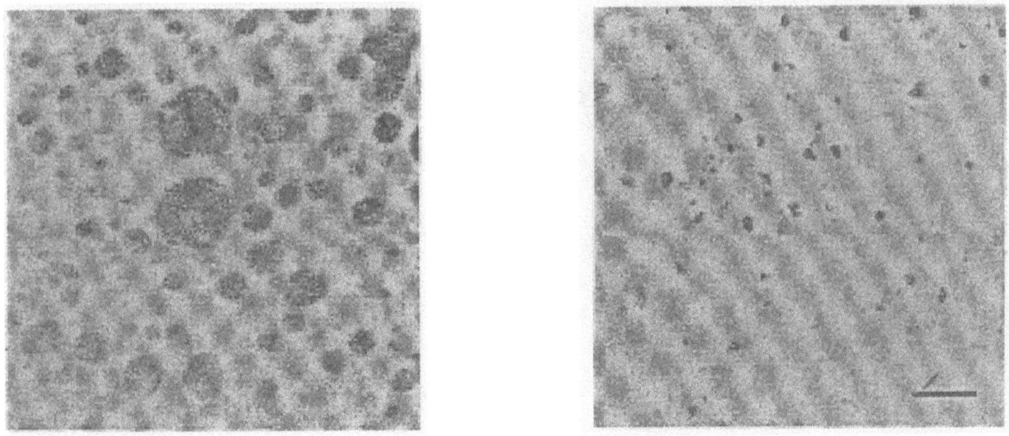

0.1 MPa. 25°C   800 MPa. 25°C   Bar = 300 nm

Figure 7–6 Transmission electron microscopy of chemically fixed casein solutions shows the effect of pressure treatment on a micellar casein solution. A pressure of 800 MPa drastically reduces the mean size of the casein micelles.

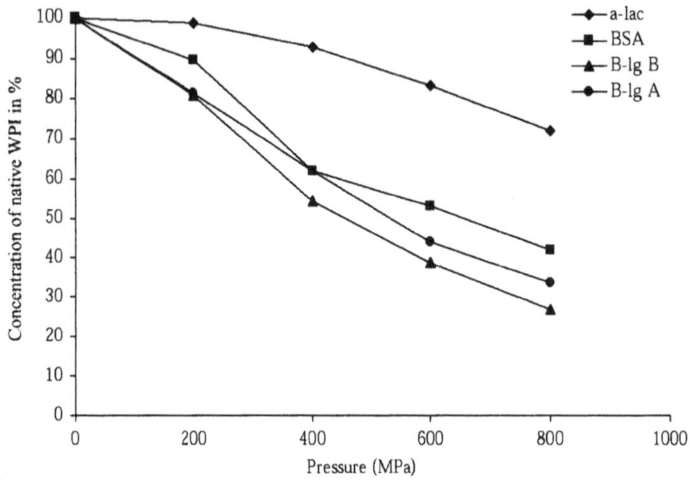

Figure 7–7 Effect of high pressure on WPI denaturation as a function of pressure and temperature (25°C) after a holding time of 10 minutes measured by high-performance liquid chromatography according to the protocol described by Resmini et al. (1989) without acidification of the WPI solution.

species. BSA is a minor component in the mixture, while the other three proteins are major components and are present in almost equal amounts in the mixture. At room temperature (25°C) protein denaturation upon pressure treatment is not complete, even at the highest pressures.

The most pressure-stable protein present in WPIs is α-lactalbumin (Nakamura et al., 1993; Felipe et al., 1997). This behavior can be confirmed by

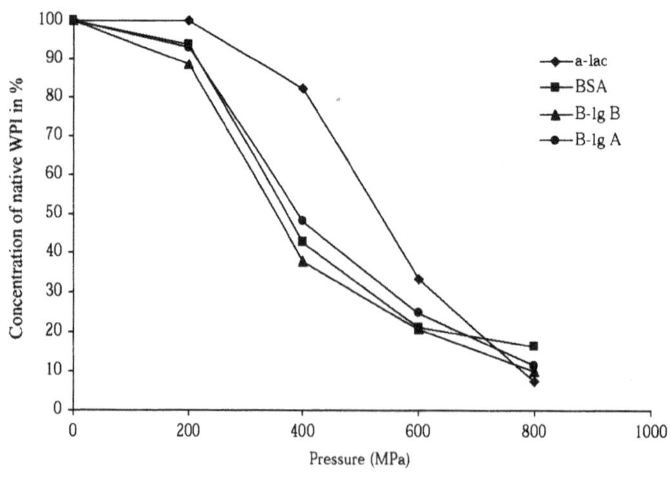

Figure 7–8 Effect of high pressure on WPI denaturation as a function of pressure and temperature (40°C) after a holding time of 10 minutes measured by high-performance liquid chromatography according to the protocol described by Resmini et al. (1989) without acidification of the WPI solution.

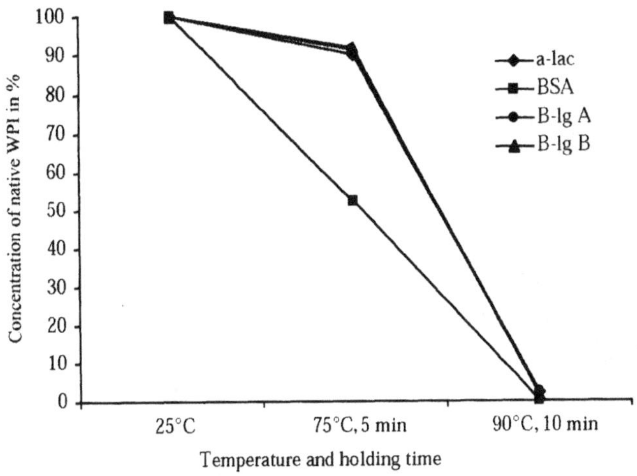

Figure 7–9 Effect of different temperature treatments (75°C for 5 minutes and 90°C for 10 minutes) on the degree of WPI denaturation as measured by high-performance liquid chromatography according to the protocol described by Resmini et al. (1989) without acidification of the WPI solution.

applying pressure to WPI solutions at room temperature. For α-lactalbumin, a denaturation of approximately 30% was achieved at the highest pressure. BSA, β-lactoglobulin B, and β-lactoglobulin A denatured at 58%, 77%, and 66%, respectively, after 10 minutes of treatment at 800 MPa. Denaturation levels of more than 50% were observed only at pressures above 400 MPa. Van Camp et al. (1997) showed that the solubility of high-pressure–treated whey protein concentrates also depends on pH. Whey proteins remain most soluble at pH 6.0 and least soluble at pH 5.0.

When the WPI samples were pressurized at 40°C (Figure 7–8), denaturation was faster. The BSA lost 75% of its solubility (from 2 mg to 0.5 mg in a 100-mg sample), but its concentration in the total blend was small (less than 5%). The other three components were almost completely denatured at 800 MPa; native proportions were 10% for β-lactoglobulin B, 11.5% for β-lactoglobulin A, and 7% for α-lactalbumin. In addition, a combination of moderate temperature (40°C) and very high pressure strongly denatures α-lactalbumin, which at room temperature is very pressure-resistant (compare Figure 7–7 with Figure 7–8). With respect to food applications, WPI is considered fully denatured after 10 minutes at 800 MPa. Denaturation at both 25°C and 40°C is very moderate at pressures up to 200 MPa and is enhanced as the pressure increases. Temperature has a more pronounced effect on denaturation than pressure (even at such moderate temperatures as 40°C). Denaturation is cooperative. Whey proteins are very susceptible to denaturation if treated for more than a few minutes above 80°C. Samples were pressure treated for only 10 minutes and at temperatures (25°C and 40°C) where at ambient pressure no denaturation occurs. These results demonstrate that the temperature effect on denaturation is enhanced in the presence of applied pressure. All the high-pressure–treated WPI solutions were clear solutions (*see* Figure 7–11).

Figure 7–9 shows the denaturation of WPI isolates as a function of temperature. Treatment at 75°C for 5 minutes is considered to be moderate denaturation conditions (below the critical denaturation temperature of 80°C), whereas treatment at 90°C for 10 minutes is considered to be severe denaturation conditions. At 75°C, BSA showed approximately 50% denaturation; all the other WPI components showed approximately 10% denaturation. This denaturation pattern is completely different from that observed after high pressure treatment, where especially BSA denaturation was much lower at the lower pressures. At 90°C, denaturation was more complete than at the strongest high pressure treatment (800 MPa at 40°C). These findings suggest that pressure treatment is more suitable than a pure heat treatment for fine-tuning the denaturation kinetics of proteins such as WPI.

Effect of High Pressure/Temperature on the Viscosity of a 12% Protein Solution

The apparent viscosity of a 12% WPI solution increases upon high pressure treatment at 25°C with a similar shear-thinning behavior in all samples, including the control (Figure 7–10 A). However, when the solution is pressure treated at 25°C, the increase in viscosity is higher at 400 MPa than at 800 MPa. A possible explanation for this somewhat surprising result is the observed differences in the relative proportions of denaturation of the four protein species at different pressures, which influence the nature of protein-protein interactions.

The results are different at 40°C. As expected, viscosity increases with increasing pressure and shear thinning is observed, but the effect is less at 800 MPa than at lower pressures (Figure 7–10 B). These results indicate that slight changes in temperature lead to differences in the rheological behavior of pressure-treated WPI solutions. In addition, pressure-treated WPI solutions are less viscous than those that receive a severe temperature treatment (compare Figures 7–10 A and B with 7–10 C). These different rheological properties might be related to differences in structures formed, as suggested by Langley et al. (1987) for thermally produced gels.

When the 12% WPI solution is treated with 0.1 MPa and 400 MPa at room temperature, the elastic modulus (G′) is equal to the viscous modulus (G″), and G′ and G″ show similar values at 0.1 MPa. At 800 MPa, G′ is greater than G″, indicating that the elastic component dominates and a gel-like system is formed (Table 7–3). The loss of the tertiary structure of globular whey proteins during denaturation leads to an increase in the volume occupied by the protein. Thus, the viscosity can be expected to increase if more denatured proteins are present. An additional viscosity increase would occur in the case of gelation of the denatured protein caused by hydrophobic interactions and the formation of intermolecular disulfide bridges. However, Fernandes and Raemy (1996) reported that gelation leading to a solidification of the 12% WPI solution did not occur with the chosen raw materials

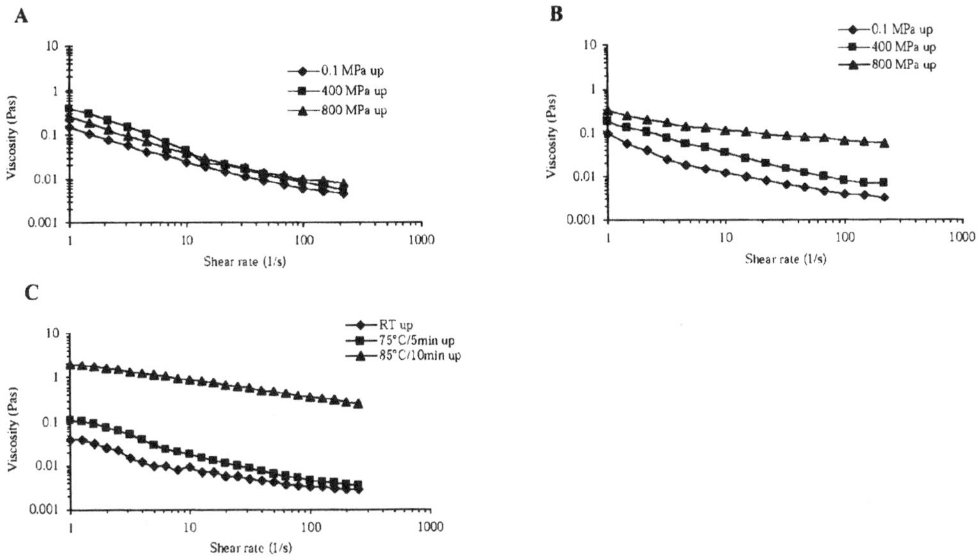

Figure 7–10 Apparent viscosity of a 12% WPI solution at 25°C (**A**), 40°C (**B**), and different heat treatments at ambient pressure (**C**). Flow curves were performed at 22°C using a 50 mm parallel plate geometry with either 250 μm, 500 μm, or 1000 μm gap size. No influence of gap size on viscosity was observed.

under the high pressure processing conditions used. Michel et al. (1998) reported that trials with another WPI source already containing 10% weight percentage of heat-induced, denatured whey protein showed solidification of a 12% WPI solution at 800 MPa. This finding indicates that the amount of denatured protein is crucial for gel formation under high pressure treatment. Furthermore, Rojas et al. (1997) showed that the gel properties strongly depend on the fractions of α- and β-lactoglobulins present.

Table 7–3 Moduli (Elastic G′ and Viscous G″) of a 12% WPI (BIPRO) Solution as a Function of Pressure and Temperature

Processing Conditions	*G′ (Pa/1.0 Hz)*	*G′ (Pa/10 Hz)*	*G″ (Pa/1.0 Hz)*	*G″ (Pa/10 Hz)*
0.1 MPa/25°C/10 min	0.103	0.267	0.098	0.292
0.1 MPa/40°C/10 min	0.004	n.d.	0.017	.085
400 MPa/25°C/10 min	0.093	0.234	0.106	0.258
400 MPa/40°C/10 min	0.055	0.217	0.053	0.163
800 MPa/25°C/10 min	0.266	0.652	1.036	0.484
800 MPa/40°C/10 min	0.170	2.952	0.716	3.769
75°C/5 min	0.045	0.077	0.091	0.228
90°C/10 min	1.949	9.676	3.597	17.559

STRUCTURE ENGINEERING BY PRESSURE IN PROTEIN-PECTIN MIXTURES

In reality, foods are always multi-component systems containing proteins, polysaccharides, and fats. Protein-polysaccharide interactions in an aqueous environment play a significant role in food structure formation and stability (for review see Tolstoguzov, 1991, 1997; Syrbe et al., 1998). The final distribution of protein, fat, and polysaccharide components determines the product's texture (rheological properties). Prediction of food protein functionality in such complex food systems is difficult because proteins interact with polysaccharides. Both proteins and polysaccharides fulfill similar structural functions and work multifunctionally and simultaneously. Therefore, food composition and processing strongly influence their functionality. Proteins and polysaccharides can act independently or can interact and phase separate (or gel). Thus, they can impart unique properties to the food's aqueous solution and, consequently, to the final food system.

Many studies have been conducted on the effect of pressure on proteins and polysaccharides at atmospheric and elevated pressures (for review see Chirife, 1996; Tolstoguzov, 1999). Less is known about the effect of pressure on the surface properties of proteins, and even less is known about its effect on proteins in the presence of other biopolymers. Only a few reports (Galazka et al., 1995; Dickinson & Pawlowsky, 1996; Galazka et al., 1996a, 1996b, 1997, 1999a, 1999b) are available on the effect of pressure on a single protein or a pressurized protein to which selected hydrocolloids have been added. Most of these studies involved hydrocolloids that were polyanionic and non-food grade.

Tolstoguzov (1998) identified three possible situations in solutions of biopolymers:

1. *Incompatibility*, which is an interaction of the biopolymers leading to segregation into two distinct immiscible aqueous phases, each phase loaded mostly with a single biopolymer species. This phenomenon is due to the limited thermodynamic compatibility of proteins and polysaccharides in aqueous media.
2. *Compact coacervation*, resulting from a net attraction between the polymers leading to a separation into two phases, with each phase containing both biopolymers. This phenomenon is attributed to the formation of an insoluble electrostatic protein-anionic polysaccharide complex.
3. *Miscibility*, resulting in a single homogeneous soluble phase, where the molecular components are either cosoluble or exist as soluble complexes.

Tolstoguzov and coworkers spent many years studying the behavior of more than 100 combinations of hydrocolloids and proteins in solvents (mostly water) and have explored their binary and ternary phase diagrams (Tolstoguzov, 1997). They found that biopolymers at the phase separation stage could generally adsorb onto other biopolymer particles (colloids).

Various phase diagrams of protein (mainly casein from skimmed milk, gelatin, soybean proteins, BSA, and ovalbumin) and polysaccharide mixtures (anionic and nonionic) have been established (Antonov et al., 1975; Zhurawskaya & Kiknadze, 1986; Tolstoguzov, 1991; Dickinson & Pawlowsky, 1996). However, the behavior of whey proteins in phase-separated systems was only recently analyzed, when Syrbe (1997) investigated the influence of heat and high pressure on the phase-separation behavior of ternary whey protein and pectin mixtures (Syrbe et al., 1995; Syrbe et al., 1998). Stabilization of the demixed whey protein phase was achieved either by thermal or high pressure denaturation (Cheftel & Dumay, 1996; Hinrichs et al., 1996; Zasypkin et al., 1996).

Native whey protein molecules show a globular tertiary structure. This is the result of polypeptide folding, leading to a decrease of the excluded volume. Globular proteins have a phase-separation threshold that is 10 times higher than those of classical random coil polymers or unfolded, unordered (denatured) proteins (Tolstoguzov et al., 1985; Tolstoguzov, 1999). The globular structure makes whey proteins miscible with anionic polysaccharides up to a certain concentration. However, the globular structure of whey proteins is lost upon denaturation, leading to a dramatic increase in excluded volume and provoking phase separation in the presence of, for example, anionic high-methoxyl pectin molecules.

According to Le Chatelier's principle, an increased concentration of incompatible biopolymers in a highly volume occupied system minimizes the excluded volume. This mechanism favors biopolymer gelation and crystallization (Tolstoguzov, 1997). Furthermore, the excluded volume is significantly greater for nonspherical macromolecules such as linear polysaccharides (Tanford, 1961). The competition for space between macromolecules determines the critical conditions for phase separation. Phase separation occurs at about 2% to 4% concentration for mixtures of gelatin or casein with linear polysaccharides, about 4% or higher for globular protein-polysaccharide mixtures, and more than 12% for mixtures of globular proteins (Tolstoguzov et al., 1985). According to Tolstoguzov (1998), effects of excluded volume include the following:

- reduction of cosolubility
- enhancement of the association of macromolecules
- enhancement of protein adsorption at oil-water and gas-water interfaces
- increase in gelation rate
- increase in elastic modulus
- reduction in the critical concentration for gelation of mixed solutions

Other typical features of phase-separated food systems include low interfacial tension, low concentrated interfacial layer between coexisting phases, adsorption of lipids and flavoring components within the interfacial layer between immiscible biopolymer phases, easy deformability of liquid-dispersed particles in flow.

Therefore, phase separation can generate a wide variety of tailor-made microstructures. By provoking solidification, polymerization, or gelation of one or both of the incompatible phases, we can kinetically "freeze-in" the phase separation before reaching thermodynamically stable conditions (e.g., a macroscopically visible two-phase system).

Figure 7–11 shows the macroscopic appearance of 1.5% high-methoxyl pectin (control), 12% whey protein (control), and 12% whey protein/1.5% pectin solutions after pressure treatments at two temperatures (25°C and 40°C) and after 1 and 7 days of aging. Differences in the samples are clearly visible. All the solutions containing pectin or whey protein alone were transparent (no turbidity, gelation, or sedimentation) before and after heat and high-pressure treatments. er visual differences were detected in these samples, and storage up to 7 .d not lead to further changes. The mixed WPI/high-methoxyl pectin solutions were transparent before treatment, but two significant differences were observed after high pressure treatments. The biopolymer aqueous mixtures treated at low pressure (0.1 and 200 MPa) and 25°C remained unchanged (transparent and nonviscous), even after aging for 7 days. Mixtures subjected to 400 to 800 MPa at 25°C exhibited a well-defined phase separation. After 7 days of storage, this phase-separation process led to an increase in size of the transparent lower layer.

The samples behaved differently when pressure treated at 40°C. Only at 400 MPa could a visible phase separation into two similarly sized layers be observed. In the samples treated at pressures below and above 400 MPa, no visible parti-

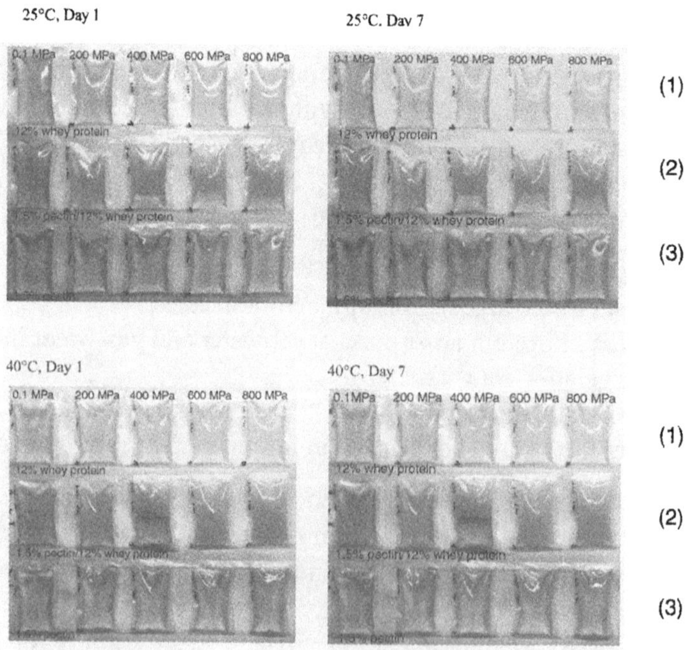

Figure 7–11 Macrophotographs of pressure-treated (1) high-methoxyl pectin (control); (2) whey protein isolate solutions; and (3) mixed whey protein isolate/high-methoxyl pectin solutions. They were treated at 25°C and 40°C. The macrophotographs were taken 1 and 7 days after treatment.

tioning into two phases occurred. No further changes in layer sizes were observed after 7 days of storage.

Influence of Pressure Treatment on the Structure of Ternary WPI/Pectin Mixtures

Structural analysis by differential interference contrast microscopy of liquid ternary WPI/JMJ-pectin mixtures at atmospheric pressure showed some protein-rich aggregates at both 25°C and 40°C (Figure 7–12). The WPI and pectin concentrations used are very close to the threshold level where phase separation occurs without any denaturation of the proteins. Thus, the observed phenomenon might correspond to a "weakly" phase-separated system.

Dilution tests were used to verify protein gelation in phase-separated systems. Phase separation without gelation, where the protein phase could be dissolved by dilution with water, was observed under all processing conditions, with the exception of 600 and 800 MPa at 40°C and heat treatment at ambient pressure at 90°C. These observations are in agreement with the viscosity data that also indicate gelation in these systems at pressures above 600 MPa and 40°C and at 90°C at ambient pressure (see viscosity results in Figure 7–13). Microscopic observation of "non-gelled, phase-separated" systems showed a wide variety of structures (Figure 7–12). This is mainly due to continuing phase separation of the well-mixed ternary solutions on the microscope slide. At the beginning of phase separation, large tubelike structures were formed (see Figure 7–12, 25°C, 400 MPa). These eventually became perfect spheres (partially visible in Figure 7–12, 40°C, 400 MPa) caused by minimization of the surface energy.

Even without any visible phase separation on a macroscopic level (see Figure 7–11, 600 and 800 MPa at 40°C), there was evidence of phase separation on a microscopic scale (see Figure 7–12, 800 MPa at 40°C). The solution can be stabilized by increasing the viscosity of the continuous pectin phase by phase-separation–induced increase of the pectin concentration in combination with protein denaturation.

Influence of Pressure Treatment on Viscosity

The apparent viscosity of ternary whey protein/pectin mixtures was always higher than the viscosity of the whey protein and pectin solutions alone (compare Figures 7–5 and 7–10 with Figure 7–13). This is mainly because mixing pectin with the whey proteins has a synergistic effect. All ternary samples (control and pressure treated) showed a shear thinning behavior, indicating structure breakup. Shear thinning was more pronounced in the gelled systems (treatments of 600 and 800 MPa at 40°C, plus heat treatment at 90°C at ambient pressure).

A drastic increase in apparent viscosity could be observed at 800 MPa and 40°C. Comparison of the heat-treated WPI/pectin mixtures at 90°C with samples treated with high pressure at 40°C showed that almost the same rheological behavior can be achieved by pure heat treatment. This is to be expected for the

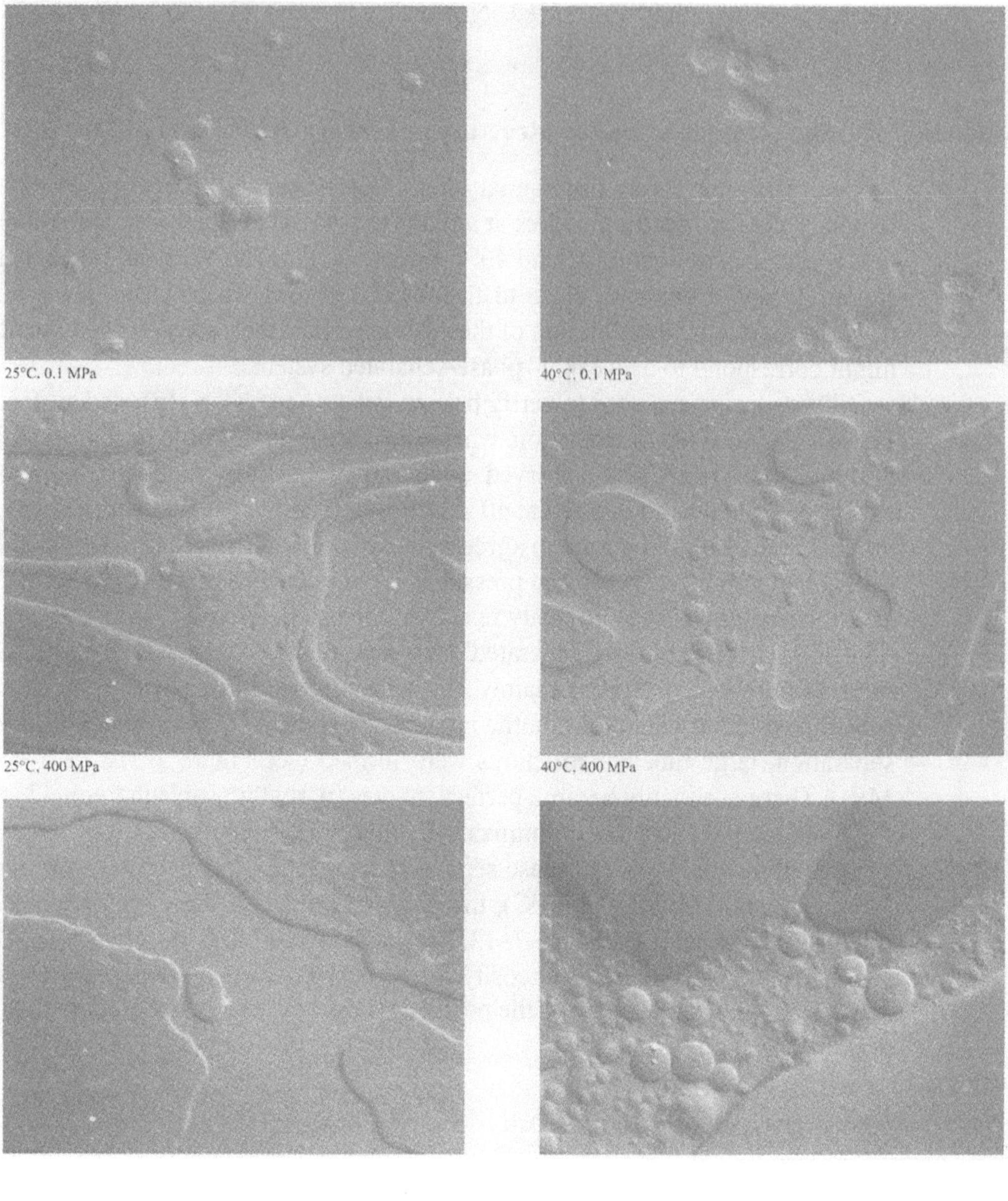

Figure 7–12 Photomicrographs of solutions after heat and pressure treatment of the phase-separated systems (0.1, 400, and 800 MPa at 25°C left and at 40°C right) as seen under differential interference contrast light microscopy.

---

systems that are almost completely denatured (90°C for 10 minutes and 800 MPa at 40°C for 10 minutes).

Analysis of the viscous modulus (G′) and elastic modulus (G″) showed that G′ is smaller than G″ in the control samples. However, in samples treated at room

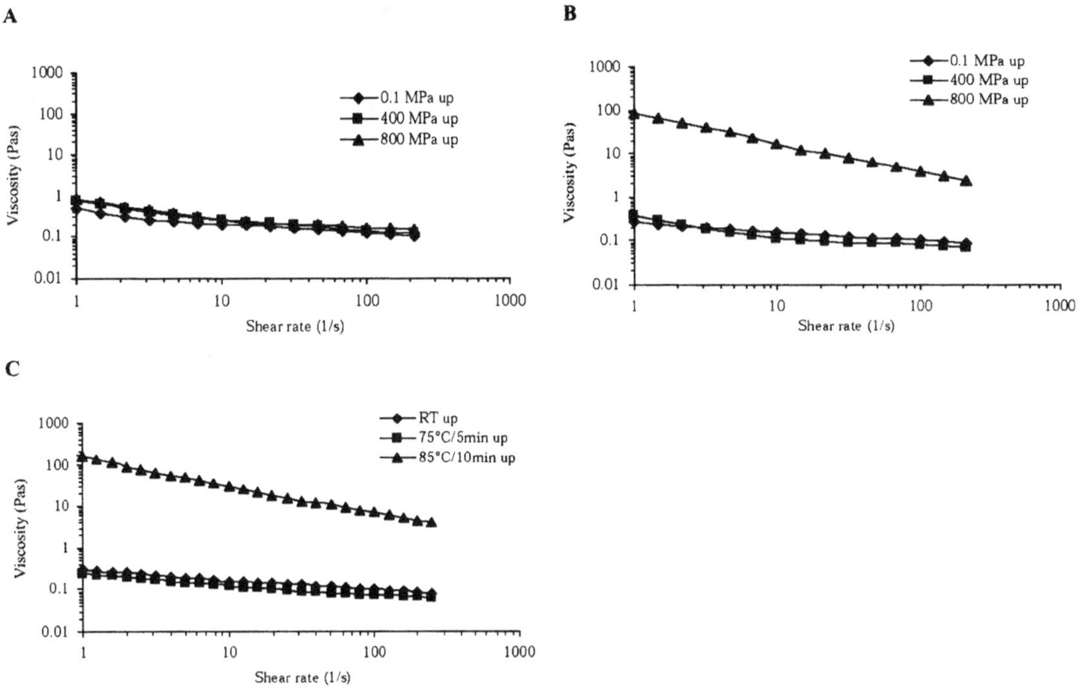

Figure 7–13 Apparent viscosity of a 12% WPI/1.5% pectin solution at 25°C (**A**), at 40°C (**B**), and at different temperatures at ambient pressure (**C**). Flow curves were performed at 22°C using a 50 mm parallel plate geometry with either 500 μm or 1000 μm gap size. No influence of gap size on apparent viscosity was observed.

temperature with pressure up to 800 MPa and those treated at 40°C with pressure up to 400 MPa, G′ and G″ increased with growing pressure (Table 7–4). In all these ternary mixtures, the viscous component dominates the system. At 800 MPa and 40°C, G′ is greater than G″. In this case, the elastic component dominates the system, indicating gelation of the protein-enriched phase. From dilution trials in combination with light microscopy, we know that WPI gels upon a pressure treatment ≥600 MPa at 40°C. The exact pressure where gelation occurs is assumed to fall between 400 and 600 MPa (10 minutes holding time).

The similar rheological behavior of the nongelled, heat-treated mixture and the pressure-treated mixtures is surprising, because there were large differences in their denaturation profiles (Figures 7–7, 7–8, and 7–9). This can be explained by the low interfacial tension in phase-separated systems (Tolstoguzov, 1998) as well as the easy deformability of liquid-dispersed particles under flow. These rheological results suggest that gelation of the protein phase is necessary to significantly influence the textural properties of ternary high-methoxyl pectin/WPI/water mixtures. However, gelation occurred only if the denaturation of all WPI species was above 60%.

Table 7–4 Moduli (Elastic G′ and Viscous G″) of a Ternary Mixture of 12% WPI (BIPRO) and 1.5% JMJ-Pectin as a Function of Pressure and Temperature

Processing Conditions	*G′ (Pa/1.0 Hz)*	*G′ (Pa/10 Hz)*	*G″ (Pa/1.0 Hz)*	*G″ (Pa/10 Hz)*
0.1 MPa/25°C/10 min	0.27	2.72	0.94	5.80
0.1 MPa/40°C/10 min	0.28	1.86	0.79	4.96
400 MPa/25°C/10 min	1.48	6.27	2.05	9.57
400 MPa/40°C/10 min	2.76	13.69	4.88	23.99
800 MPa/25°C/10 min	1.40	7.01	2.13	11.00
800 MPa/40°C/10 min	279.98	546.16	141.22	262.72
75°C/5 min	0.37	2.41	0.90	5.34
90°C/10 min	2364.54	3201.83	598.57	945.39

CONCLUSION

A combination of high hydrostatic pressure, temperature, and pressure holding time can be used to induce phase separation in miscible high-methoxyl pectin/WPI mixtures. Varying these parameters can influence the degree of denaturation of the proteins. Denaturation provokes phase separation in a ternary pectin-WPI-water system, leading to a concentration of the molecular species involved. Similar textures can be obtained either by pressure or heat treatment as indicated by rheological measurements, even if the denaturation profiles are different. Significant changes in texture can only be achieved when the protein phase is gelled. To achieve this, all WPI species must have a denaturation degree above 60%.

An intelligent combination of viscosity enhancement by phase separation and provocation of gelation caused by a sufficiently high denaturation and concentration by phase separation of the WPI upon a combined pressure/temperature or pure temperature treatment offers a potential to structure food systems in a controlled way, i.e. by provoking gelation. In order to apply the concept of structure formation by protein denaturation and subsequent phase separation in an efficient way, we must learn more about the phase behavior and kinetic data of denaturation as a function of concentration and over the whole range of processing parameters (pressure, holding time, and temperature). To establish and facilitate the creation and control of new food textures, the obtained data must be combined with a quantitative description of the resulting textures by rheology and microstructural analysis of the structures formed.

REFERENCES

Albersheim, P., Neukom, H., & Deuel, H. (1960). Splitting of pectin chain molecules in neutral solutions. *Arch. Biochem. Biophys., 90,* 46.

Anema, S.G., Lee, S.K., Schrader, K., & Buchheim, W. (1997). Effect of pH on the turbidity of pressure-treated calcium caseinate suspensions and skim milk. *Milchwissenschaft, 52,* 141–146.

Antonov, J.A., Grinberg, V.J., & Tolstoguzov, V. (1975). Phasengleichgewichte in Wasser/Eiweiss/ Polysaccharid Systemen. *Stärke, 27,* 424–431.

Autio, K. (1990). Rheological and microstructural changes of oat and barley starches during heating and cooling. *Food Struct., 9,* 297–304.

Autio, K., & Stolt, M. (1998). Effect of high pressure on the structure of starch. In Autio, K. (Ed.) *Fresh Novel Foods by High Pressure* (pp. 61–67). Espoo, Finland: Julkaisija Utgivare.

Autio, K., Stolt, M., Snauwaert, J., & Heremans, K. Pressure-induced microstructural changes in starch suspensions. *Lebensm.-Wiss. Technol.* In Press.

Balny, C., & Masson, P. (1993). Effects of high pressure on proteins. *Food Reviews International, 9,* 611–628.

Basak, S., & Ramaswamy, H.S. (1996). Ultra high pressure treatment of orange juice: A kinetic study on inactivation of pectin methyl esterase. *Food Res. Int., 29,* 601–607.

Basak, S., & Ramaswamy, H.S. (1998). Effect of high pressure processing on the texture of selected fruits and vegetables. *J. Texture Stud., 29,* 578–601.

Cheftel, J.C., & Culioli, J. (1997). Effects of high pressure on meat: A review. *Meat Science, 46,* 211–236.

Cheftel, J.C., & Dumay, E. (1996). Effects of high pressure on dairy proteins: A review. In R. Hayashi & C. Balny (Eds.), *High Pressure Bioscience and Biotechnology* (pp. 299–308). Amsterdam: Elsevier Science B.V.

Chirife, J. (1996). ISOPOW VI. International Symposium on the Properties of Water in Foods conference report. *Trends Food Sci. Technol.*

Chou, T.C., & Kokini, J.L. (1987). Rheological properties and confromation of tomato paste pectins, citrus and apple pectin. *J. Food Sci., 52,* 1658.

Crelier, S., Roberts, M.-C., & Juillerat, M.-A. (1998). Effect of high pressure treatment on the texture and enzyme activities of selected vegetables. In H. Ludwig (Ed.), *Advances in High Pressure Bioscience and Biotechnology* (pp. 413–416). Berlin and Heidelberg: Springer-Verlag.

De Vries, J.A., Rombouts, F.M., Voragen, G.J., & Pilnik, W. (1982). Comparison of the structural features of apple and citrus pectic substances. *Carbohydr. Polym., 4,* 89–101.

Dickinson, E., & Pawlowsky, K. (1996). Effect of high-pressure treatment of protein on the rheology of flocculated emulsions containing protein and polysaccharide. *J. Agric. Food Chem., 44,* 2992–3000.

Dornenburg, H., & Knorr, D. (1993). Cellular permeabilization of cultured plant tissues by high electric field pulses of ultra high pressure for the recovery of secondary metabolites. *Food Biotechnol., 7,* 35–48.

Dumay, E.M., Kalichevsky, M.T., & Cheftel, J.C. (1994). High-pressure unfolding and aggregation of betalactoglobulin and the baroprotective effects of sucrose. *J. Agric. Food Chem., 42,* 1861–1868.

Dumoulin, M., Ozawa, S., & Hayashi, R. (1998). Textural properties of pressure-induced gels of food proteins obtained under different temperatures including subzero. *J. Food Sci., 63,* 92–95.

Ezaki, H., & Hayashi, R. (1992). High pressure effects on starch: Structural change and retrogradation. In C. Balny, R. Hayashi, K. Heremans, & P. Masson (Eds.), *High Pressure and Biotechnology,* Colloque INSERM (Vol. 224, pp. 163–165). Montrouge: John Libbey Eurotext.

Felipe, X., Capellas, M., & Law, A.J.R. (1997). Comparison of the effects of high-pressure treatments and heat pasteurization on the whey proteins in goat's milk. *J. Agric. Food Chem., 45,* 627–631.

Fernandes, P.B., & Raemy, A. (1996). High pressure treatment of whey protein/polysaccharide systems. In R. Hayashi & C. Balny (Eds.), *High Pressure Bioscience and Biotechnology* (pp. 337–342). Amsterdam: Elsevier Science B.V.

Fuchigami, M., Katoh, N., & Teramoto, A. (1995). Effect of pressure shift freezing on texture, pectic composition and histological structure of carrots. In R. Hayashi & C. Balny (Eds.), *High Pressure Bioscience and Biotechnology* (pp. 379–386). Amsterdam: Elsevier Science B.V.

Funtenberger, S., Dumay, E., & Cheftel, J.C. (1995). Pressure-induced aggregation of beta-lactoglobulin in pH 7.0 buffers. *Lebensm. Wiss. Technol., 28,* 410–418.

Funtenberger, S., Dumay, E., & Cheftel, J.C. (1997). High pressure promotes beta-lactoglobulin aggregation through SH/S-S interchange reactions. *J. Agric. Food Chem., 45,* 912–921.

Galazka, V.B., Dickinson, E., & Ledward, D.A. (1996a). Effect of high pressure on the emulsifying behavior of beta-lactoglobulin. *Food Hydrocolloids, 10,* 213–219.

Galazka, V.B., Sumner, I.G., & Ledward, D.A. (1996b). Changes in protein-protein and protein-polysaccharide interactions induced by high pressure. *Food Chemistry, 57,* 393–398.

Galazka, V.B., Ledward, D.A., Dickinson, E., & Langley, K.R. (1995). High pressure effects on emulsifying behvior of whey protein concentrate. *J. Food Sci., 60,* 1341–1343.

Galazka, V.B., Ledward, D.A., Sumner, I.G., & Dickinson, E. (1997). Influence of high pressure on bovine serum albumin and its complex with dextran sulfate. *J. Agric. Food Chem., 45,* 3465–3471.

Galazka, V.B., Smith, D., Dickinson, E., & Ledward, D.A. (1999a). Effect of high pressure on protein-polysaccharide complexes. In Dickinson, E. (Ed.) *Food Emulsion and Foams: Interfaces, Interactions and Stability* (pp. 104–116). Cambridge: The Royal Society of Chemistry.

Galazka, V.B., Smith, D., Ledward, D.A., & Dickinson, E. (1999b). Interactions of ovalbumin with sulphated polysaccharides: Effects of pH, ionic strength, heat and high pressure treatment. *Food Hydrocolloids, 13,* 81–88.

Garti, N., & Reichmann, D. (1993). Hydrocolloids as emulsifiers and stabilizers. *Food Structure, 12,* 411–426.

Gaucheron, F., Famelart, M.H., Mariette, F., Raulot, K., Michel, F., & Graet, Y. (1997). Combined effects of temperature and high-pressure treatments on physicochemical characteristics of skim milk. *Food Chemistry, 59,* 439–447.

Heremans, K. (Ed.) (1997). *High Pressure Research in the Biosciences and Biotechnology.* Leuven: Leuven University Press.

Heremans, K., Meersman, F., Rubens, P., Smeller, L., Snauwaert, J., & Vermeulen, G. (1999). A comparison between pressure and temperature effects on food constituents. In F.R.A. Oliverira & J.C. Oliveira (Eds.), *Processing Foods: Quality Optimization and Process Assessment* (pp. 269–280). Boca Raton, FL: CRC Press.

Heremans, K., Van Camp, J., & Huyghebaert, A. (1997). High pressure effects on proteins. In S. Damadoran & A. Paraf (Eds.), *Food Proteins and Their Application in Foods. Part III.* New York: Marcel Dekker.

Hinrichs, J., Rademacher, B., & Kessler, H.G. (1996). Reaction kinetics of pressure-induced denaturation of whey proteins. *Milchwissenschaft, 51,* 504.

Iametti, S., Transidico, P., Bonomi, F., Vecchio, G., Pittia, P., Rovere, P., & Dall'Aglio, G. (1997). Molecular modifications of β-lactoglobulin upon exposure to high pressure. *J. Agric. Food Chem., 45,* 23–29.

Kanno, C., Tai Hau, M., Hagiwara, T., Ametani, M., & Azuma, N. (1998). Gel formation from industrial milk whey proteins under hydrostatic pressure: Effect of hydrostatic pressure and protein concentration. *J. Agric. Food Chem., 46,* 417–424.

Kasai, M., Hatae, K., & Shimada, A. (1995a). Pressure induced hardening of vegetables and its mechanisms. Poster presentation at the *International Conference on High Pressure Bioscience and Biotechnology,* Nov. 5–9, Kyoto, Japan.

Kasai, M., Hatae, K., Shimada, A., & Ibuchi, S. (1995b). Pressure pretreatment of vegetables for controlling hardness before cooking. *Nippon Shokuhin Kagaku Kaishi, 42,* 594–601.

Knorr, D. (1994). Plant cell and tissue cultures as model systems for monitoring the impact of unit operations on plant foods. *Trends Food Sci. Technol., 5,* 328–331.

Langley, K.R., Green, M.L., Brooker, B.E., & Smith, A.C. (Eds.). (1987). *Mechanical Properties of Whey Protein Gels in Relation to Composition and Microstructure.* Washington, DC: Irl Press.

Law, A.J.R. (1996). Effects of heat treatment and acidification on the dissociation of bovine casein micelles. *J. Dairy Res., 63,* 35–48.

Lee, S.K., Anema, S.G., Schrader, K., & Buchheim, W. (1996). Effect of high hydrostatic pressure on Ca-caseinate systems. *Milchwissenschaft, 51,* 17–21.

Messens, W., Van Camp, J., & Huyghebaert, A. (1997). The use of high pressure to modify the functionality of food proteins. *Trends Food Sci. Technol., 8,* 107–112.

Michel, M., Leser, M., Syrbe, A., Clerc, M.-F., Bauwens, I., von Schack, M.-L., and Watzke, H.J. (1998). Structure formation in protein-polysaccharide mixtures upon pressure treatment. In Autio, K. (Ed.), *Fresh Novel Foods by High Pressure* (pp. 69–82). Espoo, Finland: Julkaisija Utgivare.

Miles, M.J., Morris, V.J., Orford, P.D., & Ring, S.D. (1985). The roles of amylose and amylopectin in the gelation and retrogradation of starch. *Carbohydr. Res., 135,* 271–281.

Morris, V.J. (1990). Starch gelation and retrogradation. *Trends Food Sci. Technol., 1,* 2–6.

Nakamura, T., Sado, H., & Syukunobe, Y. (1993). Production of low antigenic whey protein hydrolysates by enzymatic hydrolysis and denaturation with high pressure. *Milchwissenschaft, 48,* 141–145.

Oakenfull, D., & Scott, A. (1984). Hydrophobic interaction in the gelation of high methoxyl pectins. *J. Food Sci., 49,* 1093–1098.

Oakenfull, D.G., & Scott, A.G. (1985). Gelation of high methoxypectins. *Food Technol. Austr., 37.,* 156–158.

Orford, P.D., Ring, S.G., Carroll, V., Miles, M.J., & Morrs, V.J. (1987). The effect of concentration and botanical source on the gelation and retrogradation of starch. *J. Sci. Food Agric., 39,* 169–177.

Pilnik, W. (1990). Pectin—A many splendoured thing. In G.O. Phillips & D.J. Wdlock, D.J. (Eds.), *Gums and Stabilisers for the Food Industry* (Vol. 5, pp. 209–220). New York, Tokyo: Oxford University Press.

Pittia, P., Wilde, P.J., Husband, F.A., & Clark, D.C. (1996). Functional and structural properties of beta-lactoglobulin as affected by high pressure treatment. *J. Food Sci., 61,* 1123–1128.

Prestamo, G., & Arroyo, G. (1998). High hydrostatic pressure effects on vegetable structure. *J. Food Sci., 63,* 878–881.

Relkin, P., Meylheuc, T., Launay, B., & Raynal, K. (1997). Thermal stability and sol-gel transitions of whey and egg white proteins. In P.A. Williams & G.O. Phillips (Eds.), *Gums and Stabiliser for the Food Industry* (Vol. 9, pp. 145–153). Oxford: The Royal Society of Chemistry.

Resmini, P., Pellegrino, L., Saracchi, S., & Chiodi, J. (1989). A sensitive method to determine spinacine (6-carboxy-1,2,3,4-tetrahydroimidazopyridine) in ripened cheese by HPLC. *It. J. Food Sci., 1,* 35–44.

Ring, S.G., Colonna, P., I'Anson, K.J., Kalichevsky, M.T., Miles, J.J., Morris, V.J., & Orford, P.D. (1987). The gelation and crystallisation of amylopectin. *Carbohydr. Res., 162,* 277–293.

Rojas, S.A., Goff, H.D., Vintha, S., Dalgleish, D.G., Flores, A., & Senaratne, V. (1997). Gelation of commercial fractions of beta-lactoglobulin and alpha-lactalbumin. *Int. Dairy J., 7,* 79–85.

Rolin, C., Nielsen, B.U., & Glahn, P.-E. (1998). *Pectin.* New York, Basel, Hong Kong: Marcel Dekker.

Schols, H.A., Ros, J.M., Daas, P.J.H., Bakx, D.E., & Voragen, A.G.J. (1997). Structural features of native and commercially extracted pectins. In P.A. Williams & G.O. Phillips (Eds.)., *Gums and Stabilisers for the Food Industry* (pp. 3–15). Oxford: The Royal Society of Chemistry.

Schrader, K., Buchheim, W., & Morr, C.V. (1997). High pressure effects on the colloidal calcium phosphate and the structural integrity of micellar casein in milk. Part 1: High pressure dissolution of colloidal calcium phosphate in heated milk systems. *Nahrung, 41,* 133–138.

Shannon, J.C., & Garwood, D.L. (1984). Genetics and physiology of starch development. In R.L. Whistler, J.N. Miller, & E.F. Paschal (Eds.), *Starch, Chemistry and Technology* (pp. 25–86). London: Academic Press.

Smelt, J. (1998). Recent advances in the microbiology of high pressure processing. *Trends Food Sci. Technol., 9,* 152–158.

Stolt, M., Oinonen, S., & Autio, K. (2000). Effect of high pressure on the physical properties of barley starch. *Innov. Food Sci. Technol.* (in press).

Stolt, M., Stoforos, N.G., Taoukis, P.S., & Autio, K. (1999). Evaluation and modelling of rheological properties of high pressure treated waxy maize starch dispersions. *J. Food Eng., 40,* 293–298.

Stute, R., Klingler, R.W., Boguslawski, S., Eshtiaghi, M.N., & Knorr, D. (1996). Effects of high pressures treatment on starches. *Starke, 48,* 399–408.

Syrbe, A. (1997). *Polymer Incompatibility in Aqueous Whey Protein and Polysaccharide Solution—Phase Separation and Microgel Particle Formation.* Düsseldorf: VDI Verlag GmbH.

Syrbe, A., Bauer, W.J., & Klostermeyer, H. (1998). Polymer science concepts in dairy systems—An overview of milk protein and food hydrocolloid interaction. *Int. Dairy J., 8,* 179–193.

Syrbe, A., Fernandes, P.B., Dannenberg, F., Bauer, W., & Klostermeyer, H. (1995). Whey protein-polysaccharide mixtures: Polymer incompatibility and its application. In E. Dickinson & D. Lorient (Eds.), *Food Macromolecules and Colloids* (pp. 328–339). London: The Royal Society of Chemistry.

Tanford, C. (1961). *Physical Chemistry of Macromolecules.* New York: John Wiley & Sons.

Tedford, L.A., Kelly, S.M., Price, N.C., & Schaschke, C.J. (1998). Combined effects of thermal and pressure processing on food protein structure. *Food and Bioproducts Processing, 76,* 80–86.

Tolstoguzov, V. (1991). Functional properties of food proteins and the role of protein-polysaccharide interactions. *Food Hydrocolloids, 4,* 429.

Tolstoguzov, V.B. (1997). Protein-polysaccharide interactions. In S. Damodaran & A. Paraf (Eds.), *Food Proteins and Their Applications* (Vol. 80, pp. 171–198). New York: Marcel Dekker.

Tolstoguzov, V.B. (1998). Functional properties of protein-polysaccharide mixtures. In S.E. Hill, D.A. Ledward, & J.R. Mitchell (Eds.), *Functional Properties of Food Molecules* (pp. 252–276). Gaithersburg, MD: Aspen.

Tolstoguzov, V.B. (1999). The role of water in intermolecular interactions in food. In Y.S. Roos, R.B. Leslie, & P.J. Lillford (Eds.), *Water Management in the Design and Distribution of Quality Foods* (pp. 199–233). Lancaster and Basel: Technomic.

Tolstoguzov, V.B., Grinberg, V.Y., & Gurov, A.N. (1985). Some physicochemical approaches to the problem of protein texturization. *J. Agric. Food Chem., 33,* 151–159.

Van Camp, J., & Huyghebaert, A. (1995). A comparative rheological study of heat and high pressure induced whey protein gels. *Food Chem., 54,* 357–364.

Van Camp, J., Messens, W., Clement, J., & Huyghebaert, A. (1997). Influence of pH and calcium chloride on the high-pressure-induced aggregation of a whey protein concentrate. *J. Agric. Food Chem., 45,* 1600–1607.

Yen, G.-C., & Lin, H.-T. (1998). High pressure and heat treatments effects on pectic substances in guava juice. In F. Shaidi, C.-T. Ho, & N. van Chuyen (Eds.), *Process-Induced Chemical Changes in Food* (pp. 81–90). New York: Plenum Press.

Zasypkin, D.V., Dumay, E., & Cheftel, J.C. (1996). Pressure- and heat-induced gelation of mixed β-lactoglobulin/xanthan solution. *Food Hydrocolloids, 10,* 203.

Zhurawskaya, N.A., & Kiknadze, E.V. (1986). Concentration of protein as a result of the phase separation of water-protein-polysaccharide systems. Part 1: Phase equilibria in water milk proteins-polysaccharide systems. *Nahrung, 30,* 601–613.

CHAPTER 8

Effects of High Pressure on Water-Ice Transitions in Foods

Siegfried Denys, Oliver Schlüter, Marc E.G. Hendrickx, and Dietrich Knorr

INTRODUCTION

The effects of high hydrostatic pressure on the phase transitions and the phase diagram of water (as shown in Figure 8–1) are well known (Bridgman, 1912; Wagner et al., 1994), but until recent years, little attention has been paid to possible applications in the area of processing and preservation of foodstuffs.

High pressure offers new possibilities for food processing and preservation, not only at temperatures above zero but also in the negative temperature region. According to the principle of Le Chatelier, pressure opposes reactions associated with volume increase, such as the transition from liquid water to ice I (the common ice form at atmospheric pressure). This results in a lowering of the freezing point to a minimum of −22°C at 207.5 MPa (Bridgman, 1912) (Figure 8–1), which suggests the possibility of new applications of high pressure such as pressure-shift freezing, pressure-induced thawing, pressure-assisted freezing and thawing, and preservation of foodstuffs at subzero temperatures in the liquid state (subzero storage without freezing) (Deuchi & Hayashi, 1992; Kanda & Aoki, 1993; Kalichevsky et al., 1995; Le Bail et al., 1997; Knorr et al., 1998). Pathways in the phase diagram associated with each of these processes are illustrated in Figure 8–2. Knorr et al. (1998) first introduced the terminology used here to distinguish between the different processes. *Pressure-assisted* means that the phase transition occurs under constant pressure; *pressure shift* refers to a phase transition caused by a pressure change; and *pressure-induced* means a phase transition initiated with pressure change continued at constant pressure. Unfortunately, the literature on this topic is marked by inconsistent terminology.

At certain conditions of pressure and temperature, ice polymorphs other than ice I may be formed (Figure 8–1). In contrast to ice I, these ice forms have a higher density than liquid water (at the same pressure), which explains why their

Parts of this work have been supported by the German Research Foundation (DFG-Kn 260/6-1) and by a sandwich program of the German Academic Exchange Service (DAAD).

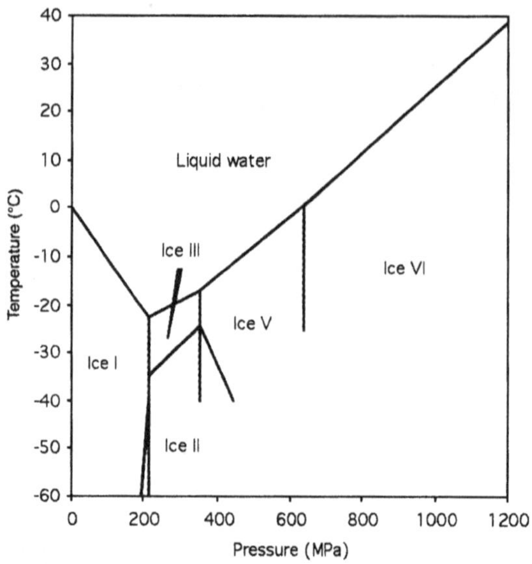

Figure 8–1 The solid–liquid phase diagram of water.

equilibrium temperature with liquid water increases at higher pressure. Ice VI may be formed at room temperature at pressures of about 900 MPa. Application of the other ice polymorphs to food processing and preservation still needs to be explored. Pressure shifting between ice polymorphs I, III, and V has been studied as a way of disrupting microorganisms (Edebo & Hedén, 1960; Hedén, 1964). However, its application to foods seems dubious because the associated volume change (greater than 17%) leads to severe tissue damage. The processes related to the transition of liquid water to ice I do not require the amount of pressure used in food pasteurization and sterilization. High pressure equipment operating under 400 MPa is less expensive and quite reliable. Therefore, these processes are likely to have commercial interest and will be considered in detail in this chapter.

THE USES OF PRESSURE IN FREEZING AND THAWING

Pressure-Shift Freezing

Principle and Advantages

Food freezing occurs in two stages: nucleation and crystal growth (Sahagian & Goff, 1996). Reduction of available liquid water as ice, combined with subzero temperatures, causes a reduction of chemical and biological processes, and renders freezing the preservation technology that guarantees the longest shelf life (Reid, 1990). As a consequence, freezing is widely used for food processing (meat and processed meats, fruits and vegetables, dairy and egg products, and cereals and baked products). At the same time, freezing and subsequent thawing of foods cause undesirable changes in their texture and sensory properties. Studies of the effects of freezing and thawing rates on food quality have led to the current trend of rapid freezing and thawing or, more economically, to the combination of slow and rapid freezing and thawing (George, 1993).

The optimum freezing rate for foods is system-dependent. Generally, slow freezing of tissue material allows osmotic transfer of water through the cell membrane and the growth of large extracellular ice crystals, causing extensive structural changes (Reid, 1990). Other problems, such as increased enzyme and microbiological activity may result from increased solute concentration. Rapid freezing, on the other hand, results in smaller ice crystals (intercellular and extracellular) and leads to better product quality. Ultra-rapid freezing (e.g., cryogenic freezing) may cause lethal intracellular ice crystallization or mechanical cracking, because of initial volume decrease (cooling) followed by volume increase (caused by transition to ice I), especially when the treated products are large. Therefore, when freezing foods, the general goal of food technologists is to create a homogeneous matrix of small ice crystals. This state arises from rapid nucleation caused by supercooling.

In this context, the effect of pressure on the water–ice I phase equilibrium has attracted attention. When a product initially is kept in the liquid state at subzero temperature combined with high pressure, sudden pressure release (pathway *acdf* in Figure 8–2) results in a very high degree of supercooling and promotes rapid ice nucleation. In theory, 22°C of supercooling can be obtained when the expan-

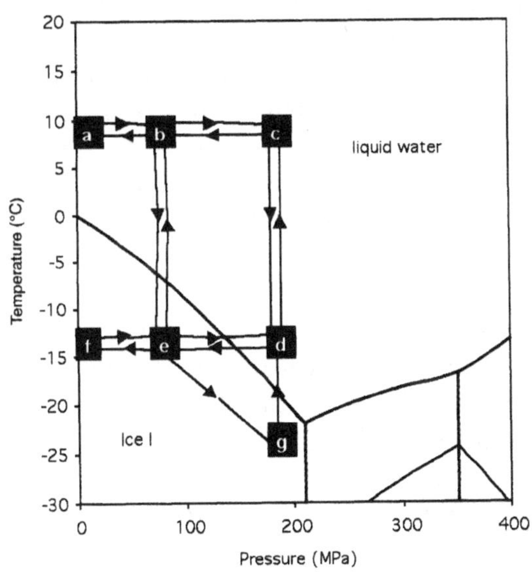

Figure 8–2 Detail of the phase diagram of water showing interesting possibilities for the use of high pressure: pressure-shift freezing (*acdf*), pressure-induced thawing (*fdca*), pressure-assisted freezing (*abef*) and thawing (*feba*), and subzero storage without freezing (*acd*).

sion is achieved very quickly. For each degree of supercooling, an increase of about tenfold in the ice nucleation rate is observed (Burke et al., 1975). Usually, conventional freezing at atmospheric pressure starts when a much smaller degree of supercooling is obtained, depending on the product and the rate of heat transfer during freezing. Furthermore, in classic freezing, the freezing kinetics have a spatial dependency: the supercooling necessary for initiating nucleation is first reached at those layers in the product that are in thermal contact with the cooling medium. In the case of pressure-shift freezing, pressure release and the corresponding supercooling are obtained uniformly and immediately throughout the sample, resulting in a uniform matrix of small ice nuclei. According to Sanz et al. (1997), about 36% of the total water content can be instantaneously converted to ice during expansion from 200 MPa to 0.1 MPa. A practical problem faced in pressure-shift freezing is that, depending on the size of the product, a large amount of latent heat must be released to complete the freezing process; thus, additional cooling is required to complete crystal growth (Kalichevsky et al., 1995). Considering the time it takes to achieve pressure release in high pressure equipment (generally 5 to 20 seconds), simultaneous removal of all latent heat is impossible, and the temperature of the product will increase to the phase transition temperature at atmospheric pressure (i.e., the pressure resulting after pressure release). Experimental data show that this happens very fast (Denys et al., 1997; Levy et al., 1999). After pressure release, the product temperature remains at this value (generally below 0°C) until all latent heat is removed. Actually, the freezing point decreases during freezing as a result of increasing solute concentration.

In summary, the primary advantage of pressure-shift freezing is the possibility of achieving a uniform and rapid nucleation, whereby a considerable amount of water is converted to ice I. The subsequent part of the process is analogous to classic freezing processes, and temperature profiles follow the same course as conventional freezing curves. To illustrate this, a typical temperature history during pressure-shift freezing, recorded in the center of a cylindrical Tylose sample (a food simulator), is shown in Figure 8–3.

Applications in Food Technology

Food applications of pressure-shift freezing are still under development, and only preliminary studies have been conducted. Most of these studies focus on the advantageous effects of pressure-shift freezing on texture and structure of products, as a result of smaller ice crystals. Japanese studies revealed that pressure-shift freezing of tofu resulted in better structure than air-blast frozen tofu. Air-blast frozen tofu developed drip, developed shape deformation, formed a rigid core in the center, and had a spongy texture when thawed (Kanda et al., 1992; Kanda & Aoki, 1993). Pressure-shift freezing produced fine ice crystals to result in lower structural damage. The tofu was restored to its original shape and homogeneous structure, and its taste and texture were the same as before freezing. Koch et al. (1996) observed that pressure-shift freezing of potato cubes resulted

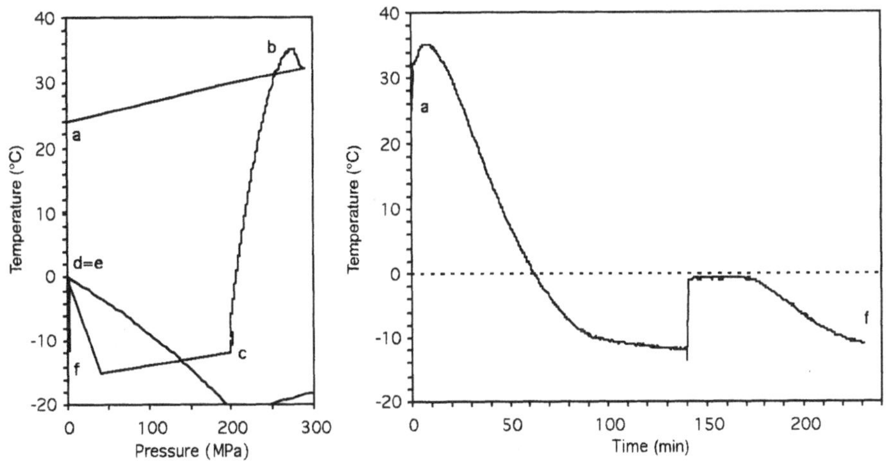

Figure 8–3 Typical temperature history during a pressure-shift freezing process, recorded in the center of a cylindrical Tylose sample. Note the temperature increase at pressure release.

in less damage of the cell structure, less drip loss, and less enzymatic browning than conventionally frozen cubes. They also emphasized the importance of subsequent storage temperature. Cooling to −30°C after pressure-shift freezing did not significantly affect the results, whereas subsequent frozen storage at −18°C resulted in quality deterioration. This could be related to the glass transition temperature of potato tissue. Fuchigami et al. (1996, 1997a, 1997b) reported on textural and histological changes in pressure-shift frozen carrots, suggesting that specific pressure-temperature conditions (e.g., 200 MPa, −20°C) are particularly suitable. Otero et al. (1998) compared damage to the microstructure of eggplants frozen by conventional air freezing and by pressure-shift freezing. Pressure-shift frozen samples had the appearance of fresh samples, and no differences between center and surface cell structure were observed (indicating uniform nucleation). Still-air and air-blast freezing caused cell separation and rupture of cell walls in the product center. Firmness decreased after freezing in descending order of magnitude: still-air frozen (75%), air-blast frozen (69%), pressure-shift frozen (56%). Similar results were observed for drip loss. Martino et al. (1998) compared the size and location of ice crystals in large pieces of pork muscle frozen by pressure-shift freezing to those obtained from pork frozen by air-blast and liquid nitrogen. In the pressure-shift frozen samples, small-sized ice crystals were observed in the center and at the surface of the treated samples. In contrast, samples from air-blast and cryogenic fluid freezing showed nonuniform ice crystal distributions.

In conclusion, pressure-shift freezing can provide an effective alternative for obtaining high-quality frozen foods. To use this technology successfully on products of large size, it is essential to remove latent heat to prevent recrystallization. Operating costs of pressure-shift freezing can probably be reduced by keeping products in the high pressure chamber only when the application of pressure is necessary. For example, products could be removed from the high pressure unit after nucleation is achieved, and the freezing process could be completed in conventional freezing equipment. This approach could potentially increase the cost-effectiveness of the pressure-shift freezing process.

Pressure-Induced Thawing

Principle and Advantages

A frozen product can be forced to the liquid area in the phase diagram by applying high pressure, thus allowing faster thawing (Deuchi & Hayashi, 1992; Kalichevsky et al., 1995; Le Bail et al., 1997; Knorr et al., 1998) (pathway *fdca* in Figure 8–2). For example, Deuchi and Hayashi (1992) found that pressure treatments of 200 MPa for 30 minutes at 5°C were sufficient to completely thaw ice prepared at temperatures from −10°C to −30°C. Reduced treatment time (20 minutes) produced a "state of sherbet" (small ice crystals in liquid water), apparently indicating the uniform nature of pressure-induced thawing. However, this may also have been caused by recrystallization in a fully thawed sample on depressurization, as a result of insufficient heating (Kalichevsky et al., 1995). In the reverse process of pressure-shift freezing, a temperature decrease is generally observed when a frozen sample is forced to the liquid region in the phase diagram. This phenomenon has also been observed when ice I is compressed (Kanda et al., 1993), and it can be attributed to the absorption of latent heat required for the melting process and the products' inability to simultaneously supply this amount of heat. The product melts at the melting temperature associated with the applied pressure, and heating is required to provide the latent heat of fusion and to prevent recrystallization on depressurization. Kanda et al. (1993) also reported partial melting of ice I upon compression, analogous to the partial freezing observed at expansion in pressure-shift freezing. A high pressure thawing process can be roughly represented by pathway *fegca* in Figure 8–2, and this explains the observed contradictory phenomenon of a sudden temperature increase in the pressure medium (caused by compression), but a fast decrease of product temperature, during pressure buildup.

As an example, a temperature profile during a pressure-induced thawing process, recorded in a Tylose cylinder, is shown in Figure 8–4. The similarity of this temperature profile to the profile during conventional thawing is notable after high pressure is reached; the difference is a shift toward lower temperature in the case of pressure-induced thawing. The increased thawing rate associated with pressure-induced thawing is mainly attributed to the larger temperature gradient between the melting product and the ambient pressure transmitting medium, resulting after compression. Faster thawing is known to reduce the loss of liquid retention properties (ice recrystallization occurring at low thawing rates causes drip) and can improve color and flavor preservation in fruit. Obviously, the effects of pressure are added to the global effect of other treatments, and this

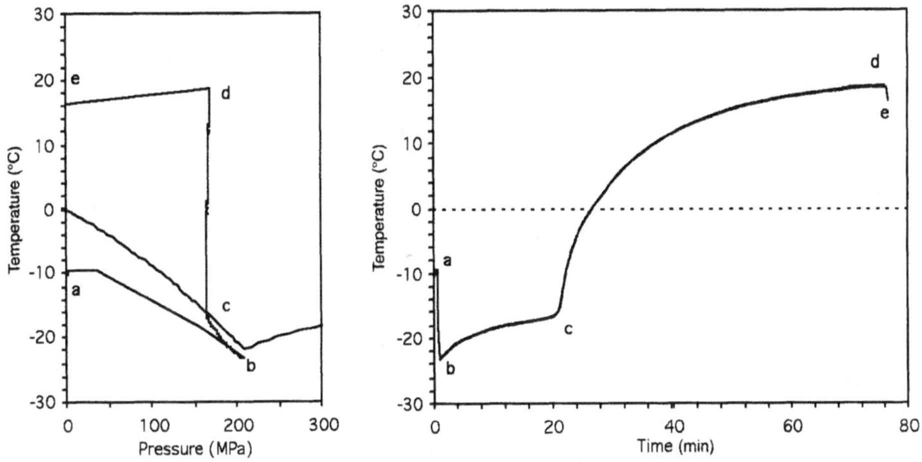

Figure 8–4 Typical temperature history during a pressure-induced thawing process, recorded in the center of a cylindrical Tylose sample. Note the temperature decrease at pressure buildup.

should be considered when comparing processes. The presence of sugar or salt reduces the amount of pressure required to thaw ice I (Deuchi & Hayashi, 1992), because of melting point depression and a shift of the water phase diagram. For surimi, the depressed melting point curve is almost parallel to that of pure water (Takai et al., 1991). This is probably the case for most real foodstuffs.

During pressure-induced thawing, unlike for pressure-shift freezing, the major part of the phase transition takes place at high pressure. Thus, the effect of pressure on the properties determining the rate of phase transition is more important than for pressure-shift freezing. Based on thermodynamic pressure-volume-temperature relations, the latent heat required for the transition from ice I to liquid water decreases with increasing pressure (Bridgman, 1912; Karino et al., 1994). Karino et al. reported a reduction from 333 kJ/kg at atmospheric pressure to 241 kJ/kg at 194.7 MPa. The reduced latent heat, as well as the large temperature gradient resulting after compressing a frozen sample, contributes to a shorter thawing time. However, because the enthalpy difference between initial state (frozen product at temperature T_i) and final state (thawed product at temperature T_f) is the same, regardless of the pathway followed, additional enthalpy must be provided to heat the thawed product to its final temperature.

Applications in Food Technology

Effects of pressure-induced thawing on food products have been studied more thoroughly than pressure-shift freezing. Most studies are on meat and fish products.

Meat products. Effects of pressure-induced thawing on meat products generally include reduction of the time required to thaw the product completely and better retention of quality factors than with conventional thawing methods. Negative effects include discoloration and whitening of the meat, depending on the pressure level applied. Deuchi and Hayashi (1991, 1992) reported that frozen beef at −20°C was thawed at 50 MPa without changing original color and with low drip. However, considering the low pressure applied in this study, it is questionable whether partial thawing occurred upon compression. Higher pressures caused denaturation of protein, which results in whitening of the meat. Thawing times for frozen beef at −10°C, thawed at 120 MPa, were one-third shorter than thawing times at atmospheric pressure, with sensory qualities comparable to those of conventionally thawed products (Makita, 1992). Again, toughening and surface whitening were observed at lower temperatures and higher pressures. Pothakamury et al. (1995) observed the most pronounced thawing time reduction; they reported that 2 kg of frozen beef at 5°C could be thawed in 80 minutes when pressurized to about 200 MPa, while thawing at atmospheric pressure at 5°C required 7 hours. The flavor and juiciness of the thawed beef were about the same for beef thawed under low humidity at 5°C, but the surface of the beef was slightly discolored.

Obviously, when comparing thawing times and quality effects, a number of factors should be considered (initial temperature of the frozen product, ambient temperature, applied pressure level, and size and thermal properties of the product). In the literature, too often these factors are not specified. In addition, effects should be compared to those observed with commercial thawing techniques, which also exhibit strong variations. In a detailed study, Zhao et al. (1996, 1998) compared the sensory properties of meat that was thawed using high pressure and conventionally thawed meat. They focused on different pressure levels (140, 210, 280, and 350 MPa), processing times (5, 15, and 30 minutes), sample diameters (55, 65, and 80 mm), and initial temperatures (−7°C, −11°C, −18°C, and −22°C) and compared properties of pressure-induced thawed beef and conventionally thawed beef (in a cooler at 3°C at atmospheric pressure). Pressures of 210 to 280 MPa provided the processing range necessary for thawing. Pressure-induced thawing occurred much faster than controls and resulted in similar texture and color. They also observed a temperature reduction in frozen beef during pressurization, resulting from the partial conversion of ice to water and the product's inability to immediately supply the latent heat necessary for thawing.

Fish products. Takai et al. (1991) reported greatly enhanced thawing rates for surimi, depending on the temperature of the sample and the medium and on the pressure used. They also noted protein denaturation and discoloration of the fish meat at higher pressures. Murakami et al. (1992, 1994) observed a decrease of drip for frozen tuna muscle that was thawed at high pressure, with no effect on the microbial quality. However, they observed that pressure had no effect on the thawing rate. Depending on the pressure level applied, they also observed a discoloration of the tissue. In addition, they observed a reduced solubility of sarcoplasmic proteins (caused by protein denaturation) with increasing thawing pressure (50 to 150 MPa). Yoshioka et al. (1996) compared pressure-induced thawed carp (100 to 300 MPa for 10 minutes) with carp thawed under running water at 15°C to 17°C. They reported that the pressure-thawed fish maintained a better quality of the fish muscle than the fish thawed with running water. Chevalier et al. (1999) studied the influence of the pressure level (0.1 to 200 MPa), the freezing rate (before thawing), the pressurization rate, and the pressure holding time for pressure-induced thawing of whiting filets in comparison with thawing at atmospheric pressure. They observed that the thawing time for the pressure-induced thawing processes was reduced by a factor of four. Freezing rate is recognized as being a significant factor in tissue damage: the higher the freezing rate, the lower drip losses. These authors also observed lower drip losses when higher pressurization rates were applied at a given pressure. As in slower thawing, ice crystal recombination apparently occurs during the partial thawing when pressurization rates are lower. However, according to Chevalier et al., (1999) the use of high pressure for thawing did not result in less thawing drip than that from atmospheric thawing. When using high pressure thawing, they obtained less drip volume than for atmospheric thawing only by prolonging the pressure holding time. They explained that this was a result of an osmotic pressure gradient and a mass counterflow diffusion (of water into the cells and solutes out of the cells) when cells are physically compressed. Indeed, a study on volume changes of yeast and blood cells submitted to high pressure showed increased permeability and mass transfer during pressure treatment (Perrier-Cornet, 1995). Probably, the reduced drip reported in earlier studies (Murakami et al., 1992; 1994) was a result of an excessive thawing time used by these authors (according to Chevalier et al., 1999). In summary, an increased thawing rate and the corresponding time saving seems to be the most obvious advantage of pressure-induced thawing of fish. Reduction of drip is another potential advantage, but this is still open for study.

Other products. Eshtiaghi and Knorr (1996) investigated pressure-induced thawing at 600 MPa (25°C and 50°C) for 15 minutes of frozen strawberries as a pretreatment in thermal processing of strawberry preparations. They observed increases in sucrose uptake and reduction of microbial counts by 2 log cycles. In contrast, they observed that commercial thawing of strawberries under atmospheric conditions resulted in fruit with lower density than the surrounding syrup; this caused the strawberries to float and caused inconsistencies in product delivery and quality. Deuchi and Hayashi (1991) observed that the texture of a 1.7% agar gel frozen at −20°C was no better after pressure-induced thawing than after thawing at ambient pressure. This probably indicates that, in this case, the structural damage occurs largely from freezing and not from thawing (Deuchi & Hayashi, 1991).

The study of packaging materials appropriate for pressure-induced thawing applications has also been initiated. Fradin et al. (1998) tested six different packaging materials on their resistance to piercing and delamination caused by residual air during pressure-induced thawing. Only one material, BB4L, containing EVA copolymer, was found to be resistant to piercing.

It is likely that pressure-induced thawing will have many applications in the food and medical fields, especially for products in which significant sample deterioration occurs during thawing. Le Bail et al. (1997) proposed an economic simulation for a pressure-induced thawing application. Based on the cost of an installation capable of treating a charge of 300 kg, maintaining a temperature of 10°C, and reaching a pressure of 100 MPa in 5 minutes, and based on the assumption that drip losses are 5% less than with conventional thawing methods, these authors calculated the time necessary to recuperate the initial investment. Recuperation times of 16 to 64 months were reported, depending on the annual tonnage treated and the primary cost of the treated product.

Pressure-Assisted Freezing and Thawing

During pressure-assisted freezing and thawing, the phase transition occurs under constant pressure (Figure 8–2). Thus, the advantageous effects of shifting the pressure along the freezing or melting line (uniform nucleation and partial freezing or melting, as in pressure-shift freezing and pressure-induced thawing) are not exploited. Haas et al. (1972) found that is was possible to retain quality when freezing fruits, vegetables, and meat products under gaseous pressure of 0.3 to 10 MPa at −20°C to −25°C. Strawberries, pineapple, and banana frozen under these conditions had a softer structure and could be eaten without thawing. Samples that were air dried after the pressure treatment showed less textural damage, larger size and less shriveling, more rapid dehydration, and more uniform rehydration than samples that were conventionally freeze-dried. Some products showed increased drip. According to the authors, these advantageous effects were the result of tiny gas bubbles visible in the product after pressure treatment. It appears that this gas remains within the tissue during drying and prevents collapse that otherwise occurs because of capillary forces. These effects are not expected in case of hydrostatic pressure.

Fuchigami et al. comprehensively studied effects of phase transitions conducted at constant pressure on the texture and histology of raw and blanched carrots (Fuchigami et al., 1996, 1997a, 1997b), kinu-tofu (Fuchigami et al., 1997c), and Chinese cabbage midribs (Fuchigami et al., 1998a). Those processes were conducted by pressurizing to the desired pressure for a certain time at −18°C to −20°C and then releasing pressure. They focused on phase transitions to ice I, ice III (340 MPa), ice V (400 MPa), and ice VI (700 MPa). In addition, they investigated the effects of compressions to 200 MPa, where no phase transition occurs at the temperature under investigation. (*Note:* In view of the freezing point depression of liquid water at pressures up to 600 MPa, and considering the relative high freezing temperatures used in these studies, it is probable that the products did not reach complete freezing and that nucleation occurred upon expansion [as in pressure-shift freezing], rather than complete phase transition at constant pressure. This could be the case for runs that used pressures of 200 to 400 MPa.) After treatment, the product was thawed at 20°C and at atmospheric conditions. The authors reported loss of texture and histological damage when pressure-assisted freezing to ice I or ice VI was conducted (Fuchigami et al.,

1997a). In contrast, treatments at 200 to 400 MPa (where pressure-shift freezing probably occurs upon pressure release) of raw carrots yielded a better texture and smaller ice crystals than conventional freezing (Fuchigami et al., 1997b). Pressure-assisted freezing to ice III and ice V also appeared to be effective in improving texture and histological structure of frozen carrots (Fuchigami et al., 1997b). They drew analogous conclusions for the texture of frozen tofu (Fuchigami et al., 1997c) and frozen Chinese cabbage (Fuchigami et al., 1998a). For Chinese cabbage, use of pressure enhanced deesterification of pectin and increased firmness and rupture strain (a "pliable" structure resulted). To eliminate additional nucleation effects caused by pressure release, Fuchigami et al. (1998b) conducted analog freezing experiments at −20°C on tofu (compressed to the same pressures as in these authors' earlier work), but pressure-assisted thawing was applied. When tofu was frozen at 200 to 500 MPa and then thawed at atmospheric pressure, the size of ice crystals increased slightly because of phase transition when pressure was reduced. However, the pore size in tofu thawed at high pressure was the same as for untreated tofu. The textural quality of tofu treated at 200, 340, and 400 MPa was better than that of tofu frozen at 100 MPa or above 500 MPa, regardless of the thawing method. Again, it should be mentioned that these samples probably did not reach complete freezing.

In conclusion, phase transitions conducted at constant pressure lack the advantages of pressure-shift freezing and pressure-induced thawing. Therefore, pressure-assisted freezing and thawing are unlikely to have the commercial potential of pressure-shift freezing and pressure-induced thawing. However, these processes can serve as a useful research tool because they allow investigators to eliminate effects caused by pressure changes (e.g., to determine the pressure dependence of the latent heat, discussed in the section below).

Preservation at Low Temperature and High Pressure (Subzero Storage without Freezing)

Because of the anomalous phase behavior of water, using moderately high hydrostatic pressure permits one to maintain liquid aqueous systems at temperatures as low as −22°C (Figure 8–2), and to take advantage of the low temperature while avoiding the disadvantageous effects of freezing. In 1977, Charm et al. (1977) studied fish quality and stability of the enzymes horseradish peroxidase and red-crab trypsin on storage at −3°C under 24 MPa (sufficient to prevent freezing). They observed that pressure inhibition of enzyme activity occurred below a certain critical temperature, depending on the enzyme. Cod fillet stored under these conditions for 36 days had a perceived storage time of 7 days at 1°C under ambient pressure compared to 6 days for a product that went through a similar period of frozen storage. Samples stored at 1°C under ambient pressure were unacceptable after 9 days. The authors pointed out that a significant energy saving could be made using pressure storage rather than freezing. Deterioration caused by freezing and thawing effects is also avoided.

Recently, investigators have studied the effects of more extreme pressures and temperatures. Deuchi and Hayashi (1990, 1992) showed that mechanical and rheological properties of agar gels could be maintained by nonfrozen storage at −20°C and 200 MPa for 12 days. In addition, they stored uncooked fruits and meats, microorganisms (baker's yeast and 10 kinds of bacteria), and freeze-sensitive enzymes at subzero temperature (−5°C to −20°C) with application of pressure (50 to 200 MPa) for a few days or weeks. Strawberries and tomatoes maintained their fresh taste, texture, and color on storage. Raw pork also could be successfully stored, avoiding drip losses after thawing. Microbial counts of most microorganisms in ground beef (coliforms, Enterobacteriaceae, Gram-positive and Gram-negative psychrophiles, enterococci, and lactic acid bacteria) were reduced by storage at 200 MPa and −20°C, and in some cases more than by freezing. Yeast and some bacteria were completely inactivated by such pressurization at subzero temperature, even when the pressurization was induced by volume expansion related to the phase transition (liquid to ice I) (Hayakawa et al., 1998). This is not the case at room temperature. Enzymes that are inactivated by freezing (catalase, β-amylase, cathepsin, and lactate dehydrogenase) generally had reduced activity after nonfrozen storage, but they were not inactivated. Ooide et al. (1994) examined the change of the contents of substances related to nucleic acids (ATP, ADP, IMP, and AMP) and drip volumes from the muscle of carp and chicken after nonfreezing preservation under subzero conditions (−8°C at 110 MPa and −15°C at 170 MPa for 50 days). Enzymatic degradation of the compounds related to nucleic acids occurred at −8°C as at 5°C refrigeration, although their rates were slower. However, no changes were observed when the product was stored at −18°C in a frozen condition. They concluded that pressure storage at low temperature could preserve meat for a long time without a decline of texture, but attention should be paid to enzyme activity, which is a potential problem under these conditions. Enzymatic degradation of compounds related to nucleic acids was only slightly slower under high pressure conditions than under refrigerated storage, whereas it was significantly reduced by freezing.

In conclusion, low-temperature, nonfrozen storage under pressure appears to be a potentially useful technology to prolong the shelf life of certain foods over that of refrigeration, while avoiding the damage caused by freezing. Enzyme activity is likely to limit potential storage times (unlike freezing). Other areas of study that benefit from the extension of the liquid phase toward low temperatures are cold denaturation of proteins (Foguel & Weber, 1995; Zang et al., 1995; Nash et al., 1996) and the formation of gels with exceptional properties such as egg gel (Plonce et al., 1996).

Specific Requirements for the Technical Equipment

The technical requirements for a high pressure system are defined by the planned working pressure, the working temperature, the required throughput, and the product. Because the high pressure thawing process uses moderate pressure with temperatures above 0°C, it has no additional requirements other than those for current high pressure system applications. Therefore, existing high pressure systems can be used for both experimental use and industrial use (Deplace, 1995; Olsson, 1995). For the investigation of pressure-supported phase change processes in the low temperature range (pressure-supported freezing and storage), the special material qualities needed for the pressure vessel, the seal systems, and the pressure-transmitting medium need to be examined and modified accordingly. Table 8–1 outlines the characteristics of several experimental high pressure units working at subzero temperatures. These characteristics, which are also of interest for industrial-scale applications, are discussed below.

The relevant working temperature of pressure-supported freezing is as low as −40°C. In this range, the steels conventionally used for high pressure vessels tend to crack because of their temperature-dependent flexibility (rigidity). Therefore, special alloys such as copper-beryllium (which has high thermal conductivity and is nonmagnetic) (Knorr et al., 1998) or certain other elastic alloys, which are not suitable for direct contact with the treated foods because of their toxicity and corrosive effects, are used in place of steel. The use of specific plastic materials is also conceivable, but the low thermal conductivity that may be expected from these materials can impose additional demands on the temperature-controlling system. Generally, the temperature-controlling system is adapted to the specific high pressure vessel.

The temperature of the vessel is usually controlled by external tubes wrapped around the vessel (possibly integrated into the vessel wall) and connected to a cryostatic unit. Vessels with a larger diameter can also be cooled by internally wound cooling tubes or by Peltier elements. However, the use of these latter techniques to improve the rate of heat transfer will reduce the usable volume of the vessel. A reduction in the necessary cooling capacity is possible when vessels of a smaller size (and diameter) are directly inserted into the coolant-bath of the cryostat. In multiple-vessel systems, a nearly equal temperature can be achieved in all of the connected high pressure vessels (Arabas et al., 1999). This is particularly important in the determination of inactivation kinetics of enzymes and microorganisms. However, to ensure accurate control of the temperature (e.g., during a pressure-shift freezing process), it is necessary to consider the following:

- influence of the starting temperature on the process duration (possible precooling)
- the temperature of the pressure-transmitting medium (needed for compression)
- effects caused by the time span from inserting the sample in the medium to attainment of the working pressure (avoiding early, partial freezing)
- influence of the pressure buildup rate and particular properties of the pressure-transmitting medium (temperature rise caused by compression)

The pressure holding time can be shortened in pressure-supported freezing processes by reducing the system temperature at the beginning of the process.

Table 8–1 Examples of Experimental High Pressure Units Working in the Subzero Temperature Range

	Subzero Units						Measurement		
System, Reference, Company	*Material*	*Temperature Range (°C)*	*Pressure Range (MPa)*	*Internal Volume (Diameter)*	*Pressure-Transmitting Medium*	*Cooling System*	*Thermo-couples Inside*	*Pressure Gauge*	*Measuring Rate (Hz)*
Single vessel, Levy et al. (1999), ACB (France)	Stainless steel, copper-beryllium stopper	≥ -30	≤ 300	1000 mL (80 mm)	Propanediol/water (55/45, v/v)	External (cooling circuit in vessel wall)	Type T (×2)	2	≤ 10
Single vessel, Schlüter et al. (1998), Uhde (Germany)	Special steel	≥ -40	≤ 360	600 mL (56 mm)	Glycol/ethanol (80/20, v/v)	External (cooling tubes around the vessel)	Type T (11×) or Pt100 (2×)	1	≤ 10
Single vessel, Denys et al. (1997) EPSI (Belgium)	Special steel	≥ -35	≤ 600	590 mL (50 mm)	Polyglycol	External (cooling tubes around the vessel)	Type K (7×)	1	< 0.04
Single vessel, Fuchigami et al. (1997a), Kobe Steel Ltd. (Japan)	Special steel	≥ -30	≤ 400	40 mL (25 mm)	Polyethylenne glycol	External (directly inserted in cooling bath)	Type n.s. (2×)	1	n.s.
Multiple vessels, Arabas et al. (1999), Unipress (Poland)	Copper-beryllium	≥ -50	≤ 700	5 × 4.1 mL (13 mm)	Pentane/hexane or silicon oil	External (directly inserted in cooling bath)	Type K (5×)	5 + 1	≤ 1

n.s. not specified

However, precooling a pressure vessel requires a sufficiently rapid pressure increase after inserting the product into the receptacle in order to avoid unwanted partial freezing before the actual pressure-shift freezing occurs. In other words, the time between inserting the product into the medium and attainment of pressure should be as low as possible. For this reason, suitable lock systems, such as frame locks (Olsson, 1995) or bayonet locks (Deplace, 1995), which allow for fast handling, should be chosen. In addition, the material of the seals must be adapted to the low temperatures. For example, suitable seal systems could be conical metal seals or polymerized epoxy (e.g., polyurethane) seals.

It is necessary to choose a pressure-transmitting medium that will not undergo an unwanted phase transition (liquid to solid). For this reason, pure water is unsuitable. Alcohol, oils or synthetic mixtures, or combinations of these (with or without water) are employed as a medium (Table 8–1). In industrial food processing, the use of edible oils as a pressure-transmitting medium is conceivable because of their thermophysical qualities. To prevent unwanted migration of the pressure-transmitting medium to the treated food, certain packaging materials (e.g., multilayer films) (Fradin et al., 1998; Amanatidou et al., 2000) must be used. However, this also affects the heat transfer coefficient.

Special challenges are associated with the measurement of temperature and pressure in experimental research on pressure-supported phase transitions. The product-specific limits can be determined via recording of pressure and temperature changes in the pressure vessel, although the course of events under high pressure can be represented more precisely with faster data recording. To assess temperature fields, a multi-thermocouple (e.g., Unipress) can be used to measure up to 11 temperatures simultaneously (the thickness of the thermocouple tips is less than 0.2 mm) in the high pressure receptacle (Schlüter et al., 1998). Because of the complexity of a high pressure–low temperature system, it should be designed specifically around the process parameters of pressure, temperature, and time to implement a controlled and efficient process.

MODELING HEAT TRANSFER DURING PROCESSES WITH PHASE TRANSITIONS AT HIGH PRESSURE

All the applications described above share the fact that they are based on the impact of high hydrostatic pressure on phase transitions between liquid water and one of the ice forms. In order to have commercial potential, applications should yield products with improved quality or produce some economic benefit, such as time saving. From an engineering point of view, theoretically based heat transfer models that allow us to predict the temperature history of products undergoing pressure-shift freezing and pressure-induced thawing would be very useful. These are the applications for which the greatest benefits in terms of time saving are expected, and heat transfer models would allow operators to quantify this advantage. In addition, prediction of freezing and thawing times is necessary to design and evaluate industrial freezing and thawing equipment and to optimize the quality of the end product (Bakal & Hayakawa, 1973; Ramaswamy & Tung, 1984; Mannapperuma & Singh, 1988; Cleland & Özilgen, 1998). Over the past three decades, many mathematical models have been proposed for conventional freezing and thawing processes (at atmospheric pressure). Available methods range from simple semitheoretical equations to sophisticated computer programs based on finite difference and finite element techniques. All of these methods have their inherent advantages and disadvantages. An overview of these methods, with critical evaluation of some of them has been given by Cleland (1985) and Cleland and Özilgen (1998). Recently, others have attempted to model heat transfer during complex processes such as pressure-shift freezing and pressure-induced thawing.

Theoretical Heat Transfer Models and Formulation of the Phase Change at Constant Pressure

Physically, phase change in foods at constant pressure can be approached as conductive heat transfer governed in a three-dimensional constant volume (V) by the following partial differential equation:

$$\rho \cdot C_p \cdot \frac{\delta T}{\delta t} = \frac{\delta}{\delta x}\left(k \cdot \frac{\delta T}{\delta x}\right) + \frac{\delta}{\delta y}\left(k \cdot \frac{\delta T}{\delta y}\right) + \frac{\delta}{\delta z}\left(k \cdot \frac{\delta T}{\delta z}\right) + Q \qquad (1)$$

where T is the temperature (°C), ρ is the density of the food (g m^{-3}), C_p its specific heat (J g^{-1} °C^{-1}), k its thermal conductivity (W m^{-1} °C^{-1}), and Q its latent heat (J). *X, y,* and *z* are cartesian coordinates. Nonlinearity arises from the temperature variable thermal properties and the complex nature of the boundary conditions applied.

Thus, approximate methods must be used or simplifying assumptions must be made to find an analytical solution. One assumption that is often made is that the phase change occurs at a unique temperature and that thermal properties undergo a step change in values at this point (Carslaw & Jaeger, 1959). This condition is true for the freezing and thawing of pure substances but not for phase changes in most solid, high-moisture foods. However, the possibility of deriving simple prediction methods by using this assumption has been explored intensively (Cleland, 1985). Actually, during freezing and thawing of foods, the phase change takes place over a wide range of temperatures caused by freezing point depression, and the properties of foods change considerably over this temperature range. Moreover, when pressure is applied, the initial freezing point is shifted toward lower temperatures, depending on the product (Figure 8–2). Pressure, like temperature, influences the properties to be used in the governing heat transfer equation and, thus, introduces additional complexity.

Considering the nonlinearity introduced by temperature and pressure effects, it is advisable to resort to numerical methods when modeling phase transitions during processes such as pressure-shift freezing or pressure-induced thawing. These have proven to be a useful alternative, allowing for complex geometries and handling nonhomogeneous initial temperature distributions, variable boundary conditions, and nonlinear and nonisotropic thermal properties. A heat transfer problem can be formulated in finite difference form either by replacing the derivatives in the governing equation by equivalent finite difference forms or by writing heat balance equations for the nodes of a finite element grid of the object. The gradual phase change can be incorporated in many ways. The two most commonly used methods when modeling heat transfer during conventional, atmospheric freezing and thawing processes are the apparent specific heat formulation and the enthalpy formulation (Mannapperuma & Singh, 1988). In the apparent specific heat formulation, the phase change is incorporated within a suitably defined "apparent specific heat" (Bonacina & Comini, 1973). In this way, the same governing equation is used for both phases by incorporating the moving boundary in the governing equation. This equation is quasi-linear, but the thermal properties become temperature dependent when the phase change range is spanned. These dependencies are nonlinear and exhibit discontinuities at the initial freezing point, which prevent an analytical solution. For example, Figure 8–5 shows the apparent specific heat and the thermal conductivity in the freezing range for Tylose.

A numerical solution is straightforward but limited by stability criteria restricting the time increment and oscillations caused by the sharp peak of the apparent specific heat function at the initial phase-change temperature. This problem can be avoided in the enthalpy formulation, where enthalpy is the primary dependent variable and temperature is the secondary dependent variable. In this way, one governing equation is applicable for both phases. Therefore, a one-to-one equation should exist between enthalpy and temperature. Dussinberre (1949, 1962)

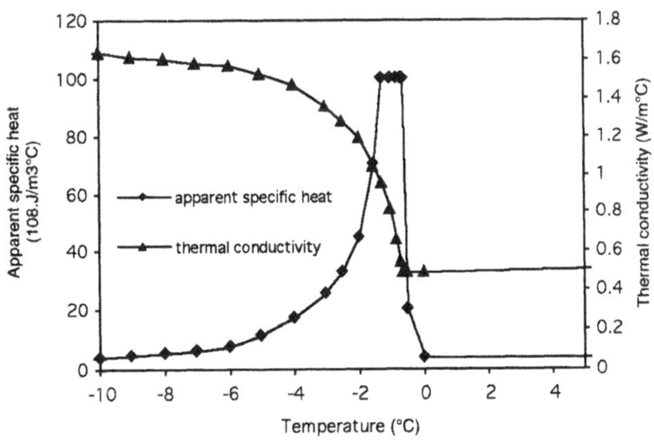

Figure 8–5 The apparent specific heat and the thermal conductivity in the freezing range for Tylose.

introduced such an equation in a qualitative manner. The temperature dependence of enthalpy and thermal conductivity are expressed as enthalpy dependences of temperature and thermal conductivity in order to use enthalpy as the primary dependent variable: $T = T(H)$ and $k = k(H)$.

State-of-the-Art Heat Transfer Models for Pressure-Shift Freezing and Pressure-Induced Thawing

Both the apparent specific heat and the enthalpy formulation approach seem very useful for numerically predicting temperature evolution during phase transition processes in which pressure is involved. Effects of pressure on thermal and physical properties can easily be incorporated in a numerical scheme. To date, the literature includes only the apparent specific heat formulation approach. Lemaire (1994) and Chourot et al. (1996, 1997) used a numerical Cranck Nicholson scheme and modeled the apparent specific heat by a triangular function spanned over 1°C to predict the temperature during pressure-induced thawing of pure water and an aqueous solution of 4.3% NaCl. Their final model considered one-dimensional heat transfer with phase change in an infinite cylinder submitted to convective heat transfer at the surface. The phase change temperature and the latent heat followed the regression equations provided by Bridgman (1912). The model accounted for the pressure effect on the latent heat by adjusting the surface of the specific heat peak at each time step according to the applied pressure. All parameters to solve the numerical system were expressed as functions of pressure and temperature but, because of lack of thermal property data, density and thermal conductivity were considered to be pressure independent. This model did not account for the effect of pressure buildup (i.e., the observed temperature decrease caused by partial melting), and this oversight introduced inaccuracy in the temperature distribution resulting after compression. Still, comparison between experimental measurements and predicted profiles showed a good level of agreement. The model allowed investigators to distinguish between effects of the decrease of the latent heat and specific heat of the liquid (10% reduction in thawing time between 0.1 and 150 MPa) and effects of the increase of the temperature difference between the thawing sample and the surrounding pressure transfer medium (according to the authors, this is the main factor). According to Chourot et al. (1996), lack of knowledge of thermal and physical properties (especially of convection in the annular gap between sample and vessel wall under high pressure) hampers accurate predictions of the temperature distribution within the samples. Recently, the model was used as a simulation tool by Boillereaux et al. (1999), who attempted to control the thawing front propagation during pressure-induced thawing of water. Controlling the thawing kinetics is known to affect the quality of foods.

Denys et al. (1997, 2000a) used an analog numerical approach (an explicit numerical algorithm with apparent specific heat formulation) to predict temperature history in Tylose (used as a food simulator), undergoing pressure-shift freezing and pressure-induced thawing processes. They included an energy balance to calculate the amount of heat released or absorbed when expanding (pressure-shift freezing) or compressing (pressure-induced thawing) a nonfrozen or a frozen sample. According to this approach, the temperature obtained after the pressure shift depends on the extent of the shift, both in the case of pressure-shift freezing and pressure-induced thawing. This was experimentally observed for Tylose (Denys et al., 1997, 2000a). The predictive capacity of the model for pressure-induced thawing was considerably improved by taking into account the pressure dependency of the latent heat of Tylose (Denys et al., 2000a). The effect of pressure on the latent heat of Tylose could experimentally be determined from a series of pressure-assisted freezing experiments conducted at different constant pressure levels. The explicit numerical algorithm (used at constant pressure) was combined with a least-sum-of-squares optimization procedure, minimizing the sum of the squared differences between experimental and predicted temperatures. The corresponding latent heat at each pressure level was estimated by adjusting the height or width of the apparent specific heat peak.

To illustrate this approach, a reduction to 90% of the specific heat values within an appropriate temperature interval and a shift of + 0.2°C of the left-handed slope of the specific heat peak are visualized in Figure 8–6A. The effect of both actions in terms of the latent heat is the same (Figure 8–6B). Figure 8–7 shows experimental and predicted temperature profiles in the center of a Tylose cylinder undergoing pressure-assisted freezing processes.

The predicted profiles were generated through the optimization program and correspond to the optimal "reduction" of the apparent specific heat peak. Both reducing the height and reducing the width of the apparent specific heat peak resulted in excellent agreement with the experimental temperature profiles for pressure-induced thawing. Pressure-induced thawing profiles collected at different positions in a Tylose cylinder (at the center, at the border, and between center and border) and predicted profiles are shown in Figure 8–8. The dotted lines in this figure represent predictions in which the latent heat effect was not taken into account. Denys et al. (2000a) demonstrated the importance of using an appropriate latent heat value, especially in the case of pressure-induced thawing where the phase transition takes place at high pressure.

Schlüter et al. (1999) fitted pressure-assisted freezing curves obtained for potato cylinders during high pressure treatment at 50, 100, and 150 MPa by using a finite difference numerical procedure (radial, symmetrical, one-dimensional heat conduction) and also reported the need to modify the literature values for the apparent specific heat and for the thermal conductivity of potato tissue. The peak value of the apparent specific heat function was markedly reduced at high pressure, but the general shape of the applied distribution (Weibull) of the specific heat could be retained. This approach was also used for fitting temperature profiles recorded during pressure-assisted thawing of potato cylinders (Schlüter et al., 1999).

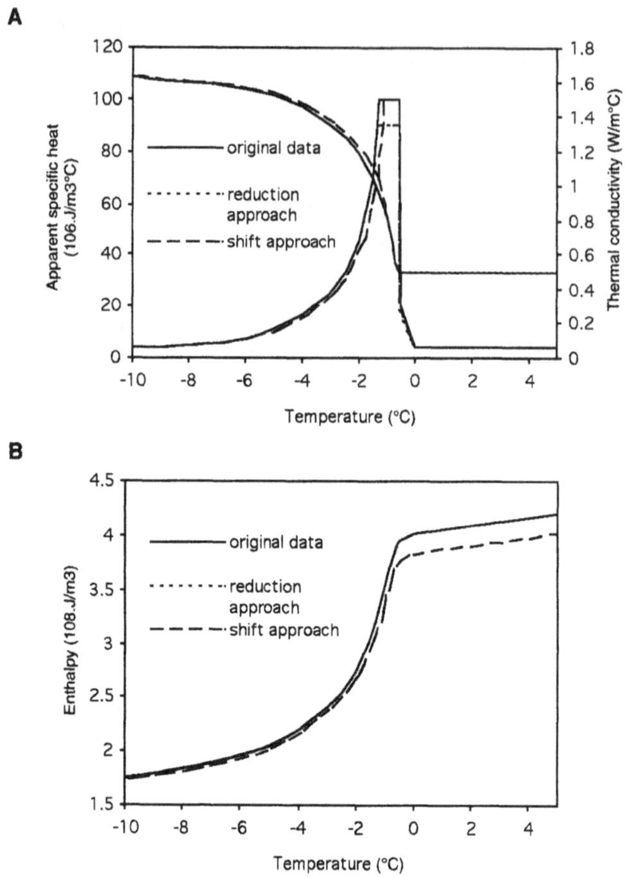

Figure 8–6 The effect of pressure on the latent heat of Tylose. (**A**) Illustration of two ways to account for pressure effects on the latent heat of Tylose: height reduction to 90% of the specific heat values within an appropriate temperature interval, and width reduction by applying a shift of + 0.2°C of the left-handed slope of the specific heat peak. (**B**) The effect of both actions in terms of the enthalpy of Tylose is the same.

Phase Transition Temperature and Equilibrium State

An essential step in the modeling of freezing and thawing processes is the determination of the real phase transition temperature, which is dependent on the specific properties of the considered material and the working pressure. At atmospheric conditions, the precise determination of the product-dependent freezing or melting points may be possible with the aid of various analytical methods (Cammenga & Epple, 1995). Compared to differential scanning calorimetry (DSC), a more feasible method of measurement is the Thermistor-cryoscope method, in which the temperature of the sample is recorded over time while cooling below the freezing point (Rahman, 1995). In this case, the crystallization temperature is derived from the relatively long temperature plateau, which follows the nucleation because of latent heat being released. If the latent heat of fusion is dissipated to a large extent, the temperature of the sample

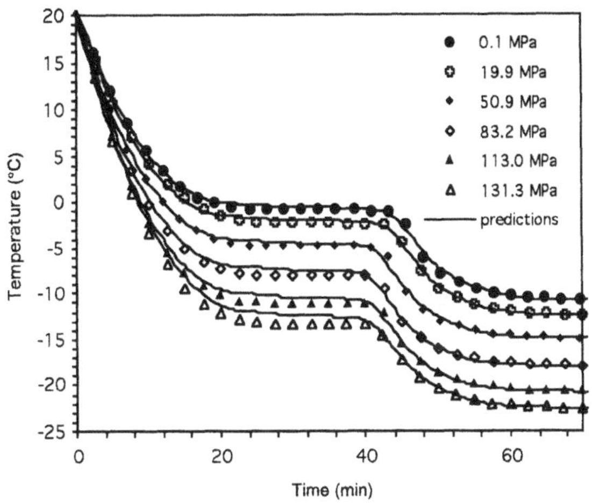

Figure 8–7 Six experimental and predicted temperature profiles recorded in the center of a Tylose cylinder undergoing pressure-assisted freezing at different pressures. The predicted profiles were generated through an optimization program and correspond to the optimal "height reduction" of the apparent specific heat peak.

decreases in accordance with the defaulted ambient temperature. In the same way, the freezing and melting point of foods can be determined at high pressure. The temperature change in the center of a sample related to the phase transition can be plotted versus time or pressure (Figure 8–9).

It is also possible to observe the phase transition by measuring the pressure change, if the sample is sufficiently large compared to the volume of the pressure vessel. After the beginning of crystallization (nucleation [N]), an increase in the specific sample volume (e.g., during pressure-assisted freezing of liquid to ice I) leads to an increase of the system pressure (freezing-induced pressurization [FIP]) and to a corresponding depression of the initial freezing point (freezing point depression [FPD]). The projection of the freezing curve into the phase diagram of water confirms the shift of the freezing point (FPD) parallel to the phase transition line. Only when the resulting latent heat of fusion is mostly removed does the temperature of the sample center decrease with deviation from the equilibrium line to the direction of the defaulted temperature level.

It is possible to represent the product-specific pressure-dependent freezing or melting curve, for example, by determining single freezing points at different pressure levels. The slope of the phase-transition line can be obtained by extrapolation. According to Figure 8–10, the slope of the phase-equilibrium line (liquid to ice I) of foods (e.g., potato) is parallel to that of pure water, starting at a comparably lower freezing point. Cooling of a sample at a pressure of 250 MPa results in freezing to ice I along the extended phase-transition line (curve B).

Because the transition to ice III cannot be obtained directly as expected (Knorr et al., 1998), it may be possible to increase the effective temperature gradient

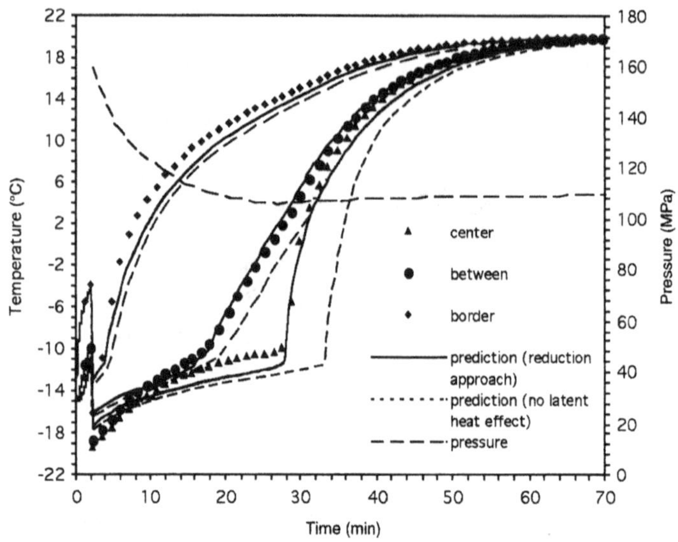

Figure 8–8 Experimental temperature profiles collected at different positions in a Tylose cylinder (at the center, at the border, and between center and border) undergoing pressure-induced thawing, and predicted temperature profiles. The dotted lines represent predictions in which latent heat effect was not taken into account.

(sample medium) during pressure-induced thawing and pressure-shift freezing of foods to levels that are above the theoretical limits accepted until now. Consequently, it may be possible to realize an additional acceleration of pressure-supported phase-transition processes.

The temperature stability at the phase boundary offers another approach to the experimental description of the equilibrium line. Using a suitable measurement system, the shift of the phase-transition temperature in the center of a sample can be recorded during manual change of the system pressure, so that a long range of the phase boundary can be determined (Figure 8–10, curve A). The almost simultaneous temperature jump in the center and at the edge of the subcooled sample indicates the regular beginning of crystallization over the sectional view of the sample (independent of radial distance) because of expansion (pressure-shift freezing). Levy et al. (1999) observed the uniform nucleation throughout an oil-in-water emulsion by the simultaneous rise of temperature with pressure shift-freezing.

Temperature Change Caused by Adiabatic Compression

Generally, researchers who are investigating high pressure prefer compression in short time. Often, however, they do not consider the increased temperature of the pressure medium and product that corresponds to increased pressure, and this oversight introduces errors in their results. Adiabatic compression or decompression is an isentropic process. The temperature increase/decrease ΔT (°C) of

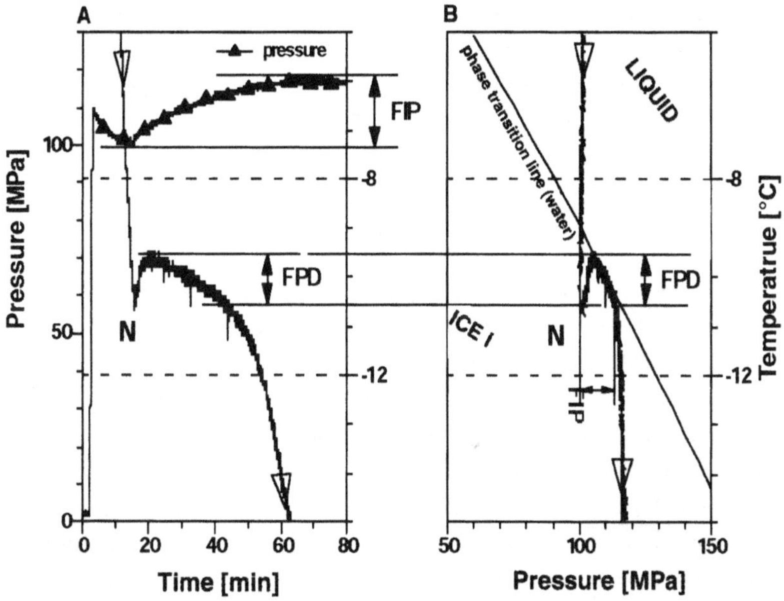

Figure 8–9 Determining the pressure-dependent freezing point by measuring the temperature at the sample center (e.g., potato cylinder) and the pressure during pressure-assisted freezing. (**A**) Plot of time-temperature and plot of time-pressure. (**B**) Plot of pressure-temperature in the phase diagram of water (N: nucleation, FIP: freezing-induced pressurization, FPD: freezing point depression).

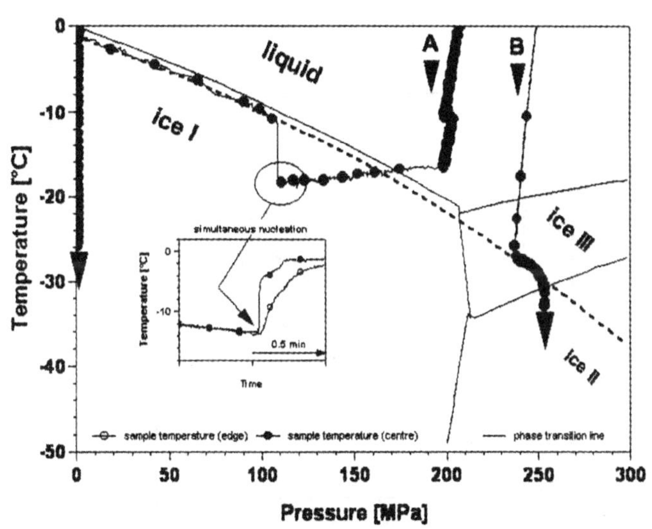

Figure 8–10 Determination of the phase-transition line of foods by extrapolation (broken line). Plot of temperature (center of a potato cylinder) versus pressure in the phase diagram of water. Curve A: freezing curve caused by pressure-shift freezing. Curve B: freezing curve caused by pressure-assisted freezing in the ice III region.

a product, subject to a pressure increase/decrease ΔP (Pa), is proportionate to the expansivity α_p (°C^{-1}) and the specific volume V (m^3 g^{-1}) of the product, and inversely proportionate to its specific heat C_p (J g^{-1} °C^{-1}) as described by the following equation (e.g., Zemansky, 1957):

$$\frac{\Delta T}{\Delta P} = \frac{T \cdot V \cdot \alpha_p}{C_p} \tag{2}$$

Makita (1992) measured the temperature rise accompanied by the compression of liquid water and observed that although the temperature of water near the wall of a high pressure vessel decreased gradually because of the high thermal conductivity of the metal, the center of the water layer reached a temperature close to the theoretical value of adiabatic compression (equation 2), when the compression rate was more than 5 MPa per second. Denys et al. (2000b) determined the expansivity of tomato paste and applesauce at pressures up to 520 MPa, based on equation 2, and used these values in a numerical heat transfer model to calculate temperature history and process uniformity in these products when subjected to compression or expansion. Figure 8–11 shows the calculated adiabatic temperature increase corresponding to compressions up to 600 MPa at three different initial temperatures (10°C, 30°C, and 50°C), for tomato paste, applesauce, and pure water. Very similar values are observed for the food products. It is likely that the temperature increase of a product is determined by its water content. When organic solvents or oils are employed as a pressure medium, the temperature rise should be several times larger than when water is used, because of their higher expansivity and lower heat capacity (Kalichevsky et al., 1995).

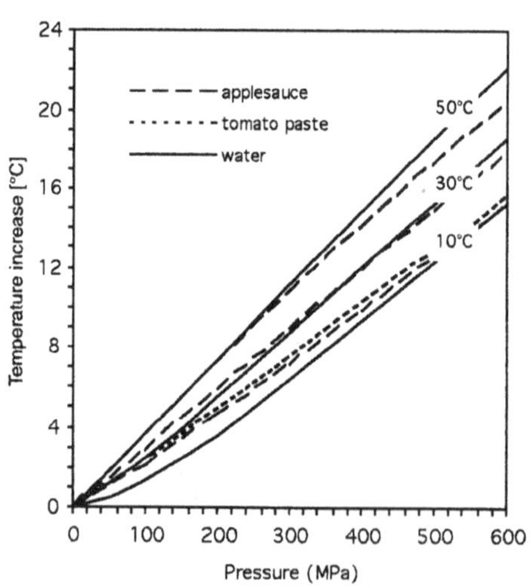

Figure 8–11 Calculated adiabatic temperature increases corresponding to compressions up to 600 MPa for applesauce, tomato paste, and pure water. Temperatures indicated in the figure are initial temperatures (before compression).

Obviously, in cases where compression or expansion is accompanied with a phase transition, such as in pressure-shift freezing and pressure-induced thawing, the temperature change caused by the pressure change is influenced by the release or absorption of the latent heat. This effect is not accounted for in the heat transfer models described above. Probably, the effect is negligible with regard to the heat released or absorbed upon freezing or thawing. Besides, determination of the expansivity in the phase-change region is expected to be difficult because of the overlapping effects. Morley (1986) used the adiabatic temperature change induced by compression up to 6 MPa to investigate physical properties of muscle tissue at temperatures from −24°C to +16°C and obtained data for the expansivity. At −1°C, the initial freezing point of muscle tissue, the expansivity becomes large and negative. On further reduction in temperature, expansivity decreases to zero and then becomes positive below −9°C. Partial melting of ice crystals when compressing a partially frozen product could explain the large negative value of the expansivity observed by Morley.

Thermal and Physical Properties Required

There has been a great deal of research, reviewed by many authors, into obtaining thermal property data required to calculate the progress of freezing or thawing in foods at atmospheric pressure. No method to predict the freezing or thawing time of foods can be accurate if the required food thermal properties cannot be obtained accurately or are not representative of the whole product. For obvious reasons, when high pressure is involved, the pressure dependency of the necessary thermal and physical properties should be determined when calculating heat transfer. In the case of freezing and thawing, these properties include density, thermal conductivity, apparent specific heat, latent heat of fusion, and initial freezing point. Properties are strongly dependent on temperature and pressure in the range where the phase change takes place. Some properties, such as density and thermal conductivity, exhibit stronger dependence on pressure than on temperature within the pressure and temperature range of interest here. In addition, when pressure is applied, other thermodynamic properties such as compressibility and expansivity attain significance. These thermodynamic properties facilitate calculation of density and temperature changes corresponding to compression and expansion of products. Collection of thermal property data is expected to be tedious and expensive because of the variability in product composition and structure and the wide range of temperature and pressure for which data are required. Data for frozen foods using conventional measuring techniques at atmospheric pressure are particularly difficult to determine because it is difficult to maintain the appropriate conditions. Furthermore, high hydrostatic pressure technology poses challenging requirements for sealing, and this obstructs the development of in situ measuring techniques.

Because water is the most important ingredient in many food products, its characteristics under pressure are very important. For ordinary liquid water, thermodynamic and other properties have been studied thoroughly in an extended temperature and pressure range and extensive data are available. For example, Ter Minassian and Pruzan (1981) reported values of the compressibility, expansivity, and specific heat of water up to 500 MPa for an extended temperature range including the supercooled region down to −40°C. Lüdemann (1992; 1994) reviewed the properties of water and aqueous solutions under high pressure. Nakahara (1990) described some properties of water relevant to the pressure treatment of food. Currently, official international formulations for all thermodynamic and transport properties in a large range of temperatures and pressures are developed and maintained by the International Association for the Properties of Water and Steam (IAPWS). This organization recently released a new international formulation for thermodynamic properties of ordinary water for general and scientific use (Wagner & Pruβ, 1999). Software implementations of these international formulations are available (e.g., NIST/ASME Steam Properties). IAPWS-reported values for some relevant water properties in the range of 0.1 to 240 MPa and at temperatures as low as −20°C are included in Figure 8–12.

Bridgman (1912) and Karino et al. (1994) studied the effect of pressure on the volumetric behavior of liquid water and the different ice forms. They both reported values of the latent heat required for the phase transition over a large range of temperature and pressure, based on pressure-temperature-volume relations (Table 8–2), thus providing a useful tool to those involved with the processing of foods at low temperatures and high pressures.

Chizhov and Nagornov (1991) described a thermodynamic equation of state for specific volumes of liquid water, ice I, and their mixtures under pressure. They determined phase boundaries of pure water and ice I, and isentropes of ice-water mixture. Based on this equation of state, Sanz et al. (1997) and Otero et al. (1997) predicted the amount of ice appearing instantaneously upon pressure release in the case of pressure-shift freezing. They integrated the thermodynamic equations of state, characterizing specific volume and specific heat of liquid water and of ice I, as a function of pressure and temperature within the interval of interest for freezing studies, and studied final volume of specific volume of frozen mixtures after adiabatic expansion along the phase-change curve. According to these authors, a high pressure-assisted freezing process may be defined as an isentropic and adiabatic process that transforms the product from an initial point on the liquid water-ice I melting curve (characterized by a steady-state low temperature, high pressure, and a definite liquid water-ice I proportion) to a final point on the same curve (with a lower pressure, a higher temperature, and higher ice content). Supercooling was not considered in the theory. They reported liquid water fractions after quasi-static adiabatic compression from 0 to 210 MPa and after quasi-static adiabatic expansion from 210 to 0 MPa. Their model predicts an instantaneous ice formation of up to 36% upon expansion along the freezing curve.

Numerous attempts have been made to model food properties mathematically and to derive calculation methods for food thermal properties at atmospheric pres-

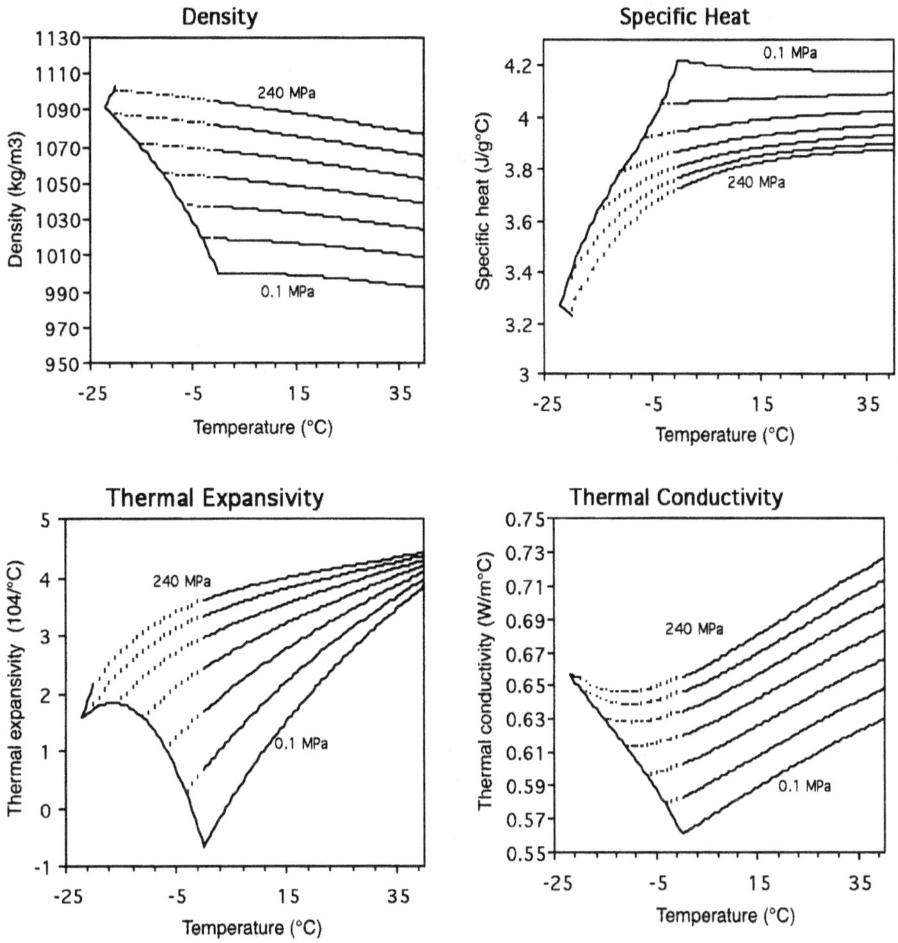

Figure 8–12 Values for the density, specific heat, thermal expansivity, and thermal conductivity of pure water in the range 0.1 to 240 MPa and at temperatures as low as −20°C (according to the NIST/ASME software). The thick lines represent thermal properties at the melting line; dotted lines are properties in the subzero range at high pressure

sure (Lind, 1991; Miles et al., 1983). Most of these models are based on the composition or the water content of the product under investigation. Probably, this approach is applicable for food properties at high pressure, and the properties of ordinary water could be very useful. A lot of work is still to be done in this area. Because of the complexity of foods and interactions between food components, the role of water in determining the thermal and physical properties of foods may not be easily identified, and in situ determination of the properties of foodstuffs at pressures above atmospheric pressure may be advisable. Denys and Hendrickx (1999) used a line heat source probe method to measure thermal conductivity in a pilot-scale high pressure unit and reported thermal conductivity values of commercially canned tomato paste and applesauce at pressures up to 400 MPa and temperatures of 30°C and 65°C. The authors also reported experimentally mea-

Table 8–2 Thermodynamic Properties Related to Phase Transitions of Water

Phase Transition	*T (°C)*	*P (MPa)*	*ΔV (cm³ g⁻¹)*	*ΔH (kJ kg⁻¹)*
Liquid-ice I	−20	193.3	+0.1313	−241
	−15	156.0	+0.1218	−262
	−10	110.9	+0.1122	−285
	−5	59.8	+0.1013	−308
	0	0.1	+0.0900	−334
Ice I-ice II	−35	212.3	−0.2177	−42.5
Ice I-ice III	−20	211.5	−0.1919	+14.6
	−25	206.3	−0.1773	+23.4
Ice II-ice III	−25	330.6	+0.0148	+68.2
Ice II-ice V	−25	350.2	+0.0401	+66.5
Ice III-ice V	−25	341.1	+0.0546	−3.64
	−20	345.5	+0.0547	−3.72
Liquid-ice III	−22	207.5	−0.0466	−213
	−20	246.2	−0.0371	−226
	−17	346.3	−0.0241	−257
Liquid-ice V	−20	308.0	−0.0828	−253
	−15	372.8	−0.0754	−265
	−10	442.4	−0.0679	−276
	−5	533.7	−0.0603	−285
	0	623.9	−0.0527	−293
Ice V-ice VI	−20	624.4	−0.0381	−0.76
	0	626.0	−0.0389	−0.83
Liquid-ice VI	−10	518.0	−0.0960	−264
	0	623.9	−0.0916	−295
	10	749.5	−0.0844	−311
	20	882.9	−0.0751	−320
	30	1038.9	−0.0663	−330

sured values of the density and expansivity of the same products at high pressure (Denys et al., 2000b). When phase transitions are involved, food engineers often consider foods to consist of two fractions: aqueous solution and solids. During the phase change, interactions between these components take place.

Miles and Morley (1978) used thermodynamic equations to estimate the effect of hydrostatic pressure on some of the thermal and physical properties (ice content, enthalpy, and density) of frozen foods. Their research allowed for internal compressions up to 6 MPa, which develop when foods are rapidly frozen (causing cracks) and when machines are used to mold food products (where high compressions force frozen pieces through a shaping channel). These authors concluded that pressure can have a marked effect on the physical properties of frozen foods, but they did not provide experimental verification. According to their equations, the proportion of ice is expected to decrease with increasing pressure at constant temperature. The ice content is little affected by pressure at low temperature but is decreased rapidly by increasing pressures near the transition temperature. The enthalpy of a frozen food product maintained at constant temperature is increased by application of pressure by a "background level" at all temperatures and pressures plus a much greater amount near the transition temperature. The partial change of density with pressure is positive at all pressures and temperatures and, as temperature increases, the change is greater. As already mentioned, measurements of thermal and physical properties in the range of phase transitions is complex because of the interaction of several effects.

CONCLUSION

The phase diagram of water has attracted a great deal of attention, and researchers are currently investigating applications such as pressure-shift freezing, pressure-induced thawing, pressure-assisted freezing and thawing, and subzero storage without freezing. Pressure-shift freezing and pressure-induced thawing, in particular, are expected to have commercial potential and deserve further study. Pressure-shift freezing can provide an effective alternative to traditional freezing methods. Pressure-induced thawing allows reduced thawing time and better retention of important quality factors than with conventional methods. In both processes, the removal or absorbance of latent heat appears to be a major factor in determining the economical potential. Pressure-assisted freezing and thawing processes lack the advantages offered by pressure-shift freezing and pressure-induced thawing, but these processes could provide information on relevant properties of foods at high pressure. Subzero storage without freezing appears to prolong the shelf life of certain foods, while avoiding the damage caused by freezing.

Similar to freezing and thawing processes at atmospheric conditions, theoretically based heat transfer models could be very useful for researchers, operators, and equipment designers. These models can be used to calculate freezing and thawing times corresponding to processes, to gain insight into freezing and thawing kinetics at high pressure, to optimize the quality of end products, and to design and evaluate industrial equipment. Research in this area has been initiated, but determination of the necessary thermal and physical properties is indispensable.

REFERENCES

Amanatidou, A., Schlüter, O., Lemkau, K., Gorris, L.G.M., Smid, E. J., & Knorr, D. (2000). Effect of combined application of high pressure treatment and modified atmospheres on shelf-life of fresh atlantic salmon. *Innov. Food Sci. Emerg. Technol., 1,* 87–98.

Arabas, J., Szczepek, J., Dmowski, L., Heinz, V., & Fronberg-Broczek, M. (1999). New technique for kinetic studies of pressure-temperature induced changes of biological materials. In H.

Ludwig (Ed.), *Advances in High Pressure Bioscience and Biotechnology* (pp. 537–540). Heidelberg: Springer-Verlag.

Bakal, A., & Hayakawa, K.I. (1973). Heat transfer during freezing and thawing of foods. *Adv. Food Res., 20,* 217–256.

Boillereaux, L., Chourot, J.-M., & Havet, M. (1999). Nonlinear trajectory control of high pressure thawing. *J. Proc. Contr., 9,* 351–356.

Bonacina, C., & Comini, G. (1973). On the solution of the nonlinear heat conduction equations by numerical methods. *Int. J. Heat Mass Transfer, 16,* 581–589.

Bridgman, P.W. (1912). Water, in the liquid and five solid forms, under pressure. *Proc. Am. Acad. Arts Sci. XLVII., 13,* 439–558.

Burke, M.J., George, M.F., & Bryant, R.G. (1975). In R.B. Duckworth (Ed.), *Water Relations of Foods (Food Science and Technology Monographs)* (pp. 111–135). London: Academic Press.

Cammenga, H.K., & Epple, M. (1995). Grundlagen der Thermischen Analysetechniken und ihre Anwendungen in der präparativen Chemie. *Angew. Chem., 107,* 1284–1301.

Carslaw, H.S., & Jaeger, J.C. (1959). *Conduction of Heat in Solids.* London: Oxford University Press (Clarendon).

Charm, S.E., Longmaid, H.E., & Carver, J. (1977). A simple system for extending refrigerated, nonfrozen preservation of biological material using pressure. *Cryobiology, 14*(5), 625–636.

Chevalier, D., Le Bail, A., Chourot, J.-M., & Chantreau, P. (1999). High pressure thawing of fish (whiting): Influence of the process parameters on drip losses. *Lebensm.-Wiss. u.-Technol., 32,* 25–31.

Chizhov, V.E., & Nagornov, O.V. (1991). Thermodynamic properties of ice, water and their mixture under high pressure. *IAHS Publ., 208,* 463–470.

Chourot, J.-M., Boillereaux, L., Havet, M., & Le Bail, A. (1997). Numerical modeling of high pressure thawing: Application to water thawing. *J. Food Eng., 34,* 63–75.

Chourot, J.M., Lemaire, R., Cornier, G., & Le Bail, A. (1996). Modeling of high pressure thawing. In R. Hayashi & C. Balny (Eds.), *High Pressure Bioscience and Biotechnology* (pp. 439–445). Amsterdam: Elsevier Science B.V.

Cleland, A.C., & Özilgen, S. (1998). Thermal design calculations for food freezing equipment—Past, present and future. *Int. J. Refrig., 21*(5), 359–371.

Cleland, D.J. (1985). Prediction of freezing and thawing times for food. Ph.D. Dissertation, Massey University, New Zealand.

Denys, S., Van Loey, A.M., & Hendrickx, M.E. (2000a). Modeling conductive heat transfer during high-pressure thawing processes and detamination of latent heat as a function of pressure. *Biotechnol. Progr., 16,* 447–455.

Denys, S., & Hendrickx, M.E. (2000b). Modeling approach for evaluating process uniformity during batch high hydrostatic pressure processing: Combination of a numerical heat transfer model and enzyme inactivation kinetics. *Innov. Food Sci. Emerg. Technol., 2,* 5–19.

Denys, S., & Hendrickx, M.E. (1999). Measurement of the thermal conductivity of foods at high pressure. *J. Food Sci., 64,* 709–713.

Denys, S., Van Loey, A.M., Hendrickx, M.E., & Tobback, P.P. (1997). Modeling heat transfer during high-pressure freezing and thawing. *Biotechnol. Progr., 13*(4), 416–423.

Deplace, G. (1995). Design of high pressure isostatic units for laboratory and industrial treatment of food products. In D.A. Ledward, D.E. Johnston, R.G. Earnshaw, & A.P.M. Hasting (Eds.), *High Pressure Processing of Foods* (pp. 137–154). Nottingham: Nottingham University Press.

Deuchi, T., & Hayashi, R. (1990). A new approach for food preservation: Use of non-freezing conditions at subzero temperature generated under moderate high pressure. In R. Hayashi (Ed.), *Pressure Processed Food: Research and Development* (pp. 37–51). Kyoto: San-Ei Suppan.

Deuchi, T., & Hayashi, R. (1991). Pressure application to thawing of frozen foods and to food preservation under subzero temperature. In R. Hayashi (Ed.), *High Pressure Science for Food* (pp. 101–110). Kyoto: San-Ei Suppan.

Deuchi, T., & Hayashi, R. (1992). High pressure treatments at subzero temperature: Application to preservation, rapid freezing and rapid thawing of foods. In C. Balny, R. Hayashi, K. Heremans, & P. Masson (Eds.), *High Pressure and Biotechnology,* Colloque INSERM (Vol. 224, pp. 353–355). Montrouge: John Libbey Eurotext.

Dussinberre, G.M. (1949). *Numerical Analysis of Heat Transfer.* New York: McGraw-Hill.

Dussinberre, G.M. (1962). *Heat Transfer Calculations by Finite Difference.* Scranton, PA: International Book.

Edebo, L., & Hedén, C.-G. (1960). Disruption of frozen bacteria as a consequence of changes in the crystal structure of ice. *J. Biochem. Microbiol. Technol. Eng., 2,* 113–120.

Eshtiaghi, M.N., & Knorr, D. (1996). High hydrostatic pressure thawing for the processing of fruit preparations from frozen strawberries. *Food Biotechnol., 10*(2), 143–148.

Foguel, D., & Weber, G. (1995). Pressure-induced dissociation and denaturation of allophycocyanin at subzero temperatures. *J. Biol. Chem., 270,* 28759–28766.

Fradin, J.F., Le Bail, A., Sanz, P.D., & Molina-García, A.D. (1998). Note. Behaviour of packaging materials during high pressure thawing. *Food Sci. Tech. Int., 4*(6), 419–424.

Fuchigami, M., Kato, N., & Teramoto, A. (1996). Effect of pressure-shift freezing on texture, pectic composition and histological structure of carrots. In R. Hayashi & C. Balny (Eds.), *High Pressure Bioscience and Biotechnology* (Vol. 13, pp. 379–386). Amsterdam: Elsevier Science B.V.

Fuchigami, M., Kato, N., & Teramoto, A. (1997a). High-pressure-freezing effects on textural quality of carrots. *J. Food Sci., 62,* 804–808.

Fuchigami, M., Miyazaki, K., Kato, N., & Teramoto, A. (1997b). Histological changes in high-pressure-frozen carrots. *J. Food Sci., 62,* 809–812.

Fuchigami, M., & Teramoto, A. (1997c). Structural and textural changes in kinu-tofu due to high-pressure-freezing. *J. Food Sci., 62,* 828–832.

Fuchigami, M., Kato, N., & Teramoto, A. (1998a). High-pressure-freezing effects on textural quality of Chinese cabbage. *J. Food Sci., 63,* 122–125.

Fuchigami, M., Teramoto, A., & Ogawa, N. (1998b). Structural and textural quality of kinu-tofu frozen-then-thawed at high-pressure. *J. Food Sci., 63,* 1054–1057.

George, R.M. (1993). Freezing processes used in the food industry. *Trends Food Sci. Technol., 4,* 134–138.

Haas, G.J., Prescott, H.E., Jr, & D'Intino, J. (1972). Pressure freezing-air drying: A new technique to reduce deterioration in drying tissue. *J. Food Sci., 37,* 430–433.

Hayakawa, K., Ueno, Y., Kawamura, S., Kado, T., Kamiyama, K., Tamaoka, Y., Morishita, M., Matsumoto, M. Kawai, S., & Hayashi, R. (1998). Microorganism inactivation using high pressure generation in sealed vessels under sub-zero temperature. In H. Ludwig (Ed.), *Advances in High Pressure Bioscience and Biotechnology* (pp. 59–63). Heidelberg: Springer-Verlag.

Hedén, C.-G. (1964). Effects of hydrostatic pressure on microbial systems. *Bacteriol. Rev., 28*(1), 14–29.

Kalichevsky, M.T., Knorr, D., & Lillford, P.J. (1995). Potential food applications of high-pressure effects on ice-water transitions. *Trends Food Sci. Technol., 6,* 253–259.

Kanda, T., Kitagawa, K., & Fujinuma, K. (1993). Behavior of ice under pressure: For the basis of high pressure sterilization under low temperature. In R. Hayashi (Ed.), *High Pressure Bioscience and Food Science* (pp. 24–26). Kyoto: San-Ei Suppan.

Kanda, Y., & Aoki, M. (1993). Development of pressure-shift freezing method, Part I: Observation of ice crystals of frozen tofu. In R. Hayashi (Ed.), *High Pressure Bioscience and Food Science* (pp 27–33). Kyoto: San-Ei Suppan.

Kanda, Y., Aoki, M., & Kosugi, T. (1992). Freezing of tofu (soybean curd) by pressure-shift freezing and its structure. *Nippon Shokuhin Gakkaishi, 39*(7), 608–614.

Karino, S., Hane, H., & Makita, T. (1994). Behavior of water and ice at low temperature and high pressure. In R. Hayashi, S. Kunugi, S. Shimada, & A. Suzuki (Eds.), *High Pressure Bioscience* (pp. 2–9). Kyoto: San-Ei Suppan.

Knorr, D., Schlueter, O., & Heinz, V. (1998). Impact of high hydrostatic pressure on phase transitions of foods. *Food Technol., 52,* 42–45.

Koch, H., Seyderhelm, I., Wille, P., Kalichevsky, M.T., & Knorr, D. (1996). Pressure-shift freezing and its influence on texture, color, microstructure and rehydration behaviour of potato cubes. *Nahrung-Food, 40*(3), 125–131.

Le Bail, A., Chourot, J.-M., Barillot, P., & Lebas, J.-M. (1997). Congélation—Décongélation a haute pression. *Rev. Gen. Froid, 972,* 51–56.

Lemaire, R. (1994). La décongélation sous haute préssion. Technical Report, DEA Report, ENSIA Massey, France.

Levy, J., Dumay, E., & Cheftel, J.C. (1999). Freezing kinetics of a model oil-in-water emulsion under high pressure or by pressure release. Impact on ice crystals and oil droplets. *Lebensm.-Wiss. u.-Technol., 32,* 396–405.

Lind, I. (1991). The measurement and prediction of thermal properties during freezing and thawing—A review with particular reference to meat and dough. *J. Food Eng., 13,* 285–319.

Lüdemann, H.-D. (1992). Water and aqueous solutions under high pressure. In C. Balny, R. Hayashi, K. Heremans, & P. Masson (Eds.), *High Pressure and Biotechnology,* Colloque INSERM (Vol. 224, pp. 371–379). Montrouge: John Libbey Eurotext.

Lüdemann, H.-D. (1994). Water and its solutions at high pressures and low temperatures. *Polish J. Chem., 68,* 1–22.

Makita, T. (1992). Application of high pressure and thermophysical properties of water to biotechnology. *Fluid Phase Equilibria, 76,* 87–95.

Mannapperuma, J.D., & Singh, R.P. (1988). Prediction of freezing and thawing times of foods using a numerical method based on enthalpy formulation. *J. Food Sci., 53,* 626–630.

Martino, M.N., Otero, L., Sanz, P.D., & Zaritzky, N.E. (1998). Size and location of ice crystals in pork frozen by high-pressure-assisted freezing as compared to classical methods. *Meat Sci., 50*(3), 303–313.

Miles, C.A., & Morley, M.J. (1978). The effect of hydrostatic pressure on the physical properties of frozen foods. *J. Phys. D: Appl. Phys., 11,* 201–207.

Miles, C.A., Van Beek, G., & Veerkamp, C.H. (1983). Calculation of thermophysical properties of foods. In R. Jowitt (Ed.), *Physical Properties of Foods* (pp. 269–312). London: Applied Science.

Morley, M.J. (1986). Derivation of physical properties of muscle tissue from adiabatic pressure-induced temperature measurements. *J. Food Technol., 21,* 269–277.

Murakami, T., Kimura, I., Miyawaka, H., Sugimoto, M., & Satake, M. (1994). High pressure thawing of frozen fish. In R. Hayashi, S. Kunugi, S. Shimada, & A. Suzuki (Eds.), *High Pressure Bioscience* (pp. 304–311). Kyoto: San-Ei Suppan.

Murakami, T., Kimura, I., Yamagishi, T., Yamashita, M., Sugimoto, M., & Satake, M. (1992). Thawing of frozen fish by hydrostatic pressure. In C. Balny, R. Hayashi, K. Heremans, & P. Masson (Eds.), *High Pressure and Biotechnology,* Colloque INSERM (Vol. 224, pp. 329–331). Montrouge: John Libbey Eurotext.

Nakahara, M. (1990). Water and ions at high pressure: Their fundamental properties relevant to the pressure treatment of food. In R. Hayashi (Ed.), *Pressure Processed Food: Research and Development* (pp. 3–21). Kyoto: San-Ei Suppan.

Nash, D., Lee, B.-S., & Jonas, J. (1996). Hydrogen-exchange kinetics in the cold denaturated state of ribonuclease A. *Biochim. Biophys. Acta, 1297,* 40–48.

Olsson, S. (1995). Production equipment for commercial use. In D.A. Ledward, D.E. Johnston, R.G. Earnshaw, & A.P.M. Hasting (Eds.), *High Pressure Processing of Foods* (pp. 167–180). Nottingham: Nottingham University Press.

Ooide, A., Kameyama, Y., Iwata, N., Uchio, R., Karino, S., & Kanyama, N. (1994). Non-freezing preservation of fresh foods under subzero temperature. In R. Hayashi, S. Kunugi, S. Shimada, & A. Suzuki (Eds.), *High Pressure Bioscience* (pp. 344–351). Kyoto: San-Ei Suppan.

Otero, L., Solas, M.T., & Sanz, P.D. (1998). Contrasting effects of high-pressure-assisted freezing and conventional air-freezing on eggplant tissue microstructure. *Z. Lebensm. Unters. Forsch., 206,* 338–342.

Otero, L., Sanz, P.D., de Elvira, C., & Carrasco, J.A. (1997). Modelling thermodynamic properties of water in the high-pressure-assisted freezing process. In K. Heremans (Ed.), *High Pressure Research in the Biosciences and Biotechnology* (pp. 347–350). Leuven: Leuven University Press.

Perrier-Cornet, J.-M. (1995). Action des hautes pressions hydrostatiques sur la morphologie et l'activité de différents types de cellules. Ph.D. Dissertation, University of Bourgogne, France.

Plonce, E., Pla, R., Capellas, M., Gervilla, R., & Guamis, B. (1996). Texture and microstructure of egg gels formed under high hydrostatic pressure at low temperature. Poster presentation at the Flair-Flow Europe Symposium "High Pressure Processing of Foods," Nov. 6–7, Cologne, Germany.

Pothakamury, U.R., Barbosa-Canovas, G.V., Swanson, B.G., & Meyer, R.S. (1995). The pressure builds for better food processing. *Chem. Eng. Progr., March,* 45–53.

Rahman, S. (1995). *Food Properties Handbook.* Boca Raton, New York, London, Tokyo: CRC Press.

Ramaswamy, H.S., & Tung, M.A. (1984). A review on predicting freezing times of foods. *J. Food Proc. Eng., 7,* 196–203.

Reid, D.S. (1990). Optimizing the quality of frozen foods. *Food Technol., 44,* 78–82.

Sahagian, M.E., & Goff, H.D. (1996). Fundamental aspects of the freezing process. In L.E. Jeremiah (Ed.), *Freezing Effects on Food Quality* (pp. 1–50). New York: Marcel Dekker.

Sanz, P.D., Otero, L., de Elvira, C., & Carrasco, J.A. (1997). Freezing processes in high-pressure domains. *Int. J. Refrig., 20*(5), 301–307.

Schlüter, O., George, S., Heinz, V., & Knorr, D. (1999). Pressure assisted thawing of potato cylinders. In H. Ludwig (Ed.), *Advances in High Pressure Bioscience and Biotechnology* (pp. 475–480). Heidelberg: Springer-Verlag.

Schlüter, O., Heinz, V., & Knorr, D. (1998). Freezing of potato cylinders during high pressure treatment. In N.S. Isaacs (Ed.), *High Pressure Food Science, Bioscience and Chemistry* (pp. 317–324). Cambridge: The Royal Society of Chemistry.

Takai, R., Kozhima, T.T., & Suzuki, T. (1991). Low temperature thawing by using high pressure. In *Les Actes du XVIIIeme Congrés International du Froid, Montreal, Québec* (pp. 1951–1955). Paris: Int. Inst. of Refrigeration.

Ter Minassian, L., & Pruzan, P. (1981). Thermodynamic properties of water under pressure up to 5 kbar and between 28 and 120°C: Estimations in the supercooled region down to −40°C. *J. Chem. Phys., 75,* 3064–3072.

Wagner, W., & Pruβ, A. (1999). The IAPWS formulation 1995 for the thermodynamic properties of ordinary water substance for general and scientific use. *J. Phys. Chem. Ref. Data.* Submitted for publication.

Wagner, W., Saul, A., & Pruβ, A. (1994). International equations for the pressure along the melting and along the sublimation curve of ordinary water substance. *J. Chem. Ref. Data, 23*(3), 515–527.

Yoshioka, K., Yamada, A., & Maki, T. (1996). Application of high pressurization to fish meat: Changes in the physical properties of carp muscle resulting from high pressure thawing. In R. Hayashi & C. Balny (Eds.), *High Pressure Bioscience and Biotechnology* (pp. 369–374). Amsterdam: Elsevier Science B.V.

Zang, J., Peng, X., Jonas, A., & Jonas, J. (1995). NMR study of the cold, heat, & pressure unfolding of ribonuclease A. *Biochemistry, 34,* 8631–8641.

Zemansky, M.W. (1957). *Heat and Thermodynamics* (pp. 248–253). New York: McGraw-Hill.

Zhao, Y., Flores, R.A., & Olson, D.G. (1996). The action of high hydrostatic pressure on the thawing of frozen meat. Poster presentation at the *1996 IFT Annual Meeting*, New Orleans, USA.

Zhao, Y., Flores, R.A., & Olson, D.G. (1998). High hydrostatic pressure effects on rapid thawing of frozen beef. *J. Food Sci., 63,* 272–275.

Part III

Food Products and Processes

CHAPTER 9

Industrial-Scale High Pressure Processing of Foods

Pierpaolo Rovere

INTRODUCTION

Since high pressure processing has been identified as a potential technology for the food industry, the production of products with extended shelf life and with fresh-like character no longer seems to be out of reach. Numerous research groups have focused on this subject, and their many studies have provided a wide range of results. This chapter provides an overview of some of these studies. Previous results of food preservation by high pressure processing at ambient temperatures led to criticisms of high pressure processing and forced food scientists to refocus their efforts on the appropriate use of combined pressure-heat processing. During the past few years, these efforts have led to novel applications of high pressure processing.

HIGH PRESSURE PROCESSING: STATE OF THE ART

High pressure processing of food dates back to the beginning of the 20th century. It was Hite (1899) who first tried to extend the shelf life of food by exposing it to high pressure. In spite of his encouraging findings, the process was forgotten until it was rediscovered and exploited industrially in the 1990s in Japan (Hayashi, 1989a).

With the aid of some public financing, Japanese manufacturers developed a first generation of products using this new processing technique in 1991 (Figure 9–1). It was difficult to see remarkable differences in quality between traditionally processed and high pressure processed foods. It was necessary for the resulting products to comply with Japanese food regulations (mandatory thermal treatment requirements), to be processed at an achievable operating pressure level, and to satisfy consumers' nutritional habits.

The author is thankful to SSICA-Italy laboratory for its great effort on high pressure processing applied research during the last years.

Figure 9–1 High pressure processed products on the Japanese market in 1995.

In 1996, a French company (Ulti) began manufacturing orange juice using high pressure processing. Today the company, which is now known as Pampryl, continues to produce orange juice using this technology. The company refers to high pressure technology in the declaration text on the packages (Figure 9–2).

The first North American company to apply this new technology was Avomex Inc., a New Mexican company that started to market various avocado products for bulk and retail use in 1996. Within a few years, guacamole became the best example of a high pressure processed product that was superior to traditionally processed products. The refrigerated shelf life of this product is 45 days without major quality losses. Its success is based on the heat sensitivity of the avocado

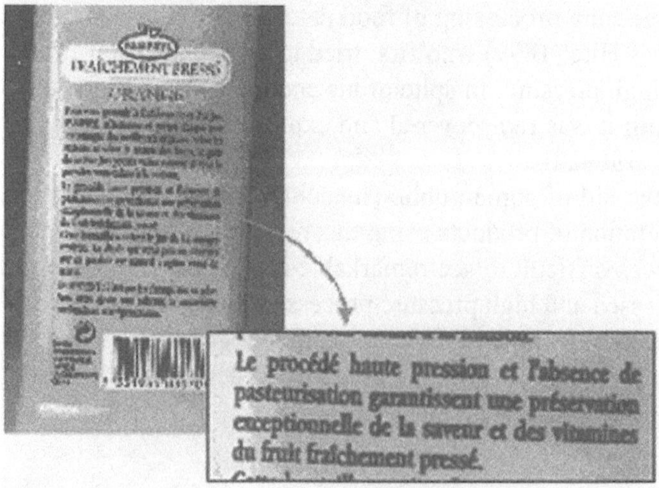

Figure 9–2 High pressure processed orange juice on the French market. Note the labeling.

pulp and the low pressure stability of the avocado polyphenoloxidase. An example of the high pressure autoclaves used by Avomex is presented in Figure 9–3.

In 1999, a Spanish manufacturer (Espuna SA Company in Olot, Spain) installed a plant that uses high pressure technology to process cooked ham. The treatment reduces the risk of recontamination during slicing and prolongs the product's commercial shelf life to approximately eight weeks (Figure 9–4).

EFFECTS OF PRESSURE ON REAL FOODS

Mechanical Effects

Surprisingly, the isostatic pressing of most food does little damage to its structure. In a packaged product immersed in a compressed liquid, the pressure is transferred homogeneously and instantaneously throughout the product. Under these conditions, the product is iso-pressed, since its internal pressure becomes equalized to the external one. In practical terms, the gaseous part of the product completely disappears while the liquids and solids remain according to their compressibilities. Because of the low volume contraction of liquids and solids, particulate products such as foods are not mechanically damaged during pressurization. The elastic capacity of many foods helps them to recover their original structure and shape (Figure 9–5).

It is important to consider the mechanical effect of pressure on packaged or bulk products. When products with compressible tissues (e.g., strawberries, mushrooms, raspberries) are processed in a container without being immersed in sufficient liquid, they could end up with mechanical damage caused by direct

A B

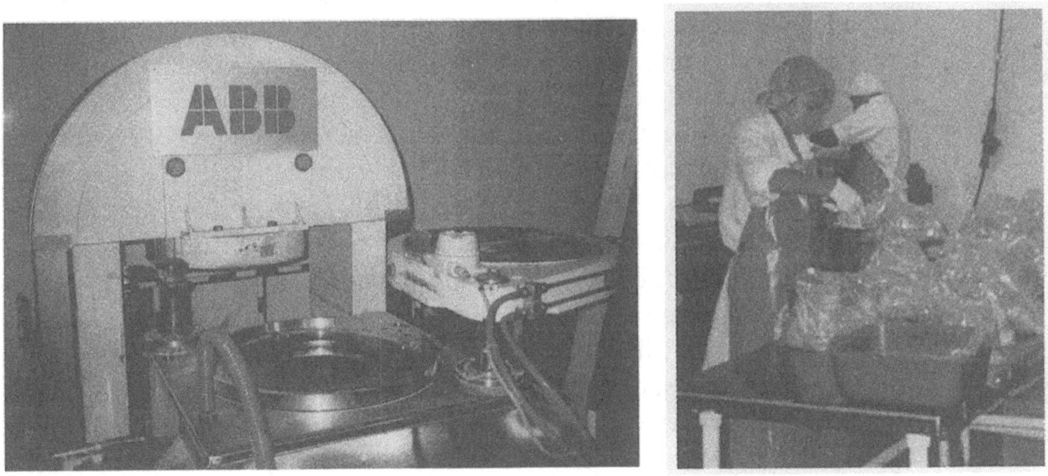

Figure 9–3 High pressure production of avocado products. (**A**) Installed high pressure autoclave at Avomex, Inc. (New Mexico). (**B**) The preparation of avocado paste at Avomex.

Figure 9–4 High pressure processing of cooked ham at Espuna SA, Co., in Olot, Spain.

contact with the wall of the container. In addition, the possibility of mechanical damage is related to the use of direct (continuous) pumping or indirect (batch) pumping. When direct pumping equipment is used, the maximum size of the processed particulates is limited by the ratings of valves and pumps (Figure 9–6).

Microbial Inactivation

The effect of pressure on various microorganisms has been reported elsewhere in this book (see Chapters 3 and 7). This chapter summarizes its practical effect on food applications. Pressure treatment simultaneously increases the temperature of any food matrix subjected to the process. The temperature itself has an effect on microbial destruction at ambient pressure. Pressure and temperature have a synergistic effect, even with only small increases in processing temperature. Thus, when processing foods, pressure, temperature, and time are affecting microbial inactivation (Figure 9–7) (Rovere et al., 1996a, 1996b; Rovere, 1998).

The effect of the surrounding environment is crucial for microbial inactivation. Moreover, water activity, pH, uniformity of the matrix, and ionic strength have a strong impact on microbial destruction (Maggi, 1993). For these reasons, it is important to study the effect of the process in real foods in order to evaluate the repeatability of results obtained from experiments in model systems. Studies of the shelf life of representative food samples inoculated with high levels of spoiling and pathogenic microorganisms are extremely important in order to evaluate the processing technology. The introduction of hurdle technologies (food processing technologies used in combination with high pressure) is of scientific interest and can help to address processing problems related to high pressure treatment (such as enzyme or spore inactivation).

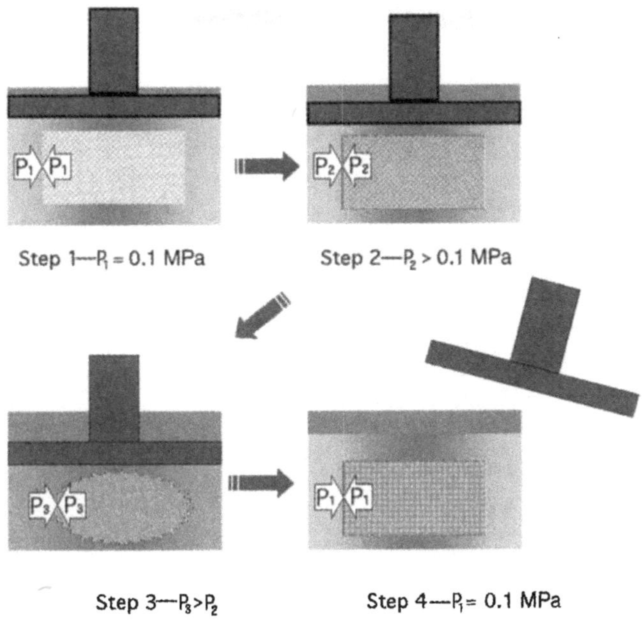

Figure 9–5 Equilibration of pressure during high pressure processing on a small food particle. Step 1—introduction of the product; step 2—increase of pressure and penetration of fluid into the product structure; step 3—pressurization and isostatic compression; step 4—pressure release and depressurization of the product.

Enzyme Inactivation

Enzyme activity is only partially affected by high pressure unless a sufficiently high temperature level is reached or a modification of the matrix produces a synergistic effect on the inactivation. In some cases, the required temperatures are

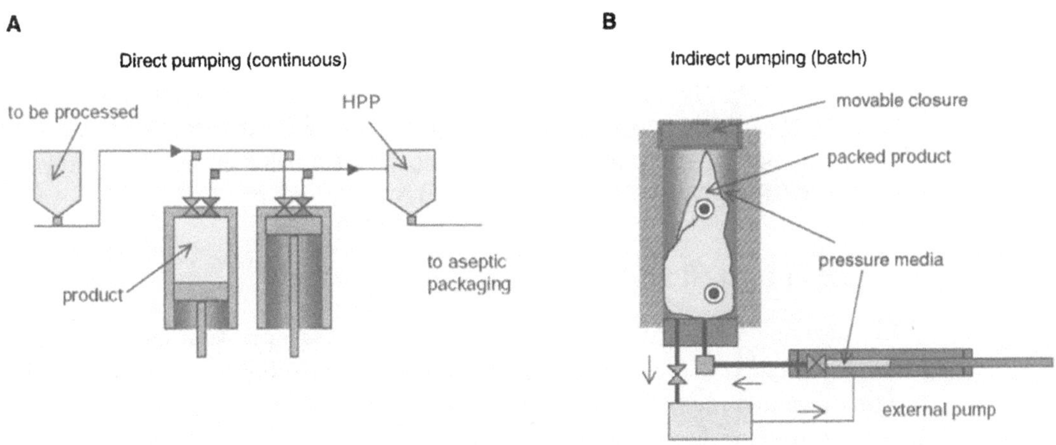

Figure 9–6 Conceptual difference between direct (**A**) and indirect high pressure equipment (**B**).

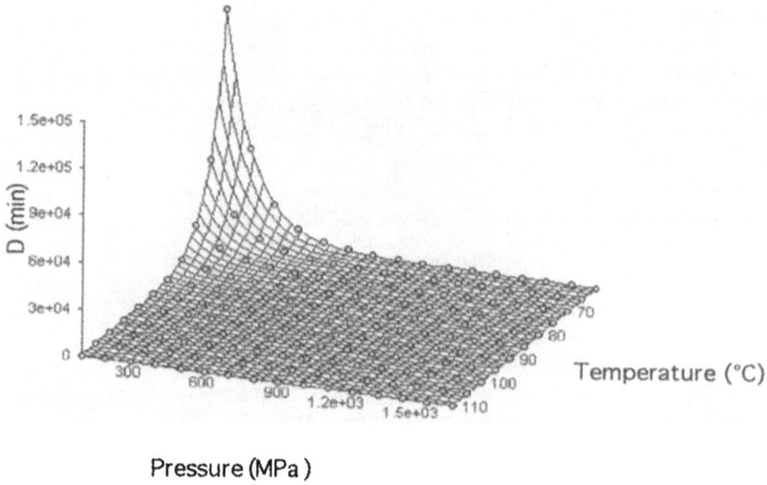

Figure 9–7 Decimal reduction time (D) changes at different pressures and temperatures in *Clostridium sporogenes* PA-3679 (ATCC 7955).

close to the thermal inactivation level at atmospheric pressure. The situation is similar for changes in medium pH. However, elevated pressure may stabilize enzymes upon thermal inactivation (Rovere et al., 1993). Examples of this behavior can be seen in the residual activity of many enzymes that are responsible for food product deterioration such as polyphenoloxidase (Figure 9–8), peroxidase, lipoxygenase, and pectinase.

The food industry uses a variety of techniques in combination with high pressure processing to reduce or control enzyme activity. For example, reduced oxygen intake, use of ascorbic acid or sulfides, active packaging, and modified atmospheres are suitable measures to reduce enzyme deterioration. The combination of hurdles seems to be the right way to introduce minimal processing to the food industry. The use of high pressure to increase activity or selectively inactivate individual enzymes may be desirable within the framework of novel process design. One example is the selective inactivation of pectinase in the presence of the nonaffected pectinmethylesterase, which favors the formation of calcium pectate in fruits and vegetables and produces a higher mechanical strength of the tissue particles and an increased viscosity of juices and purees (European patent PCT/EP97/01265; Rovere, 1997a, 1999).

Effects on Molecules of Interest for Food Technology

Because macromolecules in food are affected by high pressure treatment, high pressure also affects their mutual interactions. This also holds true for biological systems (such as membranes) that are not regarded as true macromolecules. No effects of pressure have been reported for the modification of covalent bonds in

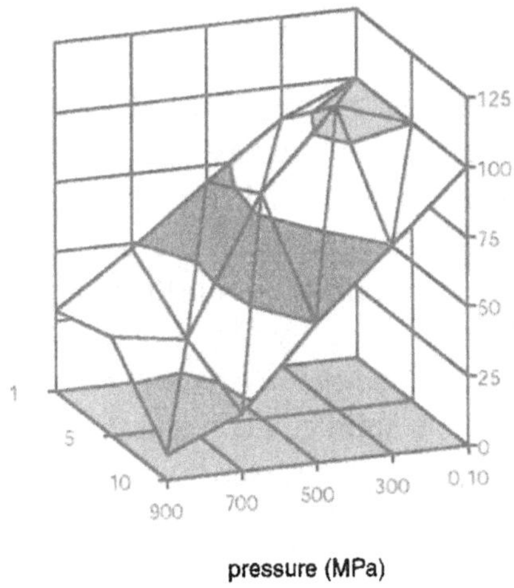

Figure 9–8 Polyphenoloxidase inactivation in apple puree at 50°C.

macromolecules (i.e., formation or breakage of covalent bonds and cross-links), with the exception of the disulfide bonds in proteins. However, water-dependent interactions that stabilize the structure of biological macromolecules are strongly affected by changes in pressure. Changes in their higher-order conformation can affect the biological functions of macromolecules. In the case of food, the nature of a given structure of macromolecules is often related to technologically relevant properties.

One class of macromolecules for which modifications are relevant from the viewpoint of food technology is proteins. The effects of high pressure on proteins have been the subject of many reviews (Balny & Masson, 1993). Pressure treatment can cause a variety of effects on proteins, inducing reversible or irreversible structural modifications leading to protein denaturation, aggregation, or gel formation. All of these effects depend on the nature and concentration of the protein, as well as on the applied pressure, temperature, treatment time, and characteristics of the surrounding environment (such as pH or ionic strength). Indeed, as pointed out in most studies, the physicochemical characteristics of the solvent and the state of hydration are very important in predicting the effect of high pressure on protein structure. In this context, food components that affect the water activity or the solvation of surface protein charges have been investigated both theoretically and experimentally (Cheftel & Dumay, 1997).

Researchers have conducted many studies on the effects of physical treatments on the structure of a variety of isolated food proteins. For many proteins, pressure treatment resulted in a modification of the secondary and tertiary structure with progressive exposition of reactive groups with increasing process intensity (Iametti et al., 1998). In several cases, these structural modifications led to reactions resulting in the formation of cross-linked aggregates, which are different from temperature-induced structures (Iametti et al., 1998; Iametti & Rovere, 1999).

Polysaccharides are another class of macromolecules that respond to high pressure. The structure of starches, pectins, and alginates is affected upon exposure to high pressure. The modification of the relationship between water and macromolecules (starch granules, for example, increase in size and react in a different way to iodine) (Figure 9–9) leads to new and interesting structures.

The water-holding capacity of gels can be improved by exposing pectic dispersions to pressure. The characteristics of the gels depend on the degree of esterification, the soluble solid content, and the pH of the dispersion medium. They are also determined by the pressure level, processing time, and temperature. This general behavior can be correlated with the ability of individual molecules to unfold and expose reactive groups to the surrounding aqueous medium.

However, despite the detailed structural information that can be derived from studies on isolated food components, it is still impossible to reproduce all the complex networks of interactions among the various constituents found in real food systems. Thus, food scientists must take care when they extrapolate results from model systems, and they must conduct experiments on real foods in order to identify the nature of the treatment-induced modifications and their influence on the properties of the food under investigation. Frequently, an understanding of the structural modifications behind a practical observation is the key that allows the replication and industrial exploitation of the process.

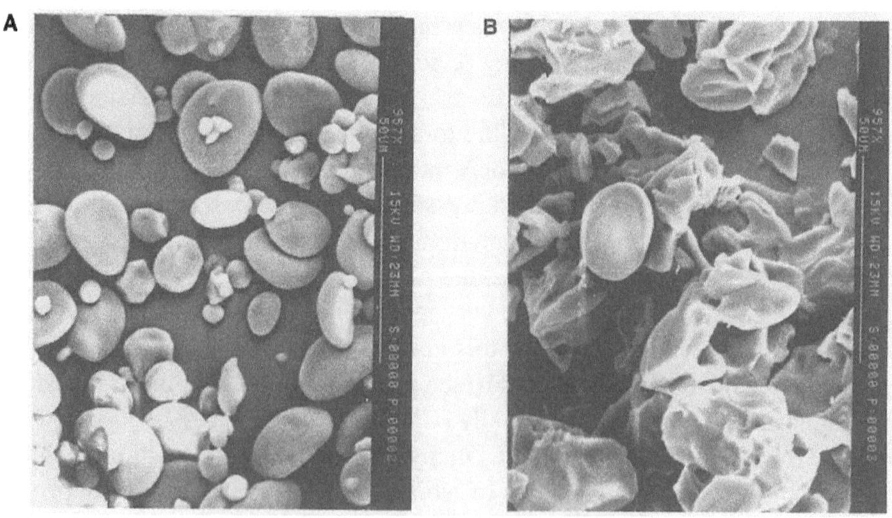

Figure 9–9 The effect of high pressure on starch granules. (**A**) Native dry wheat starch granules. (**B**) Gelatinized wheat starch after high pressure treatment.

Effects of Pressure Processing on Egg Albumen

Chicken egg albumen represents an extensively used food ingredient, mostly because of its functional properties. The gelling, emulsifying, and foaming properties of fresh albumen are fundamental for the production and determine the final properties (texture, flavor, etc.) of many food products. However, most of the functional properties of egg albumen are lost or modified after even the mildest heat treatment (e.g., pasteurization, which is normally used for the preservation of egg components).

Because high pressure processing allows liquid whole egg, albumen, or yolk to retain its natural flavor and nutritional value, this technology is a useful tool for the preservation of this sensitive product (Hayashi, 1989b). Bonomi et al. (1998) observed that liquid egg white partially coagulated when treated at pressure higher than 500 MPa, and that strong self-supporting gels were formed after exposure to high pressure. The hardness and the consistency of the gels increased with increasing pressure, but remained significantly different from those obtained by heat treatment.

Perhaps more interesting from a technological viewpoint was the observation that albumen may be kept in a liquid state after high pressure treatment in the presence of appropriate concentrations of common food ingredients such as NaCl or sucrose (Bonomi et al., 1998). As a result of protein unfolding, the treated albumen has increased viscosity, but retains its foaming (Table 9–1) and heat-gelling properties. The partial unfolding of egg albumen proteins during pressure treatment dramatically increases their susceptibility to hydrolysis by trypsin. In fresh albumen, the major protein component (ovalbumin) is completely resistant to digestion by trypsin. Pressure treatment of albumen at 400 to 600 MPa results in a strong increase in its susceptibility to trypsin. In the pres-

Table 9–1 Properties of Foams Formed by Egg Albumen Containing NaCl or Sucrose

	No Additions	*10% Sucrose*		*10% NaCl*	
Characteristics[a]	*Untreated*	*Untreated*	*Treated*[b]	*Untreated*	*Treated*[b]
A1, (g*s)[c]	1275	2267	1228	2615	1322
A2, (g*s)[d]	881	1326	872	1655	901
Consistency (A1/A2)	1.45	1.71	1.41	1.58	1.47
Foam density (g/mL)	ND[e]	0.26	0.31	0.25	0.18

[a]Treated or untreated egg albumen was beaten for 3.5 min at 25°C with a Philips Masterchef beater set at maximum speed. Foam density was measured by weight. Foam consistency was measured with a Stable Micro System Texture. Results are the average of triplicate determinations.

[b]600 MPa for 5 minutes.

[c]Data refer to the measured area during probe penetration (A1).

[d]Data refer to the measured area during probe release (A2).

[e]ND: not determined.

ence of sucrose, a 600 MPa treatment increased the extent of proteolysis by a factor of 15. In the presence of NaCl, a four-fold increase in proteolysis was observed. The S-form of ovalbumin, which is considered an index of egg aging, was not found in any of the pressure-treated samples; this suggests that pressure does not cause covalent protein aggregation. Also, the amount of ovalbumin identified by an antiserum in pressure-treated albumen was reduced to 40% of that found in untreated albumen; this phenomenon is potentially a consequence of an altered pattern of sequential epitopes in the treated ovalbumin.

These experiments demonstrate the potential benefits of high pressure processing of egg albumen. Not only does pressure treatment inactivate microorganisms, but it also helps albumen retain or improve desirable properties. Indeed, after pressure treatment protein digestibility is strongly increased, and in vitro recognition of ovalbumin epitopes by anti-ovalbumin antibodies is lowered. Sucrose and NaCl have different effects on the functionality of the pressure-treated product, a fact that could be exploited to customize the product by pressure treatment for subsequent use in different technological tasks.

Effects of Pressure Processing on Fruit Products

The preparation of fresh fruit products is of great interest for the food industry (European patent PCT/eP97/0777968A1) because of consumers' renewed interest in natural and healthy foods. Conventional thermal processing yields products that lose their typical attributes of freshness. In recent years, the use of optimized thermal treatments and the introduction of new ingredients have improved the organoleptic properties of conventional products. This has resulted in products that differ from fresh ones. The short shelf life and low safety level of fresh, unprocessed fruits are significant drawbacks. Crispiness, sharp cutting surfaces, color, and taste are attributes that any fresh product must keep. High pressure processing can help foods achieve these qualities.

The high rate of microbial inactivation in acid products may correspond to microbial stability (Figures 9–10 and 9–11). On the other hand, the low enzyme inactivation (Figure 9–12) limits the shelf life of plant products. Nevertheless, measures to control the residual enzyme activity (such as refrigerated distribution or suitable packaging materials) are in line with the high-quality standards required for "fresh, minimally processed" foods.

The remarkable increase in firmness of fruit tissue during storage is most likely due to the selective inactivation of pectic enzymes that cause the complexation of calcium ions with demethylated pectin chains. High pressure processing enhances the rate of this long-term reaction as a consequence of the pressure-induced cell disruption and the loss of compartmentalization (Rovere, 1997b). However, sometimes the process may affect the taste and flavor of the product. These changes are the result of the biochemical oxidation of fatty-acid traces, a mechanism that involves a sequence of enzymes that are not usually sensitive to pressure. Different ways of solving this problem have been suggested. The most simple measures are in agreement with good manufacturing practices: fast material flow on the pro-

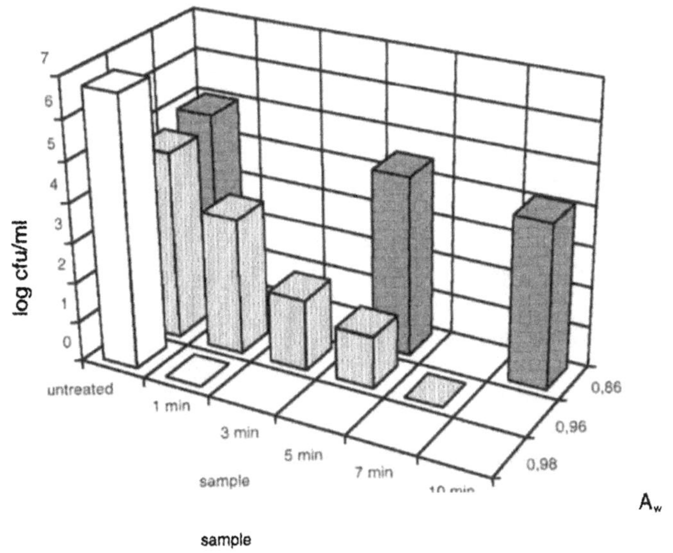

Figure 9–10 Effect of high pressure processing at room temperature on *Saccharomyces cerevisiae* strain inoculated into apricot puree with 25% and 50% of added sucrose (water activity 0.96 and 0.86, respectively).

cessing line, minimal dead times, maximum de-aeration, and refrigeration of the raw material before processing. In fact, a low starting temperature and a fast treatment slow down or suppress the lipase activity before the concentration of precursors (free fatty acids) is sufficiently high to initiate the formation of off-flavors

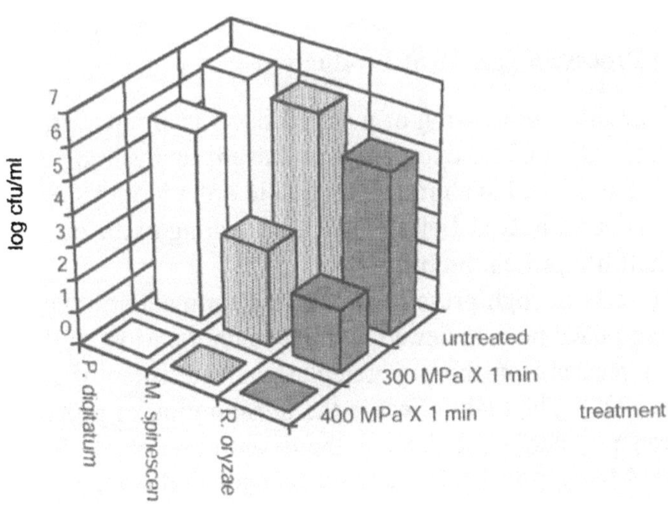

Figure 9–11 Effect of high pressure processing at room temperature on spores of *Penicillium digitatum* (M-30393), *Mucor spinescens* (GE), and *Rhizopus oryzae* (R7A) inoculated in natural apricot puree.

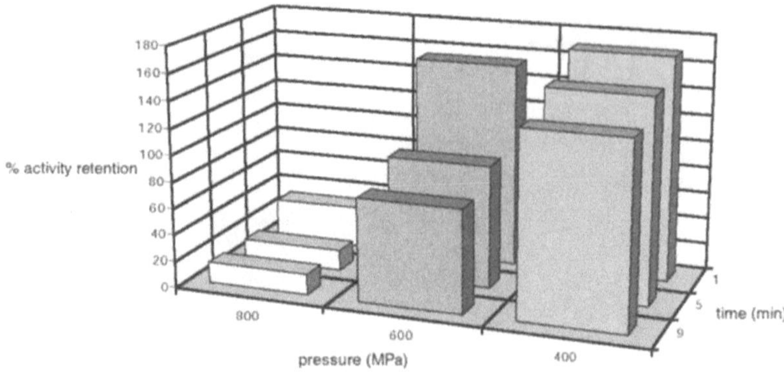

Figure 9–12 Effect of high pressure processing on pulp polyphenoloxidase in apple (Morgen Duft variety).

(Rovere, 1996). Thus, with proper techniques, natural colors, taste, and typical attributes can be retained in sensitive fruits such as kiwi, strawberries, or raspberries and related products over a comparably long period of time. The proper combination of techniques allows food manufacturers to prepare high-quality conventional foods and to develop completely new products (Figure 9–13).

Moreover, consumers are more likely to accept these products when the visual freshness is accompanied by a "natural" behavior. The fact that a fruit salad has an oxidative stability up to several hours after opening the package is important to confirm its natural characteristics and to limit the life of a product formulated without any added preservative. The use of ascorbic acid can help increase the oxidative stability of the product and reduce the technical hurdles to be overcome before packaging.

Effects of Pressure Processing on Meat Products

High pressure processing of meat products is an interesting, new application. The possibility of introducing a new hazard analysis of critical control point (HACCP) at the end of a production line in order to increase the microbial safety is a good opportunity in fields where the handling of the product and its naturally short shelf life present problems.

The effects of high pressure processing on meat and meat products are well known and have been reviewed by many authors. Most of them report a distinct effect of pressure on the organoleptic characteristics of raw meat (Cheftel & Culioli, 1997). The main effect is the denaturation of proteins when they reach pressures high enough to achieve the desired microbial effect. Pressures higher than 400 MPa can produce dramatic changes in the appearance of certain types of meat. Myofibrillar proteins are denaturated, according to processing time and operating pressure. In addition, the color changes, which is because pressure causes myoglobin to produce an increase of lightness and yellow-green compo-

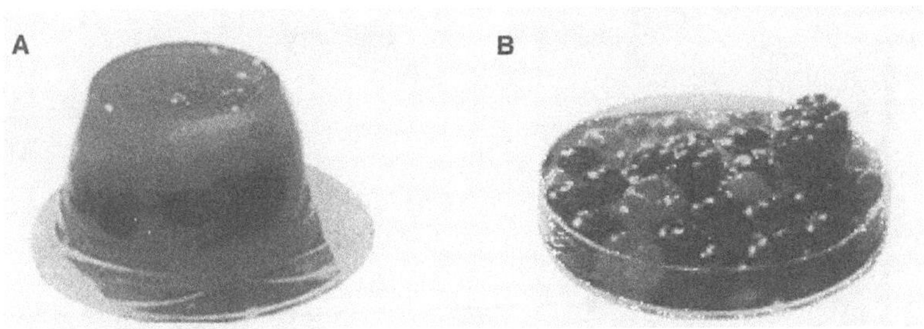

Figure 9–13 Examples of novel fruit products. (**A**) Fruit pieces in a clear juice gel; (**B**) Whole berries in smooth isotonic syrup.

nents and a decrease of red components. The texture after pressure treatment depends on the type and the structure of the meat. Dispersed, fat-rich products (minced, slurry, emulsions) prove to be less hard after pressure treatment and subsequent cooking (Carpi & Rovere, in press). Hence, high pressure applications will be limited primarily to processed meat rather than raw animal tissue. Pressure-induced inactivation of pathogenic strains of *Listeria* (Figure 9–14), *Salmonella* (Figure 9–15), and *Escherichia coli* in meat (Figure 9–16) could open up interesting possibilities for pressure treatment if food scientists can reduce the visual alterations.

One possible application is the pressure treatment of precooked meats. Pressure-treated meat can be used for the preparation of pan-covered or emulsified products where color and texture are not important because of the dominant effect of added ingredients. Emulsifying ingredients and other compounds of the

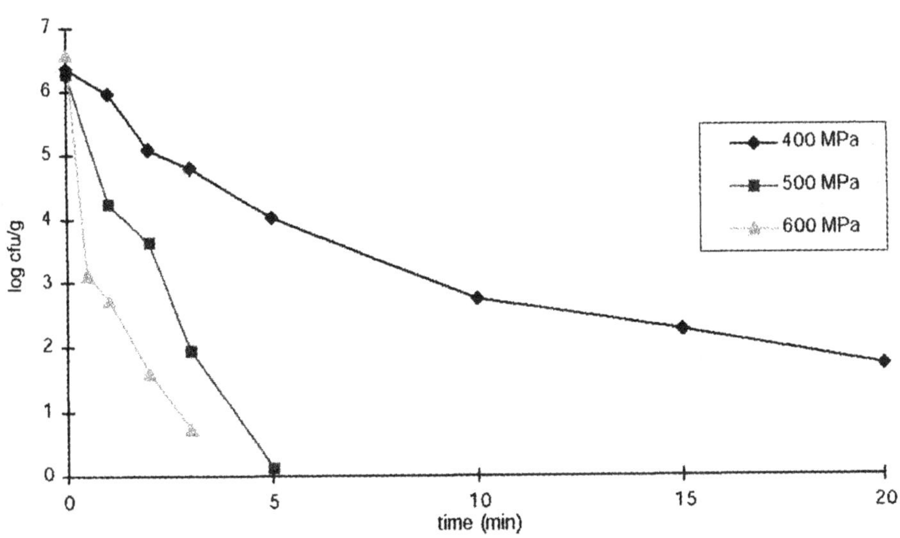

Figure 9–14 Effects of high pressure on two strains of *Listeria monocytogenes* (*Scott A* and *AL-2*) in minced chicken meat.

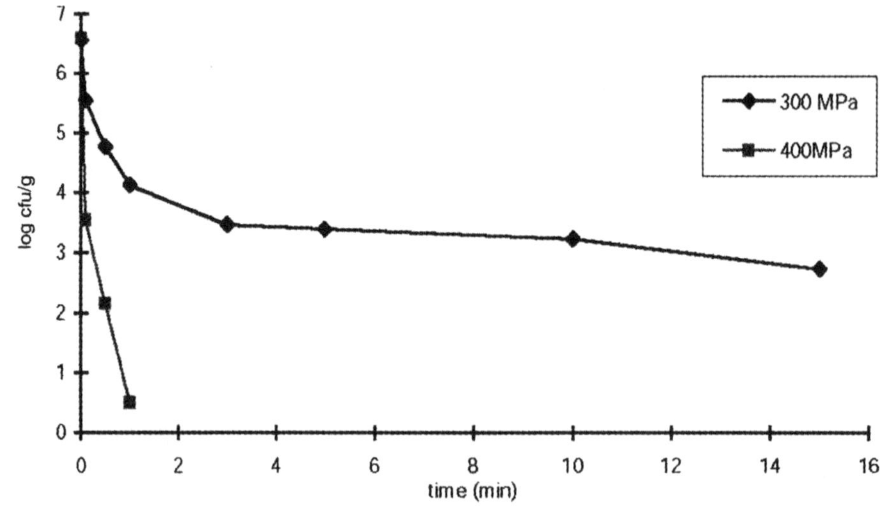

Figure 9–15 Effects of high pressure on three strains of *Salmonella* (*S. typhimurium, S. enteritidis, S. braenderup*) in minced chicken meat.

formulation can reduce color modification and water loss during cooking or high pressure processing. This opens up a new possible use in the field of ready-to-use meals. For products in which the components are prepared individually and subsequently filled into portions in a clean room environment, the safety and the duration of commercial life are key issues. The uniform and instantaneous effect

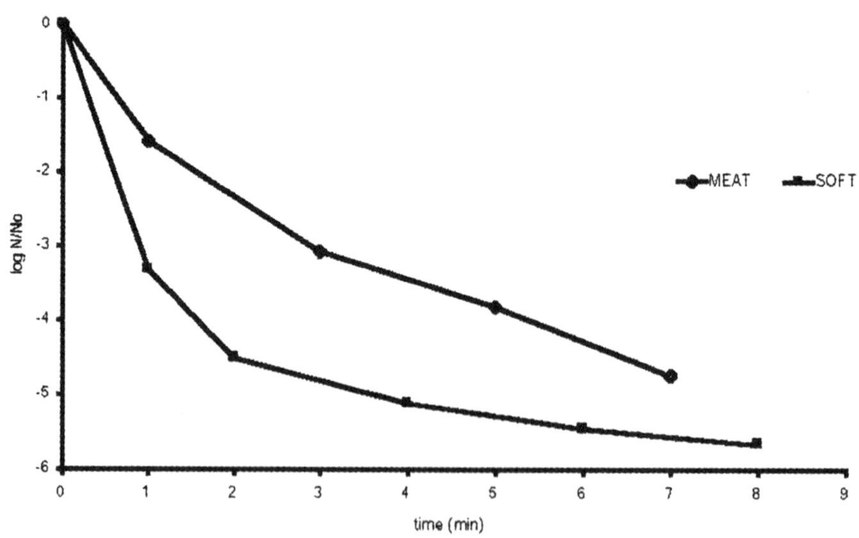

Figure 9–16 Effects of pressure treatment (600 MPa at room temperature) on *Escherichia coli* 0157 01 inoculated in soft culture broth and in meat.

of pressure can play an important role in the decontamination of such heterogeneous products. The same application can be found in sliced, cooked, or raw meat products. Cooked ham is not affected much in appearance because of the thermal denaturation of its components. In addition, dry raw meat products such as salami are not changed in color and texture because of protein denaturation that occurs during salting and curing.

THE DEVELOPMENT OF COMBINED PROCESSING

Pressure-processed products are more like unprocessed foods than are thermally processed products. Higher safety standards with a significantly increased shelf life can open up interesting new areas for high pressure processing.

It should be emphasized that high pressure processing cannot replace the thermal pasteurization of acid foods. As described, high pressure treatment can compete only with mild pasteurization of low-acid foods where it increases the quality of the product. The use of high pressure to produce low-acid preserves is the most recent possibility, which is being investigated in response to studies on the inactivation of spores by thermal treatment combined with high pressure (Figure 9–17). The products manufactured using this application will be as stable as conventionally retorted preserves. At this time, more investigations are necessary to prove the safety and the expected superior quality of this process (Table 9–2).

This process is based on the premise that pressure lowers the heat damage. Many doubts remain on the motivations of such convenience foods, which are normally positioned on the low-price retail shelf where the cost of the product determines the consumer's choice.

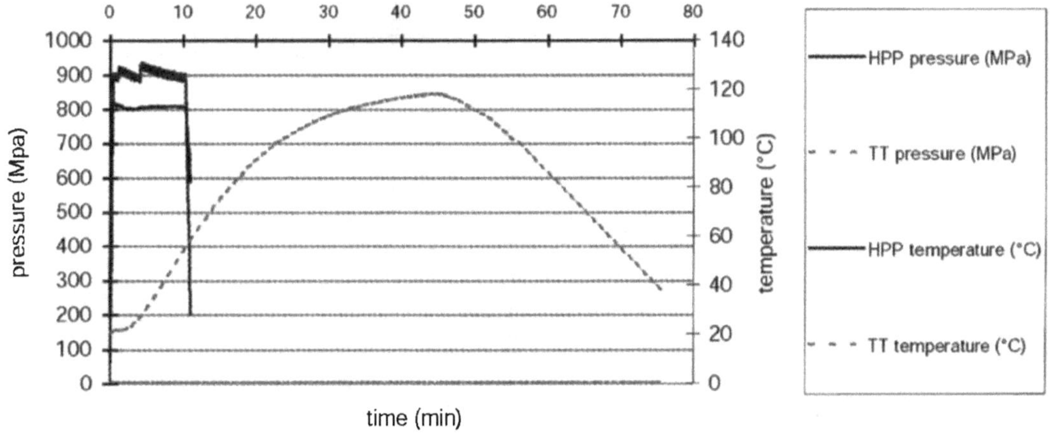

Figure 9–17 Comparison between retort (TT) and high pressure (HPP) sterilization process profiles.

Table 9–2 Comparison of Chemical Quality Markers of Retort Sterilized and High Pressure Sterilized Meat Sauce during Shelf Life

	Nontreated (Month)	*Preheated (Month)*	*High Pressure Treated (Month)*					*Retort Treated (Month)*				
Characteristic	*(0)*	*(0)*	*(1)*	*(2)*	*(3)*	*(6)*	*(12)*	*(1)*	*(2)*	*(3)*	*(6)*	*(12)*
Soluble solids (%)	16.8	18.2	18.19	18.25	18.16	18.25	18.22	20.56	20.14	20.32	20.62	20.24
pH	5.03	4.9	4.91	4.93	4.96	4.94	4.91	4.94	4.92	4.93	4.93	4.91
Total acidity (%)	0.43	0.5	0.49	0.51	0.51	0.52	0.56	0.6	0.62	0.63	0.63	0.7
Volatile acidity (%)	0	0.017	0.008	0.01	0.011	0.014	0.014	0.008	0.021	0.028	0.032	0.06
HMF (mg/kg)	<1	<1	<1	<1	<1	<1	<1	<1	<1	3	5	10

CONCLUSION

High pressure processing is opening up new and exciting possibilities for food technology in the 21st century. Many applications have yet to be discovered. The practical application of present and future research can bring new and interesting foods to the consumer's table, where functionality and safety are very important. The imagination and the curiosity of food technologists will lead to new research fields where pressure will be added as a new dimension to time and temperature.

REFERENCES

Balny, C., & Masson, P. (1993). Effects of high pressure on proteins. *Food Rev. Int., 9,* 611–628.

Bonomi, F., Donnizelli, E., Iametti, S., Pittia, P., Rovere, P., & Dall'Aglio, G. (1998). Molecular and functional properties of HP-treated egg components. In: Isaacs, N.S. (Ed.), *High Pressure Food Science, Bioscience and Chemistry* (pp. 160–165). Cambridge: The Royal Society of Chemistry.

Carpi, G., & Rovere, P. Applicazione dell'alta pressione per prolungare la vita commerciale del prosciutto cotto affettato. *Industria Conserve.* In press.

Cheftel, J.C., & Culioli, J. (1997). Effects of high pressure on meat: A review. *Meat Sci., 46,* 211–236.

Cheftel, J.C., & Dumay, E. (1997). Pressure-induced gels of betalactoglobulin: Influence of protein, polysaccharide and sucrose concentrations on gels. In: Heremans, K. (Ed.), *High Pressure Research in the Biosciences and Biotechnology* (pp. 423–426). Leuven: Leuven University Press.

European patent PCT/EP97/01265. Process for the preparation of a food product.

European patent PCT/EP97/0777968A1. Conservation de fruits.

Hayashi, R. (Ed.). (1989a). *Use of High Pressure in Food.* Kyoto: San-Ei Publishing Co.

Hayashi, R. (1989b). Application of high pressure to food processing: Pressurization of egg white and yolk, and properties of gels formed. *Agric. Biol. Chem., 53,* 2935–2939.

Hite, B. (1899). The effect of pressure on milk—A preliminary report. *Bull. West Virg. Univ. Agric. Exp. Station, 58,* 15.

Iametti, S., Donnizelli, E., Vecchio, G., Rovere, P., Gola, S., & Bonomi, F. (1998). Macroscopic and structural consequences of high-pressure treatment of ovalbumin solutions. *J. Agric. Food Chem., 46,* 3521–3527.

Iametti, S., & Rovere, P. (1999). Characterization of high-pressure treated egg albumen. *J. Agric. Food. Chem., 47,* 3611–3616.

Maggi, A. (1993). Effetti delle alte pressioni su salmonelle e altri enterobatteri in funzione del pH e del contenuto salino del brodo di cultura. *Industria Conserve, 68,* 232–235.

Rovere, P. (1994). Stabilization of apricot puree. *Prehrambeno-technol. Biotechnol. Rev., 32,* 145–150.

Rovere, P. (1996). High pressure processing induced modification in food products: The tomato. Paper presented at the meeting: *Minimal Processing Food,* Cologne Messe, GDL-EFFoST, AIR-FLOW Europe, FLAIR, Nov. 6–9.

Rovere, P. (1997a). Effects of high pressure treatment on chopped tomatoes. *Industria Conserve, 72,* 3–12.

Rovere, P. (1997b). Study of chemical and microbial stabilization of kiwifruit by high pressure treatments. *Acta Horticulturae, 444,* 671–678.

Rovere, P. (1998). Modelling and calculation of the sterilising effect in high pressure heat-treatments. *Industria Conserve, 73,* 303–313.

Rovere, P. (1999). Evaluation of different technological treatments on cherry tomatoes by means of a laboratory texturometer. *Industria Conserve, 74,* 35–39.

Rovere, P., Carpi, G., Maggi, A., & Dall'Aglio, G. (1993). Produzione e stabilizzazione di purea di mela mediante trattamenti ad alta pressione. Paper presented at 1° CISETA, Parma 18–20/10/1993.

Rovere, P., Maggi, A., Scaramuzzo, N., Gela, S., Mighioli, L., Casspi, G., & Dall'Aglio, G. (1996a). High pressure heat treatments: Evaluation of the sterilizing effect and of thermal damage. *Industria Conserve, 71,* 473–483.

Rovere, P., Tosoratti, D., & Maggi, A. (1996b). Sterilizing trials to 15,000 bar to obtain microbiological and enzymatic stability. *Industrie Alimentari, 35,* 1062–1065.

Chapter 10

High Pressure Processing of Dairy Products

Eric Needs

INTRODUCTION

It was no accident that the pioneer of high pressure research, B. H. Hite, chose milk as the medium for his studies on the use of high pressure for food preservation (Hite, 1899). The importance of food microbiology had recently become apparent following Pasteur's work on spoilage of wine and beer. Pasteur's findings had led to the controlled heat treatment of milk, especially for consumption by infants, to prevent milk-borne disease. The process was first termed pasteurization in the 1880s. For Hite, working in the 1890s, it was an obvious next step to investigate the preservation of milk by high pressure.

In the intervening 100 years, heat treatments for processing milk and cream have become universally established and accepted, to the extent that there are legal definitions for high-temperature–short-time (HTST) and ultra-high temperature (UHT) processes (Early, 1998). Given the unassailable position of heat treatment and the variable response of microorganisms to pressure, particularly the observation that milk may have a baroprotective effect for certain organisms (Patterson et al., 1995), it seems unlikely that pressure will ever replace heat for the safe production of the vast volumes of liquid milk consumed daily around the world.

Why, then, have researchers continued to investigate the effects of pressure on milk and dairy products (see recent reviews by Trujillo, et al., 1997; Balci & Wilbey, 1999; Datta & Deeth, 1999)? There may be some niche products for which heat treatment is inappropriate, and for which microbiological quality could be improved by high pressure treatment (e.g., fresh or raw milk cheeses).

I wish to thank Marta Capellas and Patricia Bland for their contributions to our research, which was supported by a grant from the European Commission, project number FAIR CT96 1113 ("High Pressure Treatment of Liquid Foods and Derived Products"). I also wish to thank Brian Brooker, who coordinated the project, and many colleagues from around Europe who have contributed to our understanding of the effects of high pressure on milk and dairy products.

But the main reason for ongoing investigation is that pressure brings about modifications to milk components (particularly proteins), which can lead to altered functionality and the possibility of novel or improved dairy products. Here, again, heat has already proved to be a useful process for the enhancement of the properties of milk proteins. However, heat and pressure have rather different effects on the structure, interactions, and properties of milk proteins. This chapter provides a brief overview of milk proteins and considers the effects of temperature before discussing the impact of pressure on milk and dairy products.

MILK PROTEINS

Dairy products are an inexpensive source of high-quality protein. The protein content of bovine milk is normally around 3.5% (goats' milk is similar and ewes' milk is higher at 5.5%). In yogurt this level is often increased to 5%, while cheese contains approximately 25% protein, depending upon variety. The major milk proteins are the caseins (approximately 80% of the total). Bovine casein contains four individual gene products, α_{s1}-, α_{s2}-, β-, and κ-caseins, which interact to form micelles. The soluble or whey proteins include β-lactoglobulin (β-lg), α-lactalbumin (α-la), serum albumin, immunoglobulins, and various enzymes. Detailed information on milk proteins can be found in Fox and McSweeney (1998).

The individual caseins have a number of important features. They are small, flexible molecules with high surface hydrophobicity. They are phosphoproteins having organic phosphorus esterified to serine residues. The number of phosphate groups varies from 1 (κ-casein) to 13 (α_{s2}-casein). α_{s2}- and κ-casein each contain two cysteine residues per mole. Of the major whey proteins, β-lg contains two cystine residues (intramolecular disulfide bonds) and one cysteine residue per mole; α-la contains four intramolecular disulfide bonds (S-S) and no free sulfhydryl groups (SH). Neither β-lg nor α-la contain phosphoserine residues. At the pH of milk, in its native state β-lg exists mainly as a dimer stabilized by hydrophobic interaction between monomers, and the free sulfhydryl group is shielded at the monomer-monomer interface.

Casein micelles are structured protein aggregates, varying in diameter between 50 and 500 nm. A model of micelle structure recently described by Horne (1998) suggests that the association of caseins is governed by a balance of attractive hydrophobic interaction and electrostatic repulsion. At the normal pH of milk and temperatures in the range 20°C to 30°C, the negative charge on the phosphoserine residues is neutralized by the colloidal calcium phosphate, which cross-links the protein and allows the hydrophobic regions of neighboring molecules to interact, thereby further binding the micelle together. An important consequence of this design is that κ-casein tends to be concentrated on the surface of the micelle. Because κ-casein effectively has a single hydrophobic domain and cannot form colloidal calcium phosphate bridges, it acts as a stop to further polymerization. For a full description and justification of this model, see Horne. The effect of high pressure treatment on milk proteins can be explained in relation to this model of the micelle.

Effect of Temperature on Milk Proteins

Before describing the effect of pressure it is worth considering the effects of the more traditional processes of heating and cooling. To reduce the growth of microorganisms and the activity of milk enzymes, raw milk is cooled to 4°C at the farm before it is transported to the dairy for pasteurization. At low temperatures hydrophobic bonds are weakened, and at 4°C significant amounts of β- and κ-casein dissociate from the micelle. This process is reversible on warming. Because casein micelles are extremely heat stable, milk can be sterilized either by UHT treatment (140°C for 5 seconds) or retorted (115°C for up to 20 minutes). Excessive heating causes dephosphorylation and some rearrangement of disulfide bonds within the micelle. However, under normal processing conditions, increasing temperature tends to strengthen hydrophobic bonds and cement micelle integrity.

Whey proteins are much less heat stable than the caseins. Heating above 70°C for 1 minute causes progressive loss of native globular conformation. Immunoglobulins and serum albumins are most susceptible. The β-lg undergoes irreversible denaturation between 80°C and 90°C. The α-la is the most heat stable; temperatures in excess of 90°C for 1 minute are required to bring about extensive denaturation (Law et al., 1994). Denaturation leads to conformational change, loss of solubility at pH 4.6, and formation of aggregates. Heating skim milk at 85°C for 20 minutes causes the formation of complexes between cysteine-containing proteins through rearrangement of the intramolecular disulfide bonds to form intermolecular bridges (Corredig & Dalgleish, 1996). The main complex formed on heating is between β-lg and κ-casein. Both contain free sulfhydryl groups, which may facilitate interaction, but it is not the only complex formed. In addition, α-la binds to the surface of the micelle. The denaturation of whey proteins and the formation of complexes with the casein micelles, as a result of heating milk, have important effects on protein functionality, which affect the quality of milk products such as yogurt and cheese. For further information on heat-induced changes in milk, see Fox (1995).

Effect of Pressure Treatment on Milk Proteins

When skim milk is pressure treated at ambient temperatures or below 25°C, there is a change in appearance. Several authors reported skim milk to appear more translucent and more greenish-yellow, while less turbid and less blue-white after pressure treatment (see Chapter 6). These subjective observations can be quantified, and Table 10–1 shows the changes in color parameters, turbidity, and transmittance with increasing pressure. Lightness (L*) values and turbidity

Table 10–1 Effect of High Pressure Treatment on the Lightness, Turbidity, and Micelle Size of Skim Milk

Pressure (MPa)	Lightness (L^*)			Turbidity		Transmittance (%)	Casein Micelle Size (Diameter μm)		
	a	*b*	*c*	*a*	*d*	*e*	*a*	*c*	*d*
Control	85	86	71	9	0.70	2	200	190	180
100	—	—	—	—	0.60	2	—	—	—
200	83	80	63	8	0.36	2	180	190	196
300	—	67	—	—	0.20	29	—	—	—
400	73	67	55	4	0.19	38	130	120	110
500	—	65	—	—	0.19	35	—	—	—
600	71	66	52	3	0.19	37	120	100	106

a Data adapted from Desobry-Banon et al. (1994); turbidity, τ (10^4NTU).
b Data from Johnston et al. (1992).
c Data adapted from Gaucheron et al. (1997).
d Data adapted from Needs et al. (2000a); turbidity, A_{320nm}.
e Data adapted from Shibauchi et al. (1992); transmittance, $T_{570\ nm}$.

decrease and transmittance increases with increasing pressure, particularly between 200 and 400 MPa (Johnston et al., 1992; Desobry-Banon et al., 1994). At pressures above 400 MPa, there is no further change.

The changes in color parameters and turbidity are due to changes in the light-scattering properties of the milk proteins (Needs et al., 2000a) as a direct result of the disruption of casein micelles. Transmission electron microscopy shows the effect of pressure treatment on the structure of the micelles (Figure 10–1).

Following treatment at 400 MPa or higher (at temperatures of 25°C or below), the micelles are broken down into small particles that appear linked into irregularly shaped aggregates. High pressure causes an increase in the exposure of hydrophobic surfaces, which has been demonstrated by an increase in the binding of fluorescent probes 8-anilino-1-napthalenesulfanate (ANS) (Johnston et al., 1992; Gaucheron et al., 1997). The increase in ANS binding is related to the decrease in particle size. This confirms that the exposed casein surfaces are hydrophobic, which may result in the micelle fragments' aggregating into irregularly shaped particles (Figure 10–1). The morphology of micelles and measurements of casein particle size in pressure-treated milk have been widely reported (Schmidt & Buchheim, 1970; Shibauchi et al., 1992; Desobry-Banon et al., 1994; Gaucheron et al., 1997; Needs et al., 2000b). Less widely studied are the changes that occur during pressure treatment. Figure 10–2 shows the changes in laser light transmission through skim milk samples under pressure (Kromkamp et al., 1996). At 300 MPa, light transmission increased rapidly during the initial 10 minutes and remained at a constant level during an additional 20 minutes. When pressure was released, there was a small instantaneous drop in light transmission, followed by a gradual decline until the end of the experiment. The final

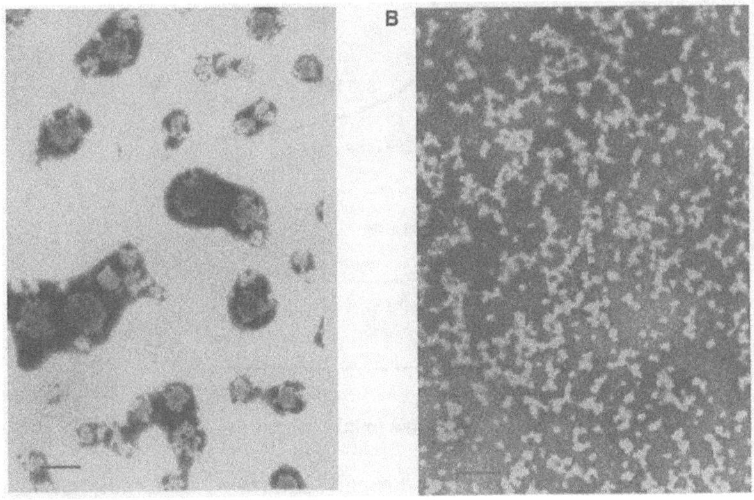

Figure 10–1 Transmission electron micrographs showing casein micelles **(A)** in untreated skim milk, and **(B)** in skim milk treated at 600 MPa for 15 minutes and held at 4°C after treatment. Micelles were visualized by negative staining; milk was spread on a grid and stained with 2% aqueous phosphotungstic acid, excess liquid was removed, and the sample was examined in a Hitachi H600 transmission electron microscope. Scale bar, 200 nm.

level of transmission remained much higher than the initial level. At 200 MPa light transmission increased but, on release of pressure, recovery appeared to be complete. These results suggest that the extent of micelle dissociation under pressure may be greater than that observed in samples following pressure treatment. Also, dissociation is not instantaneous and may be time dependent, particularly at lower pressures. The above observations account for the appearance of micelles found in cold, pressure-treated milk (Figure 10–1). However, casein micelles are very dynamic and, as explained below, warming pressure-treated milk causes these irregular aggregates to combine into new spherical micelles.

There is no evidence that disruption of casein micelles by pressures up to 600 MPa causes an increase in Ca^{++} (Johnston et al., 1992). However, a significant increase in soluble calcium has been reported (Kneifel et al., 1993; López-Fandiño et al., 1998). Anema et al. (1997) have suggested that high pressure effectively shifts the equilibrium between colloidal and soluble calcium, and this leads to micelle disintegration (Schrader et al., 1997). At ambient pressure the solubility of colloidal calcium phosphate is very pH dependent, and at the normal pH of milk the equilibrium favors the colloidal state. There may be a shift in the pH of skim milk under pressure, but the extent is unknown. Reported levels of calcium solubilization in pressure-treated skim milk are shown in Table 10–2. It appears that maximum levels of soluble calcium are reached at moderate pressures (200 to 300 MPa) and that the soluble calcium remains constant or declines at higher pressures (an exception appears to be ovine milk [López-Fandiño et al.,

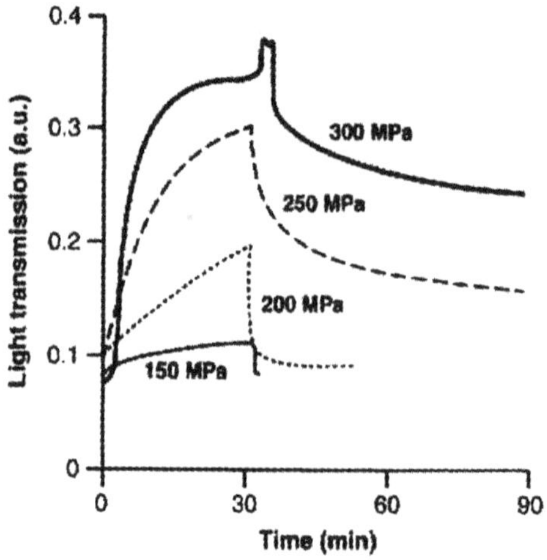

Figure 10–2 Changes in laser light transmission through samples of skim milk held under pressures up to 300 MPa and during a period up to 1 hour after pressure release. Temperature of milk before and after the experiment was 20°C. Scattering was measured with a Malvern Correlator. The wavelength of the laser light was 633 nm.

1998]). Generally the method used to measure soluble calcium was ultracentrifugation of pressure-treated milk followed by atomic absorption spectrophotometry of the supernatant. The soluble calcium may be present as calcium phosphate attached to individual casein molecules (Johnston et al., 1992) or maintaining small casein complexes that are not sedimented. The increase in soluble calcium is accompanied by a similar increase in soluble phosphorus, which

Table 10–2 Effect of Pressure on the Concentration of Calcium in Milk Sera

Pressure (MPa)	*Calcium Concentration in the Supernatant of Ultracentrifuged Pressure-Treated Milk**					*Ca^{++} Activity (mM)*
	a	*b*[1]	*b*[2]	*c*	*d*	*e*
Control	450	383	448	380	31	0.77
100	480	452	489	390	—	—
200	610	560	584	400	—	0.78
300	560	587	735	405	—	0.78
400	510	557	849	405	36	0.76
500	480	—	—	—	—	0.76
600	460	—	—	—	—	0.80

*Determined by atomic absorption spectrophotometry (*a–d*) or with an ion-sensitive electrode (*e*).
a Data adapted from Desobry-Banon et al. (1994); calcium, mg/L.
b Data from López-Fandiño et al. (1998); b[1] bovine milk, b[2] ovine milk; calcium, mg/l.
c Data adapted from Gaucheron et al. (1997); calcium, mg/kg.
d Data from Schrader et al. (1997); percent of total calcium.
e Data from Johnston et al. (1992).

would support this suggestion. The level of serum casein in pressure-treated goats' milk ultracentrifuged at 50000g also showed a maximum at 200 MPa (Law et al., 1998). Solubilization of the colloidal calcium phosphate framework as a first step leading to micelle disintegration during pressure treatment of skim milk remains a possibility, but further direct evidence is needed.

Whey proteins are also denatured by pressure. The effects on individual purified proteins (principally β-lg) on reconstituted whey protein concentrates (WPC) and on whey proteins in milk systems have been studied. Differential scanning calorimetry (DSC) thermograms of pressure-treated β-lg solutions show considerable reduction in enthalpy, which suggests loss of structure and unfolding of the molecule (Dumay et al., 1994). Circular dichroism spectra (Iametti et al., 1997) indicated that the fraction of protein that lost elements of tertiary structure was rather low (about 10%). Pressure causes β-lg dimers to dissociate and, in the presence of thiol reagents (dithiobis [2,4-nitrobenzoate]), stable monomers can be formed. The pressure required to release monomers is relatively low, and if β-lg is treated at 150 MPa the dissociation is reversible. A decrease in exposed thiol groups, following pressure treatment, occurs slowly at 10°C and ambient pressure during 20 hours (Moller et al., 1998). At higher pressures β-lg forms soluble aggregates of varying complexity. Some are noncovalent, but more than 50% are disulfide-linked oligomers (Funtenberger et al., 1997). Because β-lg contains only one free SH group per mole, the formation of oligomers must involve SH/S-S interchange reactions. As the aggregation increases with increased time of pressure treatment, the formation of intermolecular disulfide bonds is time dependent (Funtenberger et al., 1997). The extent of denaturation and aggregation is affected by environmental conditions including protein concentration, pH, and the presence of calcium (Funtenberger et al., 1995; Van Camp et al., 1997). The α-la is not readily denatured by pressure (Law et al., 1998), although it can undergo SH/S-S interchange reactions to aggregate with β-lg.

When skim milk is pressure treated, the β-lg may form intermolecular disulfide bonds with κ-casein (similar to the reaction during prolonged heating). Following pressure-treatment β-lg is precipitated with the casein fraction at pH 4.5; the protein content of cheese whey is reduced. These observations may suggest an interaction between β-lg and casein micelles. However, denatured whey protein may co-precipitate at pH 4.5 and aggregates may become entrapped in the cheese matrix. López-Fandiño et al. (1997) examined pressure-treated skim milk and protein mixtures but could find only indirect evidence of the formation of β-lg-κ-casein complexes. Protein precipitated from pressure-treated goats' milk at pH 4.6 contained aggregates too large to enter sodiumdodecylsulphate polyacrylamide gel electrophoresis (SDS-PAGE) gel under nonreducing conditions. Reduction by 2-mercaptoethanol and SDS-PAGE showed that these contained whey protein and κ-casein (Felipe et al., 1997). Using capillary electrophoresis, López-Fandiño et al. (1998) have shown that pressure treatment results in a redistribution of proteins between the soluble and micellar phases.

They examined the composition of the pellet and supernatant following ultracentrifugation of milks pressurized at 300 MPa for 30 minutes (Figure 10–3).

The levels of nonsedimentable individual caseins were related to the number of ester phosphate groups per mole capable of binding to colloidal calcium phosphate. In bovine milk, the order was β-casein, κ-casein, α_{s1}-casein, α_{s2}-casein. The nonsedimentable protein in pressure-treated goats' milk has been examined following low-speed and high-speed ultracentrifugation (Law et al., 1998). In milk treated at 20°C, β-lg content decreased linearly with increasing pressure at high speed. At low speed there was a linear increase in the individual caseins remaining in the supernatant (except for α_{s2}-casein, which increased only slightly at pressures up to 300 MPa).

Milk Protein Behavior at High Pressure: A Hypothesis

Hydrophobic interactions are favored by increasing temperature but disrupted by increasing pressure. Covalent and hydrogen bonds are generally unaffected by pressure, but disulfide bonds and ion pairs may be affected. Based upon the information described above about the response of milk protein to pressure treatment and the model for the micelle structure, it is possible to suggest a sequence of events occurring during and after pressure treatment. When cold milk is pressurized at 300 MPa and above, hydrophobic interactions rapidly dissociate. As a result micelle integrity is lost; κ-casein and much of the β-casein, which is held

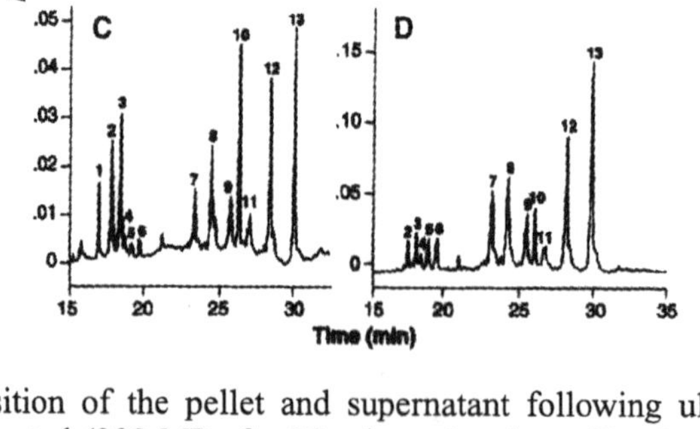

Figure 10–3 The composition of the pellet and supernatant following ultracentrifugation of raw (untreated) and pressure-treated (300 MPa for 30 minutes) ovine milk. Capillary electropherograms corresponding to **(A)** supernatant and **(B)** casein micelle pellet of raw ovine milk; **(C)** supernatant and **(D)** casein micelle pellet from pressure-treated ovine milk. 1, α-lactalbumin; 2–3, β-lactoglobulin B & A; 4–6, α_{s2}-casein; 7–9, α_{s1}-casein; 10, κ-casein; 11–13, β-casein.

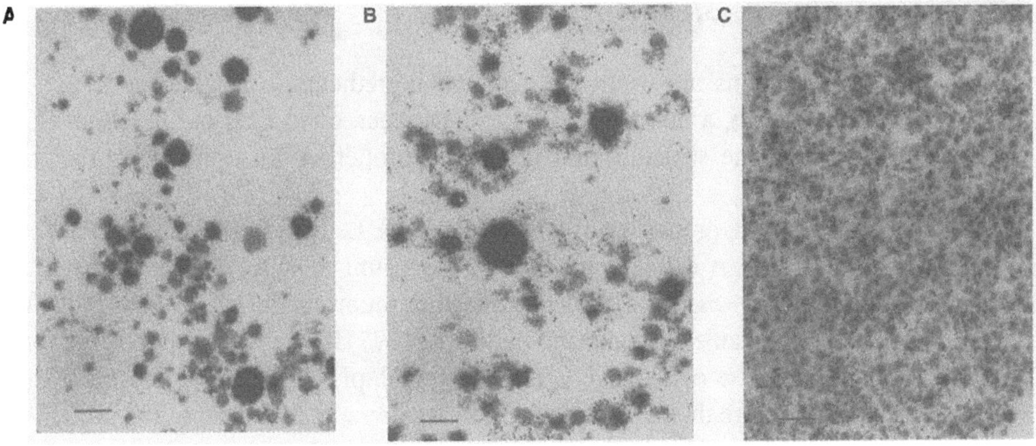

Figure 10–4 Transmission electron micrographs of micelles in milk immunolabeled with anti-bovine β-lactoglobulin. **(A)** Untreated pasteurized skim milk; very little retained immunolabel indicating absence of denatured β-lg. **(B)** Heated milk (85°C for 20 minutes); denatured β-lg adhering to micelle surfaces. **(C)** Pressure-treated milk (600 MPa for 15 minutes); intimate and assorted mixture of micelle fragments and whey protein aggregates. The β-lactoglobulin is localized by electron-dense 10 nm colloidal gold particles. For details of method, see Enright et al. (1999). Scale bar, 200 nm.

in the micelle exclusively by hydrophobic interactions, pass to the serum phase. Small clusters of α-casein, linked by colloidal calcium phosphate, may also be released. The β-lg rapidly dissociates to the monomeric form. The sulfhydryl group exposed by this dissociation is active and available to form disulfide bonds or to catalyze disulfide rearrangement. This is a slower, time-dependent process (Funtenberger et al., 1997). In heated milk, this would lead to complex formation between β-lg and κ-casein on the surface of the micelle. In milk under pressure, the β-lg may interact with κ-casein or other cystine-containing proteins present as micelle fragments or even monomeric proteins. On release of pressure, there is a rapid reformation of hydrophobic interactions, demonstrated by changes in laser light transmission (Figure 10–2) (Kromkamp et al., 1996). The composition and surface properties of these reformed micelles are very different from either untreated or heated milk micelles. Figure 10–4 shows the distribution of denatured β-lg by immunolabeling of untreated, heated, and pressure-treated skim milk. In untreated (pasteurized) milk (Figure 10–4A), immunolabeling indicates very little β-lg denaturation; the trapped gold particles are mainly independent of the micelles. In heated milk (Figure 10–4B), denaturation is extensive and the majority of the β-lg is associated with the surface of micelles. In pressure-treated milk (Figure 10–4C), the system appears much more heterogeneous and may include aggregates of β-lg, β-lg bound onto or included in micelles, and even micelles embedded in β-lg aggregates.

DAIRY FOAMS, EMULSIONS, AND GELS

Milk proteins are used as functional ingredients in a wide range of foods. For this reason, a number of studies have been conducted on the effect of high pressure on the structure and functional properties of individual milk protein preparations.

The foaming properties of β-lg (Galazka & Ledward, 1997; Pittia et al., 1997) or whey protein concentrate (WPC) (Johnston et al., 1997) solutions can be improved by pressure treatment under optimum conditions. Native β-lg foams tend to be unstable. Galazka and Ledward (1997) have shown that the time for half volume collapse of a native β-lg (0.1mg/mL, pH 7) foam was extended from 5 minutes to more than 50 minutes following pressure treatment of the protein solution at 350 MPa for 20 minutes. This was attributed to increased surface pressure (measured at a planar air-water interface by the Wihelmy plate method) and increased surface hydrophobicity (measured by ANS binding). Pittia et al. (1997) also found an increase in foam stability of a β-lg solution (0.2mg/mL, pH7) following pressure treatment. This was determined as the change in microconductivity of the foam during 5 minutes following formation. However, they also found that the initial conductivity was lower in foams from pressure-treated β-lg solutions than from the native protein (Pittia et al., 1996a). Initial conductivity is a measure of the foam liquid content. These authors argue that pressure treatment reduces the available surface-active protein and that protein aggregates would be slower to adsorb and stabilize the air-water interface. Of course, it is WPC rather than purified β-lg that is used as a food ingredient. Johnston et al. (1997) foamed WPC solutions (2.5% weight percentage) and found that pressure-treated samples had significantly higher foam expansion, foam volume stability, and foam liquid stability than did untreated samples. The optimum pressure was 400 MPa. These authors suggested that cross-linking through S-S bridges gives good viscoelastic film formation. The protein content of these samples was much higher than that used in experiments involving β-lg, and the amount of available surface-active protein would not be a limiting factor in the initial bubble formation.

β-casein is a very hydrophobic protein, and it has a random coil structure that gives high molecular flexibility. These properties result in a high rate of adsorption at the air-water interface and good foam expansion. However, β-casein foams lack stability, and drainage and collapse occur rapidly. Pittia et al. (1996b) examined the effect of pressure treatment between 300 and 900 MPa on the foaming properties of β-casein. Treatment of β-casein solution (0.2mg/mL) at 300 MPa for 30 minutes produced foams with initial conductivity similar to native β-casein solutions. However, the stability (residual conductivity at 5 minutes) of the pressure-treated sample showed a marked improvement over the native sample. Interestingly, stability was maintained in the presence of Tween 20, a competing surfactant. When the molar ratios of Tween 20 to β-casein were between 2 and 5, stability of the native β-casein foams fell by 30% to 40% while stability of foams containing pressure-treated β-casein fell by only 10%. Pittia et al. (1996b) suggest that pressure treatment facilitates an increase in thickness, rigidity, and elasticity of the adsorbed layer. This results in reduced drainage and resistance to competitive adsorption and destabilization by Tween 20.

Galazka et al. (1995; 1996) have examined the emulsifying properties of pressure-treated whey proteins. Following pressure treatments of 200 to 800 MPa for periods up to 40 minutes, emulsions were prepared from WPC or β-lg (0.4% weight percentage) with *n*-tetradecane (20% volume percentage) as the dispersed phase. The average droplet size (d_{43}) in emulsions stabilized by pressure-treated protein tended to increase with pressure and treatment time (Table 10–3). In the β-lg stabilized emulsions, droplet size increased during storage at 23°C for 10 days, indicating that stability decreased with increasing pressure and treatment time.

The anticipated increase in functional properties of whey proteins caused by pressure-induced unfolding and exposure of hydrophobic surfaces does not occur. The loss of available protein through aggregation and loss of solubility (Funtenberger et al., 1995) results in a net loss of functionality.

The effect of pressure treatment on the dispersed phase of emulsions is dependent upon the composition of the oil-water interface, the nature of the fat, and the processing conditions (especially temperature). Dumay et al. (1996) prepared emulsions containing 30% fat (glycerol trioleate or peanut oil) stabilized with 5% β-lg or sodium caseinate. These were pressure treated at 450 MPa for 15 or 30 minutes at 10°C, 25°C, or 40°C. Generally, the emulsions stabilized by sodium caseinate were unaffected by these treatments. Treatment of β-lg emulsions at 10°C or 25°C did not markedly affect droplet size. However, there were changes in the rheology of the emulsions treated at 25°C. Progressive increases in shear stress and viscosity were observed during storage at 4°C. The extent of these changes was dependent upon the choice of oil phase. Treatment at 40°C caused the emulsions to gel. Dumay et al. (1996) examined the effects of the

Table 10–3 Average Droplet Size of Emulsions Stabilized by Whey Protein Concentrate (WPC) or β-lactoglobulin

Treatment		*Droplet Average Diameter (μm)*		
			β-Lactoglobulin Emulsion[c]	
Pressure MPa	*Time (Minutes)*	*WPC Emulsion*[b]	*Fresh*	*10 day*
Untreated	—	2.1	0.86	0.86
200	20	2.1	1.4	1.9
200	40	2.6	1.5	1.9
800	20	3.7	1.5	2.7
800	40	4.8	1.6	3.6

[a]Droplet size (d_{43}) was determined immediately after production and after 10 days of storage at 23°C.
[b]Data from Galazka et al. (1995).
[c]Data from Galazka et al. (1996).

same pressure treatments applied to pasteurized or UHT-processed dairy creams (35% fat). Treating pasteurized cream at 10°C or 25°C had no important effect on fat globule size or the flow behavior. Treatment at 40°C caused fat globules to form aggregates. These flocs were easily dispersed by addition of sodiumdodecylsulfate (SDS) or by ultrasound. UHT-treated cream was more susceptible to the effects of pressure, and aggregation occurred at all processing temperatures.

When the structure of the fat in pressure-treated pasteurized cream was examined by freeze-fracture transmission electron microscopy, a large increase in fat crystallization was observed (Dumay et al., 1996). This finding is in agreement with results previously reported by Buchheim and Abou El-Nour (1992). By observing the fracture behavior of fat globules and measuring the solid fat content by nuclear magnetic resonance, they were able to quantify the extent of fat crystallization caused by pressure treatment. Even 100 MPa for 1 minute induced significant solidification. Longer times generally resulted in higher values of solid fat. There appeared to be an optimum pressure level between 300 and 350 MPa, above which the solid fat content began to fall. Importantly, the process begun under pressure continued once pressure was released and the samples were held at 23°C.

Pressure treatments of emulsions that do not cause disruption of the oil phase may have applications in dairy processing (e.g., the treatment of milk to reduce risk of microbial contamination in the production of raw milk cheese). Application of pressure to induce crystallization may be used in the aging of ice cream mixtures or the physical ripening of cream for butter making (Buchheim & Abou El-Nour, 1992).

Gelation is another functional property of whey proteins that may be induced by pressure treatment. Van Camp and Huyghebaert (1995a) examined the gelation by high pressure of a number of proteins, including WPC. The properties were compared with heat-set gels produced at 80°C for 30 minutes. Starting with protein solutions at 20°C, these authors estimated that the maximum temperature reached during pressurization to 400 MPa was 31°C. Gel strength increased in response to increases in applied pressure or treatment duration. Protein concentration was a factor affecting both gel strength and volume of nonincorporated liquid. The latter was also affected by the pH of the system. The rheology of WPC gels produced by pressure (400 MPa for 30 minutes) or heat (80°C for 30 minutes) was compared (Van Camp & Huyghebaert, 1995b) and the G' value of heat-set gels was higher than for pressure-set gels across a range of protein concentrations. Similarly, the stress-strain curves during compression to large deformation showed that the compression modulus was lower for pressure-set gels than for heat-set gels at each protein concentration. Microscopic examination showed that the pressure-induced gel had a more porous structure and a lower number of contact points. Differences in the mechanism of protein denaturation and aggregation may be responsible (Van Camp & Huyghebaert, 1995b).

Concentrated milks can form gels under pressure. Velez-Ruiz et al. (1998) concentrated milks by evaporating them to a total solid content ranging from 129 to 471 g/kg. These were treated at pressures up to 690 MPa. Measurements, which included apparent viscosity and texture profile analysis, showed that a treatment of 414 MPa caused formation of solid gels in milks containing at least 449 g/kg total solids. Evidence of gelation was apparent at lower concentrations. Some elements of the texture profile, particularly compression strength, were affected by the duration and temperature of pressure treatment; other properties such as springiness were unaffected. Lower temperature (25°C rather than 41°C) increased gel strength. From examination of the microstructure, Velez-Ruiz et al. concluded that the uniform network structure was similar to other milk gel products (yogurt and cottage cheese). They suggested that milk gels made by a combination of concentration and pressurization have attractive sensory attributes, high whiteness, good flavor, and excellent consistency.

Schwertfeger and Buccheim (1999) produced gels from skim milk using a combination of pressure and acidification. Glucono-δ-lactone was added to acidify the samples. Reaction rates and gel structure differed between gels produced under pressure and those produced at ambient pressure.

Acid-Set Gels Produced from Pressure-Treated Milk

The protein content of milk used for yogurt production is normally increased to improve textural properties. This can be achieved by concentrating the milk to the desired total solids content by ultrafiltration or evaporation. Alternatively, the milk can be fortified by additional protein in the form of skim milk powder or WPC. The protein-enriched milk is heated above 80°C (typically 90°C for 10 minutes) to induce denaturation of whey proteins as described above. Lactic acid bacteria are added, and incubation (at 42°C to 44°C) for several hours results in a drop in pH and gelation.

As an alternative to prolonged heating, the fortified milk may be pressure treated before to yogurt production. The disruption of the micelles and differences in the pattern of whey protein denaturation described above mean that the pressure-treated milk is a rather different starting material from the normal heated milk.

The effect of pressure treatment of milk below 25°C on micelle structure is shown in Figure 10–1, and differences in color parameters are shown in Table 10–1. Micelles are dynamic structures that are greatly affected by changes in the environment. Warming pressure-treated skim milk to 43°C causes the disrupted micelles to reform into spherical particles, smaller than the original micelles (Figure 10–5). This results in changes in the appearance of the skim milk. Needs et al. (2000a) have shown the changes in L*, a*, and b* that occur during the warming and acidification of skim milk fortified with whey protein. Warming caused large increases in L*, a*, and the scattering coefficient, which was consistent with reformation of small separate micelles shown by electron microscopy. Changes in color parameters during acidification resulted in a convergence in the appearance of yogurt gels prepared from pressure-treated milk and

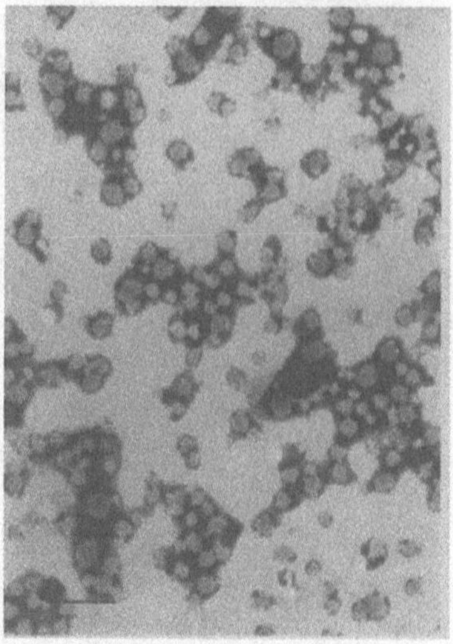

Figure 10–5 Transmission electron micrographs of casein micelles in skim milk pressure-treated at 600 MPa for 15 minutes and subsequently warmed to 43°C before yogurt production. The disrupted micelles observed in pressure-treated milk at 4°C (Figure 10–1) reform into spherical particles, smaller than the original micelles. Micelles visualized by negative staining. Scale bar, 200 nm.

control yogurts prepared from heat-treated milk (Table 10–4). Despite the similarity in the color parameters at the end of yogurt formation from pressure and heat-treated milks, there were significant differences in rheology and microstructure of these gels.

Figure 10–6 shows the change in the storage (G′) and loss (G″) moduli during the fermentation of yogurt gels from heated and pressure-treated milks. Gelation begins earlier and is more rapid in pressure-treated samples. The final value for G′ in the set yogurt gels cooled to 4°C was more than three times that of the corresponding heated milk sample. This finding suggests many more particle-particle

Table 10–4 Changes in Color Parameters as a Result of Pressure Treatment, Warming, and Fermentation during Formation of Yogurt

Treatment[a]	*Fermentation Stage*	*L**	*a**	*b**
Untreated	Milk at 4°C	81.6	−4.2	7.7
Pressure[b]	Milk at 4°C	73.6	−5.6	6.1
Pressure[b]	Milk at 43°C, pH 6.5	83.8	−4.3	6.2
Pressure[b]	Gel at 43°C, pH 4.5	89.6	−1.6	7.2
Pressure[b]	Yogurt at 4°C, pH 4.4	88.9	−2.0	8.3
Heat[c]	Yogurt at 4°C, pH 4.4	89.6	−1.7	7.9

[a]Skim milk was fortified with 19g/L of whey protein concentrate before treatment.
[b]Pressure treatment: 600 MPa for 15 minutes.
[c]Heat treatment: 85°C for 20 minutes.

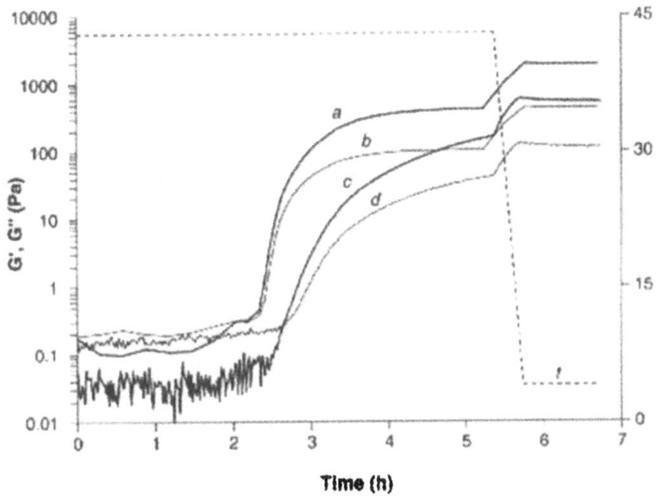

Figure 10–6 Changes in the storage (G′) and loss (G″) moduli during fermentation and formation of yogurt gels from heated (85°C for 20 minutes) and pressure-treated (600 MPa for 15 minutes) milks. The symbols *a* and *b* represent G′ and G″ values for pressure-treated samples; *c* and *d* represent G′ and G″ values for heated samples. The temperature (*t*) was held at 43°C for 5.5 hours before cooling to 4°C at the end of fermentation. Small deformation oscillatory measurements were performed on a controlled-stress Bohlin (CS) rheometer.

interactions because of the presence of an increased micelle surface area in the pressure-treated samples. However, these interactions appear to be structurally different from those formed in heated milk gels.

Figure 10–7 shows that in the heated milk gel the discrete micelles are connected, over relatively large distances, by more amorphous material. Immunolabeling shows that this gel is rich in β-lg (Needs et al., 2000a). In the pressure-treated milk gel, there appears to be more direct contact between micelles. When subjected to large deformations, these gels appear less firm and cohesive than gels of heat-treated milk. Scanning electron micrographs of these gels (Figure 10–8) reveal that the heated milk gel has a much denser network structure, which is more continuous and has smaller interstitial spaces than the pressure-treated milk gel. These differences in microstructure may well account for the differences in gel strength at large deformation.

Gels produced from pressure-treated milks or pressure-set WPC gels have different structures and rheological properties than their heat-processed counterparts. The potential to produce gels with different textures should be explored as a possible commercial application of high pressure to dairy products.

APPLICATION OF HIGH PRESSURE IN CHEESE PRODUCTION

The production of most cheeses involves formation of curd due to destabilization of casein micelles by the enzymatic cleavage of κ-casein by rennet. At the

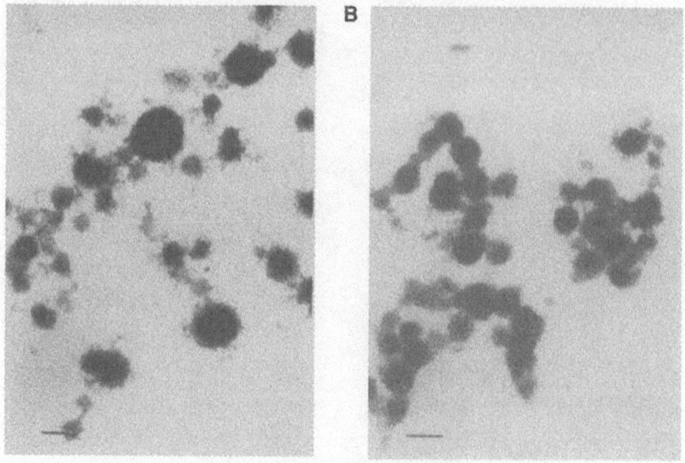

Figure 10–7 Transmission electron micrographs of yogurt. **(A)** From heated milk (85°C for 20 minutes); and **(B)** from pressure-treated milk (600 MPa for 15 minutes). Sections through the gels show structural differences. In the heated milk gel, long-range interactions between micelles are established through amorphous material (denatured whey protein). More direct micelle-micelle interactions are involved in the pressure-treated gel network. Scale bar, 200nm.

surface of the micelle, the hydrophilic portion of κ-casein projects into the serum phase forming the "hairy coat." This portion of κ-casein is negatively charged and is responsible for electrostatic repulsion between micelles. Rennet coagulation is a two-stage process. In the primary stage, rennet cleaves κ-casein very

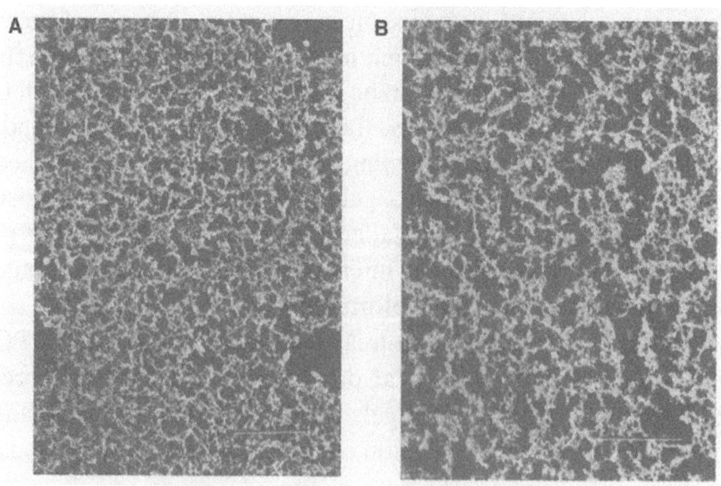

Figure 10–8 Scanning electron micrographs of milk yogurt gel. **(A)** heated milk yogurt gel, which has a much denser and more continuous network structure with smaller interstitial spaces than **(B)** pressure-treated milk yogurt gel. Scale bar, 10μm

specifically between Phe_{105}–Met_{106}, releasing the hydrophilic glycomacropeptide into the serum phase and leaving para-κ-casein attached to the micelle. The zeta potential of the micelle is reduced from approximately −20 to −10mV. In the secondary stage, at temperatures above 20°C, the destabilized micelles coagulate to form a gel or curd.

The great number of cheese varieties results largely from differences in the processing of the curd. However, the properties of the curd are fundamental to cheese quality. Important parameters of curd formation are rennet clotting time (RCT), curd firmness, syneresis, yield, and protein composition. RCT is the time between addition of rennet and the first appearance of coagulated particles in the milk. Curd firmness is normally measured at the maximum of gelation by a penetration test. In the production of many cheeses, the curd is cut before maximum strength is reached to encourage syneresis, which is the expulsion or draining of whey from the curd. Researchers have investigated the use of high pressure to facilitate the following improvements in cheese manufacturing:

- retaining whey protein to increase cheese yield, improve nutritional quality, and reduce effluent
- reducing the risk of contamination by spoilage organisms and pathogens
- improving product stability and appeal by modifying texture, flavor, salt distribution, etc.
- shortening ripening time

High-Pressure–Treated Milk for Cheese Production

The denaturation of whey protein by heating has been explored as a method of incorporating these proteins into cheese. Although cheese yield is increased, the product is generally inferior. Pressure denatures β-lg and may offer an alternative method to increase cheese yield. β-Lg denaturation increases with increasing pressure treatment of milk (600 MPa for 15 minutes results in denaturation greater than 90%). As a direct result, the protein concentration in the whey declines (Figure 10–9) and the curd weight increases (López-Fandiño et al., 1996; Needs et al., 2000b).

RCT is reduced by pressure treatment below 200 MPa. However, just as β-lg denaturation increases at higher pressures, so does RCT (López-Fandiño et al., 1996). The increase does not occur if the milk has been pressure treated in the presence of sulfhydryl blocking agent, *N*-ethylmaleimide (Needs et al., 2000b). This clearly suggests that the formation of intermolecular disulfide bonds contributes to the increase in RCT. (Note that an increase in RCT was not observed following pressure treatment of milk prepared from reconstituted skim milk powder [Shibauchi et al., 1992; Desobry-Banon et al., 1994]). Needs et al. reported that the presence of denatured β-lg only affects the second, aggregation stage of rennet coagulation and does not interfere with the enzymatic cleavage of κ-casein. Others have observed a reduction in the rate of glycomacropeptide release

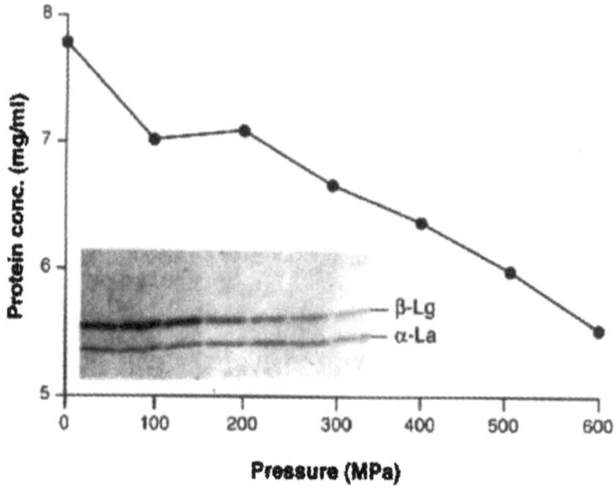

Figure 10–9 The soluble protein concentration in whey drained from cheese curds prepared from milks treated at pressures up to 600 MPa. Insert shows the protein composition by PAGE of the whey released for each pressure treatment (left to right: untreated to sample treated at 600 MPa).

(López-Fandiño et al., 1997; López-Fandiño & Olano, 1998a) but no correlation with RCT. The S-S between β-lg and κ-casein forms in the para-κ-casein region (i.e., following enzymatic cleavage of κ-casein any S-S bound β-lg would remain attached to the micelle). In heated milk, even low levels of β-lg denaturation increased RCT, and above 75% coagulation failed to occur (van Hooydonk et al., 1987). RCT of milk treated at 400 or 600 MPa was similar to that of untreated milk (Needs et al., 2000b). This suggests that, even at very high levels of β-lg denaturation in pressure-treated milk, the proportion directly associated with the micelle surface may be small. The majority of β-lg–κ-casein complexes formed may involve free κ-casein released from the micelle under pressure. The micelles would then be depleted in κ-casein, which would account for the rapid coagulation in the presence of *N*-ethylmaleimide. The presence of β-lg aggregates in the serum phase may interfere with the coagulation of the para-κ-casein micelles, causing RCT to increase.

The structure of the gel formed by rennet coagulation is affected by the treatment of the milk. Table 10–5 shows the storage modulus (G′) at the point of curd cutting for milks treated at different pressures.

All pressure treatments resulted in G′ values much higher than in the untreated control. Aggregation of the pressure-disrupted micelles produced a gel of fine protein strands combined into a continuous network. This contrasts with the gel from untreated milk, in which the strands were much courser and there were many more breaks or discontinuities in the network (Needs et al., 2000b). G′ is a measure of the number of bonds or interactions formed within a gel, and the microstructure of the gel from pressure-treated milk explains the high G′ value. Park et al. (1996) found that a heat treatment of 80°C for 30 minutes reduced

Table 10–5 The Complex Modulus (G*) and Syneresis of Cheese Curds Made from Pressure-Treated Milks

Pressure (MPa)	*G*[a] (n=5)*	*Whey Volume[b] (n=6)*
Control	93 ± 26	42 ± 2.5
200	150 ± 21	42 ± 3.9
400	190 ± 51	37 ± 5.7
600	179 ± 64	29 ± 4.5

[a]G* was measured at the point of curd cutting, and syneresis was the volume of whey drained from the curd cut in 1 hour.

[b]Volume of whey drained from cut curd as a percentage of the total volume.

maximum G′ to less than 40% of the unheated control. The level of β-lg denaturation under these conditions is similar to a treatment of 600 MPa for 20 minutes (greater than 90%), but differences in the protein-protein interactions through disulfide rearrangements occurring during pressure and heat treatments result in opposite effects on the rheological properties of the cheese curds.

A consequence of retention of whey protein in cheese curd is a reduction in syneresis or increased water retention (Marshall, 1986; Needs et al., 2000b). López-Fandiño et al. (1996) reported a 13% increase in cheese yield, but β-lg was only a small proportion. The bulk of the increase was retained water. Table 10–5 shows the volume of whey drained from the unpressed curds during 1 hour following cutting. The retention of water may be due to increased hydration of the protein surfaces (Gaucheron et al., 1997) or simply more water trapped in the finer gel network. Figure 10–10 shows the structure of pressed cheese curds made from untreated and pressure-treated milk. Conventional Cheddar cheese curd (Figure 10–10A) after pressing shows a virtually continuous protein matrix with entrapped small, irregularly shaped pockets of water and larger, spherical fat globules. The pressed curds made from pressure-treated milk have an apparently discontinuous protein matrix with more water, probably forming interconnecting channels. A product that benefits from increased water retention is fresh cheese (Trujillo et al., 1999). However, for a hard cheese such as Cheddar, retention of too much water could have adverse effects on texture and keeping quality.

Drake et al. (1997) have assessed the texture and sensory qualities of Cheddar cheese made from pressure-treated milk. A cycled pressure treatment was applied (586 MPa for 1 minute, repeated three times). RCT was short (only 60% of control RCT), which suggests that this treatment probably did not allow sufficient time for development of extensive disulfide rearrangements. There was a small increase in moisture content and yield. A trained taste panel gave a flavor score that was not significantly different from that of untreated cheese, but the texture score was lower and the defect was attributed to excess moisture. However, these were young cheeses and the treatment was designed to improve microbiological quality rather than texture.

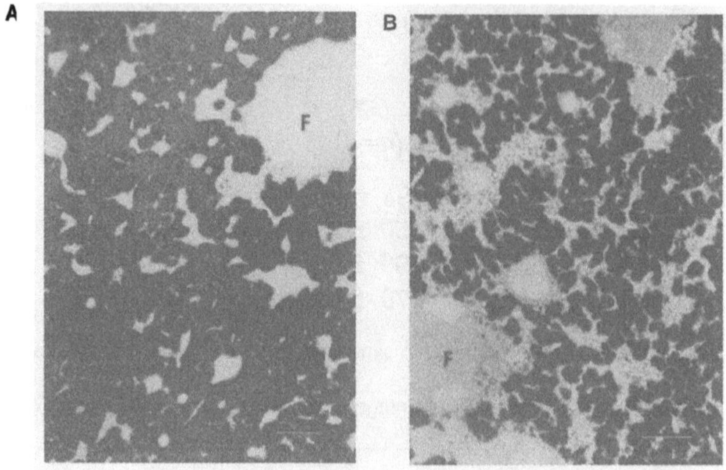

Figure 10–10 Transmission electron micrographs showing structure of miniature cheeses. **(A)** Cheese made from untreated pasteurized whole milk. **(B)** Cheese made from the same milk pressure treated at 600 MPa for 15 minutes. The conventional cheese has a solid protein matrix with small, irregularly shaped pockets of water. The cheese made from pressure-treated milk has a semicontinuous matrix with channels of entrapped aqueous phase, contributing to the increased protein content and cheese yield. F, fat. Scale bar, 1 μm.

Pressure treatment of raw ovine and caprine milks may have beneficial effects for cheese production from these specialty milks (López-Fandiño & Olano, 1998b). Treatment of ovine milk (with higher β-lg levels than bovine milk) at 200 MPa caused a significant increase in yield. Treatment of caprine milk at 300 MPa for up to 30 minutes led to improved curd firmness.

The application of pressure at elevated temperature has been investigated. The approach may appear to offer improved microbiological quality. Treatment of ewes' milk (350 MPa at 50°C) inoculated with *Listeria innocua* reduced the population by more than 7 log cycles, whereas the same pressure at 25°C resulted in only a 1 log cycle reduction (Gervilla et al., 1997). The extent of whey protein denaturation may be increased (particularly α-la) at combinations of moderate temperature and pressure. However, the effects on casein micelles may render the approach inappropriate for most dairy applications. When goats' milk is pressure treated between 300 and 350 MPa at 45°C, very large micelles (in excess of 500 nm) were formed to the exclusion of smaller micelles (Law et al., 1998). These large micelles are devoid of κ-casein and unstable, precipitating during cold storage for 24 hours. Gaucheron et al. (1997) observed a similar effect on micelle size when they treated reconstituted skim (bovine) milk at 300 MPa and 40°C; they reported no instability in this case. The combined treatment had very little impact on the stability of α-la. At the conditions described, more than 90% to 95% of α-la remained soluble and would not be included in the cheese curd. López-Fandiño and Olano (1998a) showed that, at 400 MPa and 60°C for 15 minutes, the denaturation of α-la in bovine milk approached 60%. However, they also reported that pressurizing at temperatures in excess of 40°C adversely affected, or even prevented, milk coagulation.

Before pressure treatment of milk for cheese production is adopted as a commercial process, a number of questions still need to be addressed. The extent of water retention must be optimized to match increased yield with acceptable sensory attributes. Proteolysis and flavor development need to be studied further, as these will be affected as a result of changes in the protein substrate and the activity of proteolytic enzymes. Novel or niche products may be more suitable for this technology than traditional cheeses. The effect on cheeses with specific functionality (e.g., melting and elastic properties) may be worth exploring.

High Pressure Treatment of Cheese

Brining of cheese is a lengthy process, and it may take several weeks before the salt is uniformly distributed within the cheese. Messens et al. (1999) have examined the effect of high pressure treatment during the brining of Gouda cheese. Experimental cheeses were brined for 5 hours at 400 MPa followed by 19 hours at ambient pressure. The salt and water distribution following brining and during ripening were compared with cheeses brined for 24 hours at ambient pressure. The initial salt content was identical in both types of cheese, and more than 90% of the salt was present near the rind. During 27 days of ripening, diffusion resulted in an even salt distribution throughout the cheeses. These authors concluded that high pressure brining had no effect on salt uptake or diffusion and that it was not a method for reducing brining time or the time required to reach salt equilibrium.

Accelerated Ripening of Cheese

In a separate investigation, Messens et al. (1998) examined the effect of high pressure brining on protein distribution and hydrolysis during ripening. They brined 15-g portions of Gouda cheese curd for a total of 50 minutes. During this period some samples were pressure treated for 30 minutes at 300 MPa. The total nitrogen content of the cheese serum at 1 and 14 days of ripening was higher in the samples that had received pressure treatment. This was attributed initially to the release of β-casein from the paracasein gel network, because of disruption of hydrophobic interactions. The action of plasmin hydrolyzed the β-casein and led to an accumulation of γ-caseins. As the total nitrogen in the cheese serum of pressure-treated samples increased during ripening relative to the untreated samples, these authors suggested that proteolytic enzyme activity was enhanced by the pressure-induced disruption of the gel network. This indicates that pressure treatment of the cheese may induce increased proteolysis and reduce maturation time.

Kolakowski et al. (1995) examined the effect of pressure on the proteolysis in Gouda and Camembert cheeses. They reported no influence of pressure on protein degradation in Gouda but did observe a substantial increase in pH (also reported by Messens et al. [1998]). The optimum pressure treatment for Camembert was 50 MPa for 4 hours. This led to increases in peptide and free amino acid (FAA) of 40% and 32%, respectively, compared with the untreated control.

A patent application from the Fuji Oil Company (Yokoyama et al., 1991) claimed that "although it conventionally takes 6 months to ripen Cheddar cheese, three days can bring about equivalent ripening according to the present invention." The treatment involves holding the cheese for 3 days at 50 MPa and 25°C. These investigators reported that the FAA content of the treated cheese was almost twice that of the untreated control. However, the cheese-making method was very different from that used for conventional Cheddar, especially in the initial level of starter bacteria. The amount added was at least 10 times higher than conventional inoculation rates. As a result, both the control and pressure-treated cheeses had very high levels of FAA after 3 days (Beresford et al., 1998). When the high pressure treatment described in the Fuji Oil Company patent was applied during the production of commercial Irish Cheddar (O'Reilly et al., 1998; 2000), the rate of proteolysis was similarly found to be approximately twice that of the control. But because the addition of starter was at the normal level for Irish Cheddar, the FAA concentration was well below that of a mature Cheddar. The addition of large amounts of starter culture is normally avoided, because it can cause excessive acidity or the production of bitter peptides. Casal and Gomez (1999) reported an interesting approach. They prepared cultures of lactic acid bacteria that were harvested in early stationary growth phase and resuspended in buffer (pH 7). Skim milk was inoculated with these suspensions to a final concentration of approximately 8 $\log_{10}$ cfu/mL. Samples were then treated at pressures up to 400 MPa to attenuate the lactic acid production but not the peptidolytic activity of the starter organisms. The results were complex and strain dependent. However, these authors concluded that treatments of 300 or 350 MPa would greatly reduce the acidification while retaining high aminopeptidase and dipeptidase activities. Further work is in progress to study the mechanisms involved and test the effectiveness in complex food systems such as cheese (Casal & Gomez).

No degradation of β-casein or accumulation of γ-casein was observed following the application of pressure to Cheddar cheese at 2 days (O'Reilly et al., 1998). However there was a significant increase in the cleavage of α_{s1}-casein and formation of $\alpha_{s1 1}$-casein. This primary proteolysis (caused by the residual rennet activity), in conjunction with starter proteinase and peptidases, leads to the formation of small molecular weight peptides and the release of FAA. O'Reilly et al. described the peptide profile of the water-soluble extracts of control and pressure-treated Cheddar cheeses determined by reversed phase high-performance liquid chromatography. Qualitatively the peptide composition was very similar, although the peptides were more abundant in the pressure-treated samples. These authors concluded that high pressure had an effect on both primary and secondary proteolysis, but it was the rate and not the proteolytic pathway that was affected. This is important because it suggests that the flavor chemistry (dependent upon compounds derived from the further breakdown of peptides and FAA) will be similar for high-pressure–treated and conventional cheeses. In addition, the absence of new products of proteolysis means that high-pressure–treated cheese may be the "substantial equivalent" to conventional cheese.

Butz et al. (2000) examined the evolution of volatile compounds during the ripening of Gouda cheese. The cheeses had received pressure treatments of 50 to 600 MPa for 1 hour applied 3 days after brining. A maximum of 140 substances were observed following simultaneous distillation-extraction and gas chromatography/mass spectroscopy; no new compounds were found in the pressure-treated samples. Of the 14 major components, only two differed significantly as a result of the pressure treatments. Acetonin was detected in decreasing amounts as the pressure treatment increased. Acetonin is a reduction product of diacetyl, and pressure may affect the enzymes responsible for diacetyl production (mainly NADH oxidase and diacetyl reductase). Butanoic acid, the other compound reduced by pressure, is a product of lactose fermentation. These authors suggested that this may be a direct result of lysis of the starter bacteria.

At present, the application of high pressure for accelerated cheese ripening remains an academic interest. This is because the increased rate of proteolysis occurs only while pressure is being applied. The mechanism is poorly understood. Possible explanations include increased autolysis of the starter bacteria, or conformational change of the protein to increase substrate availability or enhance enzyme-substrate interactions. A better understanding of the mechanisms may make it possible to identify conditions for the application of high pressure during cheese ripening that are of commercial interest.

MILK ENZYMES

Raw milk contains a number of enzymes. Alkaline phosphate activity is often used to assess the effectiveness of heat treatment since the enzyme is inactivated by pasteurization. In order to identify internal process markers for high pressure treatment, Rademacher et al. (1998, 1999) have examined the kinetics of milk enzyme deactivation. Alkaline phosphates were found to be pressure resistant up to 500 MPa with complete inactivation not occurring below 800 MPa (treatment time of 8 minutes). Two other enzymes, γ-glutamyltransferase and phosphohexose isomerase, were both more susceptible to pressure than alkaline phosphates. Rademacher et al. (1998; 1999) suggest that a combination of these three enzymes may be suitable to assess the extent of pressure treatment applied to raw milk, but the enzyme inactivation must first be correlated with the destruction of pathogens such as *Mycobacterium tuberculosis*.

Needs (1997) also examined alkaline phosphate activity and found that less than 20% of the activity was lost after a treatment of 400 MPa for 15 minutes. The native enzyme is associated with the milk fat globule membrane and has no activity against phosphoserine residues in casein micelles in milk. To confirm that high pressure did not change this situation, raw milk was treated at selected pressures and then incubated at 37°C for 2 and 18 hours. PAGE confirmed that no dephosphorylated caseins were produced.

Milk lipoprotein lipase is a heat-sensitive enzyme that is inactivated by pasteurization. However, it can have a significant effect on the level of FFA if the raw milk is mishandled, causing damage to the fat globules and inducing lipolysis (Needs et al., 1986). Lipases appear to be relatively pressure resistant. Goats' milk treated at 500 MPa for 15 minutes had residual enzyme activity and significantly more FFA than a pasteurized control milk (Trujillo et al., 1999). A commercial lipase preparation (origin unspecified) treated at 600 MPa for 10 minutes in Tris buffer (pH 7) lost only 40% of its activity. Treatment of 700 MPa at 45°C was required for complete inactivation (Seyderhelm et al., 1996).

Argyriou (1998) purified milk lipoprotein lipase from raw milk by affinity chromatography and measured the residual enzyme activity, after pressure treatment, using *p*-nitrophenyl butyrate as the substrate. Under the mildest conditions used (200 MPa for 5 minutes), the residual enzyme activity was reduced by 45%. However, 20% of the activity was unaffected by treatment of 800 MPa for 30 minutes (the most severe conditions tested). Residual lipase activity in pressure-treated raw milk could be important in the development of flavor in "raw" milk cheeses.

Lipolysis can be induced in raw milk by processes that damage the milk fat globule membrane and allow the enzyme access to its substrate. Argyriou (1998) measured FFA levels in pressure-treated raw milk immediately after treatment and after 18 hours of storage at 4°C. No increase in FFA was observed, and these authors concluded that the process did not induce lipolysis.

Plasmin, the principal milk proteinase, occurs mainly as the inactive precursor, plasminogen. Enzyme activity is controlled by a complex system of activators and inhibitors. Scollard et al. (2000) studied the effect of pressure on this system. They found that pressures greater than 400 MPa were required to cause any reduction in plasmin activity and, as with lipoprotein lipase (described above), even after prolonged treatment at 800 MPa approximately 20% of the activity remained. Pressure did not appear to cause any shift from plasminogen to plasmin. However, pressure treatments in the range of 300 to 400 MPa did cause an increase in proteolysis. This was thought to be an effect of pressure on casein micelle structure (i.e., an increase in availability of substrate rather than a direct effect on the enzyme).

CONCLUSION

To ensure safety and prolong shelf life, milk is processed by a variety of precisely defined and controlled heat treatments. Except for a few niche-market products, where heat treatment is undesirable, it is not likely that high pressure will replace heat for the purpose of routine microbial inactivation. However, heat can also be applied to milk to change the functional properties of milk proteins. In this area, pressure treatment can offer novel protein conformations, structures, and functional properties. Several authors have shown effects on dairy products (Trujillo, 1997; Needs et al., 2000a; López-Fandiño & Olano, 1998b). What is needed now is a systematic study for the rates of casein disaggregation and whey protein denaturation at selected pressures. An indication of the interesting possibilities that such research may reveal can be seen in Figure 10–2, the preliminary investigation reported by Kromkamp et al. (1996). The application of pressure treatment during cheese production has met with mixed success. However, there are several interesting avenues that could be developed further, e.g., raw milk cheeses, fresh cheese, and accelerated ripening of soft or mold-ripened cheeses.

REFERENCES

Anema, S.G., Lee, S.K., Schrader, K., & Buchheim, W. (1997). Effect of pH on the turbidity of pressure-treated calcium caseinate suspensions and skim milk. *Milchwissenschaft, 52,* 141–146.

Argyriou, A. (1998). Effects of high pressure treatment on the bovine milk lipoprotein lipase. MSc. Thesis, Department of Food Science and Technology, University of Reading, Reading, UK.

Balci, A.T., & Wilbey, R.A. (1999). High pressure processing of milk—The first 100 years in the development of a new technology. *Int. J. Dairy Technol., 52,* 149–155.

Beresford, T.P, O'Reilly, C.E., O'Connor, P., Murphy, P.M., & Kelly, A.L. (1998). Acceleration of cheese ripening: Current technologies, potential use of high pressure. In K. Autio (Ed.), *Fresh Novel Foods by High Pressure* (pp. 103–114). Espoo, Finland: Julkaisija Utgivare.

Buchheim, W., & Abou El-Nour, A.M. (1992). Induction of milkfat crystallization in the emulsified state by high hydrostatic-pressure. *Fett Wiss. Technol.-Fat Sci. Technol., 94,* 369–373.

Butz, P., Fernández-Garcia, A., Koller, W.B., & Tauscher, B. (2000). Effects of high pressure treatment on fermentation processes during ripening of Gouda cheese. *High Pressure Research, 19,* 37–41.

Casal, V., & Gomez, R. (1999). Effect of high pressure on the viability and enzymatic activity of mesophilic lactic acid bacteria isolated from caprine cheese. *J. Dairy Sci., 82,* 1092–1098.

Corredig, M., & Dalgleish, D.G. (1996). Effect of temperature and pH on the interactions of whey proteins with casein micelles in skim milk. *Food Res. Int., 29,* 49–55.

Datta, N., & Deeth, H.C. (1999). High pressure processing of milk and dairy products. *Austr. J. Dairy Technol., 54,* 41–48.

Desobry-Banon, S., Richard, F., & Hardy, J. (1994). Study of acid and rennet coagulation of high pressurized milk. *J. Dairy Sci., 77,* 3267–3274.

Drake, M.A., Harrison, S.L., Asplund, M., Barbosa-Canovas, G., & Swanson, B.G. (1997). High pressure treatment of milk and effects on microbiological and sensory quality of cheddar cheese. *J. Food Sci., 62,* 843.

Dumay, E., Lambert, C., Funtenberger, S., & Cheftel, J.-C. (1996). Effects of high-pressure on the physicochemical characteristics of dairy creams and model oil/water emulsions. *Food Sci. Technol.-Lebensm. Wiss. Technol., 29,* 606–625.

Dumay, E.M., Kalichevsky, M.T., & Cheftel, J.-C. (1994). High-pressure unfolding and aggregation of β-lactoglobulin and the baroprotective effects of sucrose. *J. Agric. Food Chem., 42,* 1861–1868.

Early, R. (1998). Liquid milk and cream. In R. Early (Ed.), *The Technology of Dairy Products* (pp. 13–19). London: Blackie Academic & Professional.

Enright, E., Bland, A.P., Needs, E.C., & Kelly, A.L. (1999). Proteolysis and physicochemical changes in milk on storage as affected by UHT treatment, plasmin activity and potassium iodate addition. *Int. Dairy J., 9,* 581–591.

Felipe, X., Capellas, M., & Law, A.J.R. (1997). Comparison of the effects of high-pressure treatments and heat pasteurization on the whey proteins in goat's milk. *J. Agric. Food Chem., 45,* 627–631.

Fox, P.F. (Ed.). (1995). *Heat-Induced Changes in Milk.* Brussels: International Dairy Federation, IDF—ref. S.I. 9501.

Fox, P.F., & McSweeney P.L.H. (1998). Milk proteins. In *Dairy Chemistry and Biochemistry.* London: Blackie Academic & Professional.

Funtenberger, S., Dumay, E., & Cheftel, J.-C. (1995). Pressure-induced aggregation of β-lactoglobulin in pH 7.0 buffers. *Food Sci. Technol.-Lebensm. Wiss. Technol., 28,* 410–418.

Funtenberger, S., Dumay, E., & Cheftel, J.-C. (1997). High pressure promotes β-lactoglobulin aggregation through SH/S-S interchange reactions. *J. Agric. Food Chem., 45,* 912–921.

Galazka, V.B., Dickinson, E., & Ledward, D.A. (1996). Effect of high-pressure on the emulsifying behaviour of β-lactoglobulin. *Food Hydrocolloids, 10,* 213–219.

Galazka, V.B., & Ledward, D.A. (1997). High pressure effects on the functional properties of bovine serum albumin and β-lactoglobulin. In K. Heremans (Ed.), *High Pressure Research in the Biosciences and Biotechnology* (pp. 375–378). Leuven: Leuven University Press.

Galazka, V.B., Ledward, D.A., Dickinson, E., & Langley, K.R. (1995). High-pressure effects on emulsifying behavior of whey-protein concentrate. *J. Food Sci., 60,* 1341–1343.

Gaucheron, F., Famelart, M.H., Mariette, F., Raulot, K., Michel, F., & Legraet, Y. (1997). Combined effects of temperature and high-pressure treatments on physicochemical characteristics of skim milk. *Food Chem., 59,* 439–447.

Gervilla, R., Capellas, M., Ferragut, V., & Guamis, B. (1997). Effect of high hydrostatic pressure on *Listeria innocua* 910 CECT inoculated into ewe's milk. *J. Food Prot., 60,* 33–37.

Hite, B.H. (1899). The effect of pressure in the preservation of milk. *Bull. West Virg. Univ. Agric. Exp. Station, 58,* 15–35.

Horne, D.S. (1998). Casein interactions: Casting light on the black boxes, the structure in dairy products. *Int. Dairy J., 8,* 171–177.

Iametti, S., Transidico, P., Bonomi, F., Vecchio, G., Pittia, P., Rovere, P., & Dall'Aglio, G. (1997). Molecular modifications of β-lactoglobulin upon exposure to high pressure. *J. Agric. Food Chem., 45,* 23–29.

Johnston, D.E., Austin, B.A., & Murphy, R.J. (1992). Effects of high hydrostatic-pressure on milk. *Milchwissenschaft, 47,* 760–763.

Johnston, D.E., McKinney, S.M., & McCreedy, R.W. (1997). The effects of high pressure on some functional properties of whey proteins. In K. Heremans (Ed.), *High Pressure Research in the Biosciences and Biotechnology* (pp. 403–406). Leuven: Leuven University Press.

Kneifel, W., Seiler, A., Dewit, J.N., & Kindstedt, P.S. (1993). Water-holding properties of milk protein products—A review. *Food Structure, 12*(3), 297–308.

Kolakowski, P., Reps, A., Babuchowski, A., Zmujdzian, L., Podlasin, S., & Porowski, S. (1995). Influence of high pressure on changes of cheese characteristics. *Proceedings of the XXXIIIth Meeting of the European High Pressure Research Group,* Warsaw, Poland.

Kromkamp, J., Moreira, R.M., Langeveld, L.P.M., & van Mil, P.J.J.M. (1996). Micro-organisms in milk and yoghurt: Selective inactivation by high hydrostatic pressure. *Proceedings of the IDF Symposium on Heat Treatments: 7 Alternative Methods* (pp. 267–271). Brussels: International Dairy Federation.

Law, A.J.R., Banks, J.M., Horne, D.S., Leaver, J., & West, I.G. (1994). Denaturation of the whey proteins in heated milk and their incorporation into cheddar cheese. *Milchwissenschaft, 49,* 63–67.

Law, A.J.R., Leaver, J., Felipe, X., Ferragut, V., Pla, R., & Guamis, B. (1998). Comparison of the effects of high pressure and thermal treatments on the casein micelles in goat's milk. *J. Agric. Food Chem., 46,* 2523–2530.

López-Fandiño, R., Carrascosa, A.V., & Olano, A. (1996). The effects of high-pressure on whey-protein denaturation and cheese-making properties of raw-milk. *J. Dairy Sci., 79,* 929–936.

López-Fandiño, R., Ramos, M., & Olano, A. (1997). Rennet coagulation of milk subjected to high pressures. *J. Agric. Food Chem., 45,* 3233–3237.

López-Fandiño, R., De La Fuente, A.M., Ramos, M., & Olano, B. (1998). Distribution of minerals and proteins between the soluble and colloidal phases of pressurized milks from different species. *J. Dairy Res., 65,* 69–78.

López-Fandiño, R., & Olano, A. (1998a). Effects of high pressures combined with moderate temperatures on the rennet coagulation properties of milk. *Int. Dairy J., 8,* 623–627.

López-Fandiño, R., & Olano, A. (1998b). Cheese-making properties of ovine and caprine milks submitted to high pressures. *Lait, 78,* 341–350.

Marshall, R.J. (1986). Increasing cheese yields by high heat-treatment of milk. *J. Dairy Res., 53,* 313–322.

Messens, W., Dewettinck, K., & Huyghebaert, A. (1999). Transport of sodium chloride and water in Gouda cheese as affected by high pressure brining. *Int. Dairy J., 9,* 569–576.

Messens, W., Dewettinck, K., Van Camp, J., & Huyghebaert, A. (1998). High pressure brining of Gouda cheese and its effect on the cheese serum. *Food Sci. Technol.-Lebensm. Wiss. Technol., 31,* 552–558.

Moller, R.E., Stapelfeldt, H., & Skibsted, L.H. (1998). Thiol reactivity in pressure-unfolded β-lactoglobulin: Antioxidative properties and thermal refolding. *J. Agric. Food Chem., 46,* 425–430.

Needs, E. (1997). [Effect of pressure on milk enzymes]. Unpublished data.

Needs, E.C., Anderson, M., & Morant, S.V. (1986). Interaction of factors which influence the extent of lipolysis during milking and storage of raw milk. *J. Dairy Res., 53,* 203–210.

Needs, E.C., Capellas, M., Bland, A.P., Manoj, P., MacDougal, D., & Paul, G. (2000a). Comparison of heat and pressure treatments of skimmed milk, fortified with whey protein concentrate, for set yogurt preparation: Effects on milk proteins and gel structure. *J. Dairy Res., 67,* 329–348.

Needs, E.C., Stenning, R.A. Gill, A.L., Ferragut, V., & Rich, G.T. (2000b). High pressure treatment of milk: Effects on casein micelle structure and on enzymatic coagulation. *J. Dairy Res., 67,* 31–42.

O'Reilly, C.E., O'Connor, P.M., Kelly, A.L., Murphy, P.M., & Beresford, T.P. (1998). Effect of high pressure on proteolysis during ripening of cheddar cheese. *Proceedings International Dairy Federation Congress,* Aarhus, Denmark.

O'Reilly, C.E., O'Connor, P.M., Murphy, P.M., Kelly, A.L., & Beresford, T.P. (2000). The effect of pressure of 50 MPa on cheddar cheese ripening. *Innovative Food Science and Emerging Technologies, 1*(2), 108–117.

Park, S.Y., Nakamura, K., & Niki, R. (1996). Effects of β-lactoglobulin on the rheological properties of casein micelle rennet gels. *J. Dairy Sci., 79,* 2137–2145.

Patterson, M.F., Quinn, M., Simpson, R., & Gilmour, A. (1995). Sensitivity of vegetative pathogens to high hydrostatic-pressure. *J. Food Prot., 58,* 524–529.

Pittia, P., Wilde, P.J., Husband, F., & Clark, D.C. (1996a). Functional and structural properties of β-lactoglobulin as affected by high pressure treatment. *J. Food Sci., 61,* 1123–1128.

Pittia, P., Wilde, P.J., & Clark, D.C. (1996b). The foaming properties of native and pressure-treated β-casein. *Food Hydrocolloids, 10*(3), 335–342.

Pittia, P., Wilde, P.J., & Clark, D.C. (1997). The effects of high pressure treatment on functional properties of β-casein and β-lactoglobulin. In K. Heremans (Ed.), *High Pressure Research in the Biosciences and Biotechnology* (pp. 355–358). Leuven: Leuven University Press.

Rademacher, B., Pfeiffer, B., & Kessler, H.G. (1998). Inactivation of micro-organisms and enzymes in pressure-treated raw milk. In N.S. Isaacs (Ed.), *High Pressure Food Science, Bioscience and Chemistry* (pp. 145–152). Cambridge: The Royal Society of Chemistry.

Rademacher, B., Pfeiffer, B., Brinker, C., Lechner, E., & Kessler, H.G. (1999). Kinetics of the deactivation of milk enzymes by ultrahigh pressure. *Chemie Ingenieur Technik, 71,* 392–396.

Schmidt, D.G., & Buchheim, W. (1970). Elektronenmikroskopische untersuchung der feinstruktur von caseinmicellen in kuhmilch. *Milchwissenschaft, 25,* 596–600.

Schrader, K., Buchheim, W., & Morr, C.V. (1997). High pressure effects on the colloidal calcium phosphate and the structural integrity of micellar casein in milk: 1. High pressure dissolution of colloidal calcium phosphate in heated milk systems. *Nahrung-Food, 41*(3), 133–138.

Schwertfeger, M., & Buchheim, W. (1999). Coagulation of skim milk under high hydrostatic pressure with acidification by glucono-δ-lactone. *Int. Dairy J., 9,* 487–492.

Scollard, P.G., Beresford, T.P., Needs, E.C., Murphy, P.M., & Kelly, A.L. (2001). Plasmin activity and proteolysis in high pressure-treated milk. *International Dairy Journal.* In press.

Seyderhelm, I., Boguslawski, S., Michaelis, G., & Knorr, D. (1996). Pressure-induced inactivation of selected food enzymes. *J. Food Sci., 61,* 308–310.

Shibauchi, Y., Yamamoto, H., & Sagara, Y. (1992). Conformational change of casein micelles by high pressure treatment. In C. Balny, R. Hayashi, K. Heremans, & P. Masson (Eds.), *High Pressure and Biotechnology,* Colloque INSERM (Vol. 224 pp. 239–242). Montrouge: John Libbey Eurotext.

Trujillo, A.J., Ferragut, V., Gervilla, R., Capellas, M., & Guamis, B. (1997). High hydrostatic pressure effects on milk and milk products. *Recent Res. Dev. Agric. Food Chem., 1,* 137–159.

Trujillo, A.J., Royo, C., Guamis, B., & Ferragut, V. (1999). Influence of pressurization on goat milk and cheese composition and yield. *Milchwissenschaft, 54,* 197–199.

Van Camp, J., & Huyghebaert, A. (1995a). High pressure-induced gel formation of a whey-protein and haemoglobin protein-concentrate. *Food Sci. Technol.-Lebensm. Wiss. Technol., 28,* 111–117.

Van Camp, J., & Huyghebaert, A. (1995b). A comparative rheological study of heat and high-pressure-induced whey-protein gels. *Food Chem., 54*(4), 357–364.

Van Camp, J., Messens, W., Clement, J., & Huyghebaert, A. (1997). Influence of pH and calcium chloride on the high-pressure-induced aggregation of a whey protein concentrate. *J. Agric. Food Chem., 45*(5), 1600–1607.

van Hooydonk, A.C.M., Dekoster, P.G., & Boerrigter, I.J. (1987). The renneting properties of heated milk. *Neth. Milk Dairy J., 41,* 3–18.

Velez-Ruiz, J.F., Swanson, B.G., & Barbosa-Canovas, G.V. (1998). Flow and viscoelastic properties of concentrated milk treated by high hydrostatic pressure. *Food Sci. Technol.-Lebensm. Wiss. Technol., 31,* 182–195.

Yokoyama, H., Sawamura, N., & Motobayashi, M. (1991). Method for accelerating cheese ripening. European Patent Office Application No: 91306976.1, Publication No: 0 469 857 A1, Date of Filing: 30-7-1991. Applicant: Fuji Oil Company, Ltd., Osaka, Japan.

CHAPTER 11

High Pressure Equipment Designs for Food Processing Applications

Robert W. van den Berg, H. Hoogland, H.L.M. Lelieveld, and Ludo Van Schepdael

INTRODUCTION

Although the effects of isostatic pressure on microorganism inactivation and on protein and polysaccharide denaturation have been known for more than a century (Hite, 1899), the food processing industry did not become interested in applying this technology until the past decade. This interest has been triggered by results of research at universities and institutes and by the efforts of equipment suppliers who recognize a promising new market in high pressure technology. A novel technology such as high pressure processing is only interesting when it reduces costs or creates new opportunities to improve food quality and safety. Although high pressure technology is currently more expensive than traditional processing technologies (e.g., high-temperature sterilization), the use of high pressure offers new opportunities for the food industry to respond to consumers' wishes. Consumers demand high-quality, convenient food products and are concerned with food safety. Nutrients and flavors are hardly affected by the high pressures at which microorganisms and enzymes are inactivated. High pressure processing allows food manufacturers to produce high-quality food products with an extended shelf life. Consumers may be willing to pay extra costs for superior products. A company that has rapidly introduced pressure-processed food products is Avomex, a North American company with a plant in Mexico that uses high pressure to treat fresh guacamole (avocado paste). After treatment, these products can be stored for several weeks instead of a few days, which has created profitable retail opportunities (Waugh, 2000).

Most of the companies that have started developing high pressure equipment for food processing had expertise in other applications of high pressure such as water-jet cutting and cold isostatic pressing (e.g., Flow Systems and Kobe Steel). The latter technology is used mainly in the metal and ceramics industries, and the equipment used closely resembles the equipment needed for food treatment. In the

The authors wish to thank The Netherlands Department of Economic Affairs (Programmabureau E.E.T.) for financially supporting the research to produce this chapter.

early 1990s, these companies started to develop high pressure equipment based on their own expertise, and a number of articles were published about these developments (Farr, 1990; Ishii, 1991; Deplace & Mertens, 1992; Hori et al., 1992; Kanda et al., 1992; Träff & Bergman, 1992; Mertens & Deplace, 1993; Zimmerman & Bergman, 1993; Mertens, 1994; Westerlund, 1994). Today, suppliers focus primarily on further optimization of these techniques, and developments are less pathbreaking than they were 10 years ago. Although virtually no articles concerning high pressure equipment have been published since then, several important developments have occurred. In 1999, Flow International acquired ABB Pressure Systems. The combination of Flow's hydraulics and ABB's wire-wound pressure vessels makes Flow one of the leading manufacturers of high pressure equipment for the food industry. In addition, manufacturers developed semicontinuous high pressure processing systems for processing of liquids, which is a new alternative to conventional batch systems. Meanwhile, equipment developers are working to reduce costs. Because of the high depreciation costs of high pressure equipment, the cost per liter of product is still higher than other processing techniques (such as high-temperature sterilization). Since 1997, a large Dutch project has been working to develop competitive high pressure equipment. This chapter presents an overview of state-of-the-art high pressure processing equipment and the major manufacturers of this equipment.

EQUIPMENT FOR HIGH PRESSURE PROCESSING

There are two major types of high pressure processing of food products: the (conventional) batch systems, derived from cold isostatic processing, and the semicontinuous systems. Batch systems can process both liquid and solid products, but these have to be prepacked. In-line systems can be applied only to pumpable products (e.g., orange juice). The product is pumped into the pressure vessel and pressurized using a floating piston, which separates the product from the pressure medium.

Batch Systems

Vessels and Yoke

Equipment for high pressure batch processing (Deplace & Mertens, 1992; Mertens & Deplace, 1993; Mertens, 1994; Tonello et al., 1994) consists of three main parts: the vessel, its surrounding yoke, and the hydraulics. The actual pressure treatment takes place in the pressure vessel, which is considered to be the heart of the equipment. The volume of the vessel may vary from less than a liter (for laboratory-scale applications operating at 1000 MPa) up to 500 liters (for processing units operating at 600 MPa).

In many cases, the pressure vessel is a forged cylinder constructed in low-alloy steel of high-tensile strength. The wall thickness of these (monobloc) vessels is determined by the maximum working pressure, the vessel diameter, and the number of cycles for which the vessel is designed. The use of industrial-scale monobloc vessels is limited to working pressures up to 600 MPa. Above these pressures, stresses at the inner core approach the yielding stress of the material, resulting in a diminishing fatigue resistance. To overcome this problem, pressure vessels with residual compressive stress have been developed that result in a higher yield strength. However, the costs of manufacturing these vessels are significantly higher than for monobloc vessels that are forged and processed in one single piece. Therefore, prestressed vessels are used only for high pressure processes where monoblocs cannot be used. Three techniques can be used to apply a tangential compression on the inner core of pressure vessels.

1. *Autofrettage.* The vessel is subjected to an internal pressure exceeding the operating pressure. The maximum stress in the autofrettage zone exceeds the yield stress of the material, and plastic deformation occurs. Because tangential stresses decrease rapidly in a radial direction, plastic deformation takes place at the inner core only. After depressurization, the plastically deformed region is a compressive tangential state, thus opposing the tensile stresses resulting from high pressure processing. Consequently, a higher working load can be achieved.
2. *Multilayer vessels.* By building vessels in multiple layers, the inner layers can be prestressed by giving the outer layer a negative tolerance to the inner layer. Mounting can be performed after cooling the inner cylinder and heating the outer cylinder. After cooling, the inner cylinder is under a state of compression.
3. *Wire-wound and strip-wound vessels.* This is the process Flow Pressure Systems and Kobe Steel apply to their vessels. When a rectangular spring steel wire is wound around a steel cylinder, the cylinder becomes prestressed. Spring steel wire has a yielding strength of 1600 N/mm^2 and a rupture strength of 1900 N/mm^2, which is three times stronger than normal forged high-strength steels. The repetitive mechanical deformation of the wire (cold rolling) strengthens the spring steel wire. The winding process is computer controlled. To manufacture a production-scale vessel, up to a few hundred kilometers of wire is needed. Because a wire-wound vessel has a higher yield strength and the ability to withstand the optimum stress at every diameter, it can have halve the weight of other designs, thus reducing size of the enclosing frame as well. Kobe Steel indicates that a unit containing a multilayer vessel with an inner capacity of 100 liters weighs about 29 tons; a unit containing a wire-wound vessel weighs about 10 tons. UHDE (a German company) uses a similar prestressing technique. However, instead of wires, strips compress the inner core of the vessel. In this case, the vessel is constructed of prestressed elements that are fit onto a conical liner.

To cope with the high axial forces acting on the vessel and its end closures, the vessel is surrounded by a yoke. This yoke consists of either a number of steel plates (for instance, the ACB Pressure System-Alstom Hyperbar, Figure 11–1) or a wire-wound steel frame (Flow *Quintus* System). The position of the vessel in the yoke—either horizontal or vertical—depends on the type of processing. The vessel in the Flow Quintus System is positioned vertically, whereas the ACB Pressure System-Alstom Hyperbar has a horizontally placed pressure vessel. The main advantage of the horizontal placement is that charging and discharging can take place quickly (the untreated packages may push out the treated packages). ACB Pressure System-Alstom developed a quick opening-closing system for its Hyperbar system to speed up processing (Lebas, personal communication, June 1998; February 2000). Furthermore, it is technically preferable to scale up in axial direction, which is easier for horizontal constructions. An advantage of vertical construction is that the pressure medium is reused in the vessel after discharging the product. This lowers the total cycle time, since refilling of the vessel is less time consuming.

Hydraulics

The pressure vessel is pressurized by means of a pressure medium, which should be a suitable liquid for food application processes. Typically, water mixed with a small percentage of soluble oil for pump lubrication is used. For reasons of food safety, the pressure medium must be nontoxic and hence no mineral oils should be used. The liquid is pressurized using hydraulic pumps, which have a typical power of 100 to 200 kW for production-scale units. After the vessel is loaded with a batch of products and mounted in its yoke, it is filled with pressure medium using a low-pressure, high-volume pump. During this first step, air may be removed through a de-aeration valve. In the second step, the system is pressurized to its operating pressure, either through internal or external compression.

In the case of internal compression, the system is pressurized by a piston, driven at its larger diameter by a low-pressure pump. The operating pressure in the vessel equals the pressure at the low-pressure side multiplied by the ratio between the surfaces of the piston. The advantage of internal compression is that

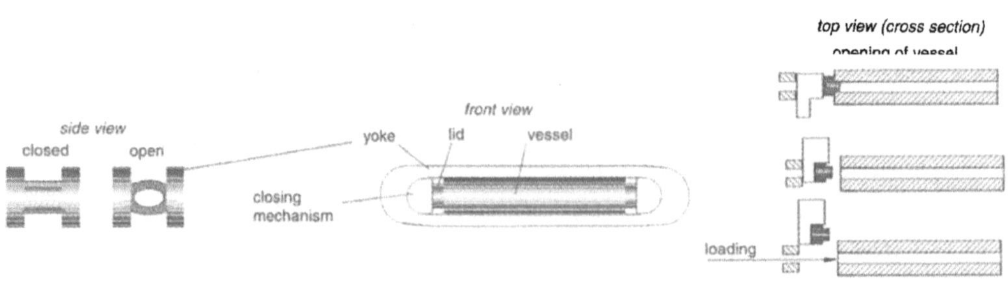

Figure 11–1 The ACB Pressure System-Alstom Hyperbar.

it achieves fast pressurization rates. Its main disadvantage is that it requires a dynamic high pressure seal between piston and inner core. Another disadvantage is that it is a single-shot operation; in case of leakage, pressure compensation is not possible.

Pressurization by means of external compression takes place through an external high pressure intensifier that pumps liquid into the vessel under high pressure. This intensifier operates through the same principle of pressure multiplication as the plunger described above. The main advantage of this technique is that static seals can be used. In practice, for operating pressures up to 600 MPa, external pressurization is common. For higher pressures, internal pressurization is more efficient than external pressurization. Because of the rapidly increasing viscosity and relatively low-diameter tubing, pressurization through an external pump is a very slow process at high pressures. Furthermore, all external components in contact with high pressure (intensifier, valves, and tubing) must be well designed to ensure an effective number of cycles (Deplace & Mertens, 1992; Mertens, 1994).

For operational efficiency, it is recommended to couple two smaller pressure pumps with half the capacity needed instead of using one large pump. If one of the pumps is out of operation because of malfunction or maintenance, the system can still be operated.

For all high pressure processing equipment, seals are the main bottleneck. In all designs, either static or dynamic seals have to be applied. Dynamic seals are especially sensitive to wear and tear and should be replaced after a relatively short number of cycles, depending on pressure. The external intensifier pump is preferable, because the dynamic seal is placed in the external pump on a small diameter. Small-diameter seals have a longer lifetime than large-diameter seals and are easy to replace. In 1998, ABB advised purchasers of their internal intensifier units to replace seals monthly when operating at 400 MPa, and weekly when operating at 600 MPa. Thus, seals may be the main factor in maintenance costs. Recent findings suggest that new materials will do a better job and reduce maintenance costs significantly.

New Developments in Batch Systems

The Agrotechnological Research Institute (ATO), Unilever Research Vlaardingen, Stork Food Processing Group, and The Netherlands Department of Economic Affairs have jointly developed the cost-efficient Barawing system for high pressure treatment of food products (Bartels, 1996; van den Berg et al., 1998). The system is based on the conclusion of recent microbiological research that efficient, high-quality sterilization of food products can only be achieved at pressures of 700 MPa or higher. The Barawing system has three main features: an internal intensifier, a composite vessel, and an axial cylinder.

Internal intensifier. A cost-competitive operation requires short cycle times, and therefore fast pressurizing. The viscosity of pressure fluids that can be used in food applications increases at high pressures. At high pressures and relatively low fluid speed, a substantial pressure drop has been recorded over a high pressure line (length 5 meters, diameter 1.6 mm, come-up time 150 seconds), as shown in Figure 11–2. For this reason, the Barawing system uses an internal intensifier design, which allows short pressurizing time, low external pressure, and high energy efficiency. In addition, it offers the option of using an accumulator for even faster pressurizing.

Composite vessel. The Barawing system adopts a vessel that is reinforced with high-strength synthetic fibers. In order to optimize the advantages of this construction, a prestressing technique has been developed. The advantages of this system over conventional high pressure processing systems are

- higher working pressure
- high number of cycles due to better fatigue resistance
- lower weight (density of composites is about 20% of the density of steel)
- faster handling
- low heat transfer (no cold spots under adiabatic temperature rise).

Figure 11–3 shows a composite test vessel designed for operation at 1100 MPa. Because of the properties of these composites, in the future they may also be used for yokes. This would yield a significant reduction in mass for the complete unit.

Axial cylinder. The Barawing system uses an axial cylinder for fast closing and opening of the vessel, thus reducing total cycle time. The design of the axial cylinder has several advantages. It reduces shear stresses in the composite material; it reduces tensile stress on the inner fiber caused by transverse Poisson

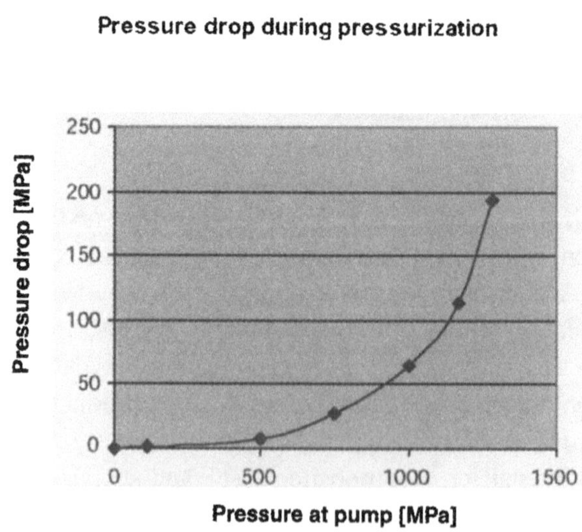

Figure 11–2 Pressure drop over a high pressure line (length 5 meters, diameter 1.6 mm, come-up time 150 seconds).

Figure 11–3 Composite pressure vessel ready to be tested. Pressure is supplied by means of an external intensifier. During pressurizing, the vessel is compressed axially by means of a hydraulic press, positioned at the bottom of the vessel.

effects; and it allows an easier liner and seal design because extrusion of the liner is prevented.

In the first phase of the project, small-scale high pressure vessels constructed of various composite materials have been built and extensively tested. Recently, participants in the joint project started a second three-year phase of the project with the goal of further commercialization of the concept. In May 2000, a pilot unit became operational. Commercialization of this new high pressure system is planned by Stork in association with Fokker Special Products.

Continuous Processing: New Developments

A semicontinuous high pressure processing system for processing of liquids has been developed by Flow, ACB Pressure System-Alstom, and Kobe Steel. Figure 11–4 presents an overview of the process. A low-pressure transport pump pumps liquids into the pressure vessel. After closing the valves, the vessel is pressurized by means of the high pressure pump. A floating piston separates the pressure transfer medium and the product. After treatment, the product is transferred to a surge vessel. By coupling a number of pressure vessels, the energy stored in a pressurized vessel can be used to pressurize a second vessel at ambient pressure, thus saving energy and process time.

Unilever patent WO97/43914 describes a continuous high pressure processing system (Agterof et al., 1997). In this system (Figure 11–5), the product is conducted in a steady flow through an open-end tube while the pressure difference

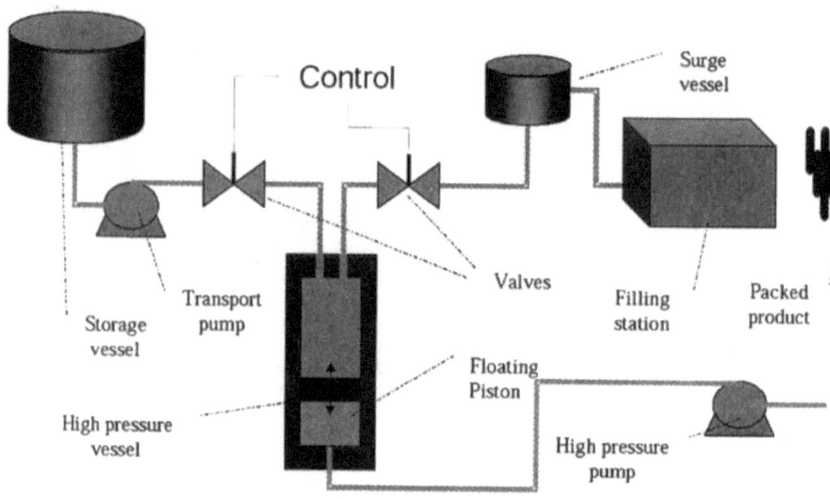

Figure 11–4 The semicontinuous high pressure processing system for processing of liquids.

between the entrance and the exit of the tube is maintained at 100 MPa or more using a high pressure pump. The pressure difference is the result of the viscous friction in the system, especially because of the exponential pressure-dependent viscosity behavior. Viscosity may increase by a factor of 10^3 to 10^4 at increased pressure. The pressure buildup in an open tube is possible with a long tube (50 to 100 m); the ratio of length to average diameter of this tube should be at least 1000 and preferably 10,000. The residence time can easily be adjusted by inserting a chamber at the upstream of the tube. Unilever suggests that an effective preservation technique might be obtained if this system is combined with pulsating electrical fields.

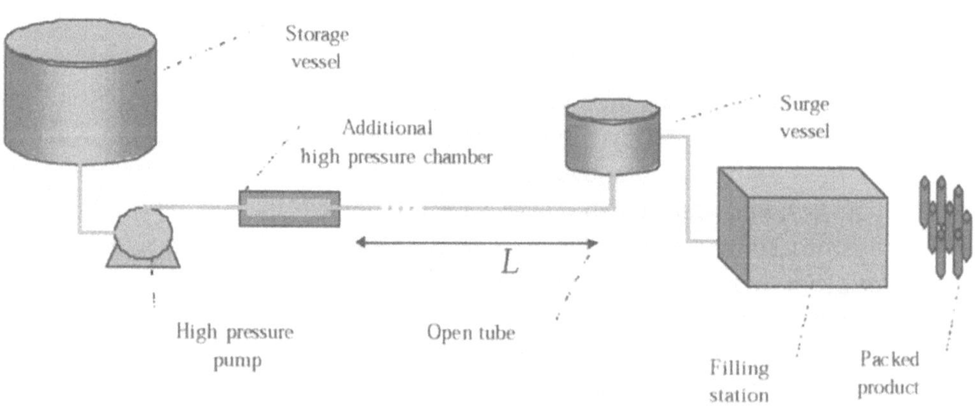

Figure 11–5 Continuous high pressure processing system for processing of liquids.

At Unilever Vlaardingen and at ATO, continuous high pressure processing laboratory-scale systems combined with pulsating electrical fields have been developed. The effects of several combinations of pressure and electrical field strength on spores and microorganisms are being currently evaluated.

Codes and Rules for Construction

For a company to use high pressure equipment, an inspection and certification by a local (national) inspection institute is required. For inspection and certification, a code describing requirements and criteria is to be used. Portions of several codes could be used for the certification of high pressure equipment, but the code that is most specifically adapted for this application is the American Society of Mechanical Engineers (ASME) Boiler and Pressure Vessel Code, Section VIII, Division 3 (Alternative rules for high pressure vessels, 1997). The rules of Division 3 include requirements for the design, construction, inspection, and overpressure protection of metallic pressure vessels with design pressures generally above 10,000 psi (approximately 69 MPa). It consists of several parts.

- *General requirements.* Scope, jurisdiction, responsibilities, duties, and general rules for inspection.
- *Material requirements (for metals).* Test requirements, requirements for bolting, and material design data.
- *Design requirements*:
  - the basic design
  - fatigue evaluation, including design fatigue curves
  - fracture mechanics
  - design rules for autofrettage
  - design for openings, closures, heads, boltings, seals, and supports
  - special design requirements for layered vessels
  - special design requirements for wire-wound vessels and frames
  - special design requirements for strip-wound vessels
  - design requirements for welded vessels
  - experimental design verification.
- *Fabrication requirements.* Welding requirements; protective linings; heat treatment of weldments; and several additional fabrication requirements for autofrettaged vessels, quenched and tempered steels, layered vessels, wire-wound vessels, and strip-wound vessels.
- *Pressure-relief devices.* Requirements for rupture disk device, pressure-relief valves, and flow capacity of relief valves.
- *Examination requirements.* Examination procedures (radiographic, ultrasonic, etc.), examination of material and welds, and acceptance criteria.
- *Testing requirements.* Impact test, hydrostatic test, gages, and transducers.
- *Marking, stamping, reports, and records.*

MAJOR MANUFACTURERS OF HIGH PRESSURE PROCESSING EQUIPMENT

Worldwide, a number of manufactures have developed ultra-high pressure equipment, each of them starting from their own area of expertise. Three major players dominate the market: Flow International Corporation, ACB Pressure System-Alstom, and Kobe Steel. Manufacturers started developing this equipment as early as 40 years ago, mainly focusing on processes such as hot and cold isostatic presses. A decade ago, equipment manufacturers started to develop high pressure equipment dedicated to the treatment of food products. A typical example of the scale of these units is the Avomex high pressure processing facility installed by Flow, which has a capacity of 215 liters operating at 600 MPa. Whereas the large equipment manufacturers focus mainly on industrial-scale equipment for the food industry, a large number of smaller companies develop laboratory-scale and pilot units for use at research institutes and universities (Waugh, 2000).

Flow International Corporation (United States and Sweden)

Flow International has its roots in ultra-high pressure waterjet cutting and has become a leader in this area. To date, the company has delivered more than 5000 units worldwide. In addition, the company is involved in isostatic presses used for the production of advanced materials and for treatment of foods, sheet-metal forming presses, and waterjet cutting and cleaning.

In 1993 Flow started developing high pressure processing systems for the food industry. In collaboration with Oregon State University, Flow studied the effects of high pressure treatment on food products. In 1999, Flow took over ABB Pressure Systems. This company was started in the early 1950s to manufacture diamonds (which require a pressure of about 8000 MPa). Later, ABB expanded its high pressure technology to other areas. In the early 1990s, ABB started its research on the effects of high pressure processing on foods together with Tetra Laval and SIK (a Swedish research institute) and later with SSICA (an Italian research institute). Currently, ABB Pressure Systems has been converted into a unit within Flow International and is called Flow Pressure Systems.

Flow Pressure Systems has developed technology for wire winding of vessels and yokes, reducing the weight by 50% and leading to an infinite lifetime of vessel and yoke (trade name Quintus). In addition, Flow International's experience gained from manufacturing water jet cutters is a sound basis for the design of intensifier pumps.

Equipment developed or manufactured by Flow International includes the following:

- At Flow Sweden a 1500 MPa unit is available for testing at temperatures up to 115°C with 1.5 liter volume.
- Approximately 20 laboratory units have been delivered to the United States, Europe, Japan, and Korea. Most of these units have a pressure of 900 MPa and a volume of 1.4 liters.

- One 22-liter, one 50-liter, and one 215-liter press have been delivered to Avomex in the United States for treatment of avocado, salsa, etc. All systems are operated at nominal pressures of 600 MPa or higher.
- Two units of 35 liter, 600 MPa have been delivered to Japan and Europe.
- Flow has delivered a number of batch presses in the range 25 to 45 liters and continuous systems for treatment of liquid food.
- In 2000, six units of 215 liters or larger with pressures from 400 to 600 MPa were delivered, among others for Avomex (Träff, personal communication, June 1998, February 2000).

ACB Pressure System-Alstom (France)

ACB Pressure System-Alstom has an extensive expertise in the development of food-processing equipment such as continuous sterilizers and canneries, and in the design and manufacture of high pressure equipment for different industrial sectors. Some of the main activities of ACB Pressure System-Alstom are high-temperature–short-time continuous sterilizers for the dairy, beverage, and pharmaceutical industry; hydraulic presses (stretch forming, super elastic forming); and cold and hot isostatic presses and high pressure equipment up to 1500 MPa.

Since 1992, the company has designed new high pressure equipment specially adapted for the treatment of food products. These high pressure processing systems are known as Hyperbar and characterized by the horizontal position of vessel and yoke. Each end is fitted with a patented quick closing-opening device to enable automatic axial loading of the vessel.

Since 1992, ACB Pressure System-Alstom has been involved in a number of European research and development programs concerning high pressure food processing. At ACB Pressure System-Alstom, several systems are available for tests including a pilot unit (700 MPa, 3 liter), a semicontinuous unit (600 MPa, 5 liters), and a mobile pilot on a trailer (500 MPa, 50 liters). About 15 laboratory units have been delivered to universities and institutes. For industrial applications, ACB Pressure System-Alstom delivered the following:

- 700 MPa (50 liter) unit to HP technologies
- several 400 MPa units to Ulti for the treatment of orange juice
- 500 MPa (50 liter) and 400 MPa (320 liter) units to Espuna for treatment of prepacked, sliced, cooked ham
- Two 500 MPa (420 liter) units scheduled to be delivered in 2000/2001 (Lebas, personal communication, June 1998, February 2000)

Kobe Steel (Japan)

With more than 30 years of experience, Kobe Steel has become a leading supplier of high pressure equipment in Japan and has sold approximately 700 units, including the largest cold isostatic processing-unit, which can handle a volume of 14,700 liters at 98 MPa. Kobe Steel is involved in the following activities:

- hot and cold isostatic presses
- isostatic extrusion
- high pressure crystallization
- super critical gas extraction
- rubber isostatic presses
- compound semiconductor single-crystal growth system

Kobe Steel has in-house capabilities to design and manufacture wire-wound vessels and yokes as well as intensifiers. In addition, Kobe Steel has developed equipment to control product temperature during ultra-high pressure. Recently, the company installed a twin vessel unit with 300-liter vessels for treatment of rice operating at 400 MPa.

Engineered Pressure Systems International (Belgium)/Engineered Pressure Systems Incorporated (United States) (Subsidiary of Belgium Headquarters)

Engineered Pressure Systems International (EPSI), formerly known as National Forge, has more than 35 years of experience in manufacturing high pressure vessels. From its roots in the forging industry, EPSI has a broad metallurgical expertise. This company has developed many types of high pressure vessels, including vessels with threaded closures (breach or continuous); vessels with unthreaded closures with yoke frame (plate or wire-wound design); and autofrettaged, shrink-fit lined, and wire-wound vessels.

EPSI primarily manufactures cold and hot isostatic presses. In addition, the company manufactures drybag cold isostatic presses (in collaboration with FREY GmbH, Germany), hydrothermal quartz growing units, deep-sea and deep-well simulators, and electrohydraulic intensifiers (standard for 100 to 700 MPa, special executions up to 900 MPa).

In the past EPSI collaborated with FMC, a Belgian subsidiary of the U.S. company, Food Machinery Corporation, supplier of food processing equipment. For the food research, EPSI constructed about 16 laboratory presses. An industrial 22-liter unit has been delivered to Avomex (Mermelstein, 1998).

UHDE Hochdrucktechnik (Germany)

UHDE was founded in 1930. It initially manufactured high pressure equipment for ammonia plants. Today, the company not only delivers equipment and complete plants but also conducts feasibility studies, process development, and product trials. UHDE manufacturers or designs the following types of equipment:

- polyethylene processes (tube reactors, heat exchangers, piping, valves)
- waterjet cutting units
- tobacco treatment plants
- urea production processes
- equipment for processes using supercritical gases (high pressure extraction

from plants, dyeing of synthetic fibers, and crystallization)
- isostatic presses

In the field of preservation and pharmacy, UHDE collaborates with the Berlin University of Technology and the University of Heidelberg.

UHDE manufactures intensifier pumps for a variety of applications:

- 300 MPa intensifier pumps for waterjet cutting
- 400 MPa intensifier pumps for injection of catalysts in polyethylene production
- 700 MPa intensifier pumps for preservation units
- 1400 MPa intensifier pumps for pressure testing and autofrettage

Vessels for operation at pressures in excess of 600 MPa are made by using strip winding, which is similar to wire winding except that the vessel is made of separate strip-wound cylinders (Vardag, personal communication, August 1998, February 2000). UHDE has manufactured several laboratory units for preservation processes and is currently working on a production-scale unit.

Other Manufacturers

A number of other companies have developed or designed high pressure equipment including the following:

- *Elmhurst Research*, a U.S. company that designs isostatic vessels by using old cannon tubes, such as 8-inch M201 howitzers.
- *Stansted Fluid Power*, a family-owned company based in the United Kingdom that is mainly involved in development of laboratory equipment, jet cutting, high pressure homogenization, and consulting.
- *Resato International*, a company in The Netherlands that is involved in development of laboratory equipment and jet cutting.
- UNIPRESS, a company in Poland associated with the High Pressure Institute of the Polish Academy of Sciences has been designing sophisticated high pressure units for various research institutes (i.e., Berlin University of Technology).

ECONOMICS OF HIGH PRESSURE PROCESSING

An important criterion in the evaluation of preservation technologies is the cost per amount of treated product (Ting & Farkas, 1995). Although high pressure processing is a very promising technique for increasing product shelf life, it is more expensive than conventional methods such as high temperature sterilization. For this reason, high pressure processing is mainly used for niche products such as fresh fruit juices, seafoods, and guacamole. Currently, costs of the high pressure processing range from 10 to 20 cents per liter, whereas costs for high-temperature treatment may be as low as 2 to 4 cents per liter.

Cost reduction of the high pressure process can be achieved by careful consideration of the process. Important determinants of costs are capacity of the system (the number of liters processed per unit of time); the depreciation costs of the system; and variable costs such as labor, occupancy, and energy. A typical breakdown scheme for the total costs of the high pressure process is presented as follows:

- area—1–2%
- labor—5–10%
- utilities, energy—2–4%
- maintenance—5–10%
- capital—75–80%

Area comprises the facility in which the processing unit is operated. Personnel includes the operators of the unit; the number of operators varies between 1 and 5, depending on the type of equipment. Energy costs are low and comprise less than 1% of total costs. Maintenance, which is mainly related to replacement of seals, is typically 4% of the investment costs per year. Thus, it is clear that capital costs are the main expense. A typical investment price for a high pressure processing system (400 to 600 MPa) is $1.5 to $2.5 million. The capital costs of installing a high pressure processing system may be distributed as follows:

- high pressure vessel, closures, and yoke—50–60%
- pump and high-pressure circuit—30–35%
- control—10–15%

Thus, the pressure vessel is the most crucial component of the unit.

In summary, cost reduction can be obtained only when at least one of the following objectives is achieved:

- Capacity is increased. This can be accomplished in two ways: (1) Upscaling (Increasing the volume of the vessel decreases the cost price per liter. Kobelco reports that total costs per liter of product for a system containing a 500-liter vessel are 50% of the costs for a unit of 70 liters. Obviously, investment costs increase with vessel volume, but not linearly.); (2) reduction of the cycle time.
- Investment costs are reduced. For this reason, the Barawing concept (van den Berg et al., 1998) focuses on reduction of pressure vessel costs in combination with a reduction of cycle time.

Reduction of Cycle Time

Capacity of the system is closely related to the cycle time of the process. For batch systems, the cycle time may be broken down into a number of steps:

1. Fill vessel with the product.
2. Close the vessel.

3. Pressurize the system.
4. Holding time.
5. Release pressure.
6. Open the vessel.
7. Empty the vessel.

The time needed for each of these steps is related to mechanical limitations, except step 4, which is related to microbiological and food quality considerations. In order to reduce holding time, an effective setting of temperature and pressure must be applied. In practice, a combination of high temperature and high pressure may reduce holding time significantly. A disadvantage is that higher working pressures result in higher demands to the high pressure system, thus increasing costs. A second disadvantage is that at higher temperatures product quality is sacrificed. Therefore, for each product a specific combination should be determined that balances quality, product safety, and economics. Recently researchers have learned that holding time can be reduced significantly by an effective combination of pressure, temperature, and two consecutive treatments.

Filling, closing, opening, and emptying of the vessel can be speeded up by automation of the system. Units such as the ACB Pressure System-Alstom Hyperbar system feature a quick closing system that can be opened and closed automatically within 20 seconds. Because the locking system displaces the lid aside from the vessel, it is possible to load and unload the vessel in a horizontal straight line (see Figure 11–1). Automated loading and discharging will decrease the overall cycle time but will also add investment costs (Lebas, personal communication, June 1998, February 2000).

The time needed for pressurizing the vessel is closely related to the power of the high pressure generator (typically, 100 to 200 kW is needed for a 200- to 300-liter system operating at 400 to 600 MPa), the final pressure required, and the type of hydraulics (internal/external intensifier). In practice, the time needed for this step may vary from seconds for laboratory-scale systems, up to 3 to 5 minutes for units of commercial use (200 to 400 liters), again depending on pressure. Increasing pumping power may increase capacity, it also increases investment costs. For processes with a relatively short holding time, it may be useful to reduce time of pressurization by applying high-power pressure pumps or accumulators. Depressurization can be quickened by using specially designed pressure valves capable of handling high discharge flow rates without erosion or cavitation damage. Typical times for this process are 20 to 30 seconds for commercial units (Träff, personal communication, June 1998, February 2000; Lebas, personal communication, June 1998, February 2000).

Obviously, capacity is related to the time that the equipment is operational. Because of the need to maintain seals, equipment is out of operation about 1 day per month. For instance, the Avomex unit is operated in three shifts a day and down on Sunday. During the down time, maintenance takes place. Effectively, the unit is in use 86% of the time (Waugh, 2000).

Also related to capacity are the *packing* efficiency and the effective *volume* of the vessel. Because of the shape of the packages and the way these are designed and stacked, effective volume may vary from 50% to 80%: the space between the packages is waste. Cubical or hexagonal packages yield a higher effective volume than cylindrical bottles. For the latter, volume utilization is most effective when the bottle diameter approaches the diameter of the vessel core, or when the diameter of the bottle is one third of the diameter of the vessel core, thus covering the cross-section of the vessel with seven bottles (78% efficiency). To achieve a maximum efficiency, ACB Pressure System-Alstom standardized its pressure vessels to two internal diameters, 90 and 280 mm. In the 90-mm vessel, cylindrical 1-liter bottles with a similar diameter are processed; the 280-mm unit contains packages of seven of these bottles. In considering the effective volume, one should take into account enclosed air and the volume of the packaging material as well (Lebas, personal communication, June 1998, February 2000).

CONCLUSION

During the past 10 years high pressure food processing has matured from laboratory-scale technology into a promising, economically feasible preservation method. High pressure technology has been introduced in the food industry, and the first products manufactured using these new techniques have emerged on the market. The costs of high pressure processing are decreasing, mainly because of new development in hardware and methods of treatment. Equipment manufacturers will further optimize their designs to reduce equipment costs and increase efficiency. The application of composites enables operation at higher pressures and lower costs. The method of treatment as introduced by Meyer (2000) may facilitate sterilization of products (activation of bacterial spores), which appeared very difficult using the conventional high pressure treatment. Thus, high pressure processing has the potential to develop into a preservation technique that is applied on a large scale in the food industry, in particular for products where retention of flavors and nutrients is desired.

REFERENCES

Agterof, W., Lelieveld, H.L.M., Reichelt, T., & Smelt, J.P.M. (1997). Method for preservation under pressure. Patent WO97/43914.

ASME Boiler and Pressure Vessel Code, Section VIII, Divison 3—Alternative Rules for High-Pressure Vessels (1997).

Bartels, P.V. (1996). High pressure reactor. Patent EP0842696.

Deplace, G., & Mertens, B. (1992). The commercial application of high pressure technology for the food industry. In C. Balny, R. Hayashi, K. Heremans, & P. Masson (Eds.), *High Pressure and Biotechnology,* Colloque INSERM (Vol. 224, pp. 469–479). Montrouge: John Libbey Eurotext.

Farr, D. (1990). High pressure technology in the food industry. *Trends Food Sci. Technol., 1,* 14–16.

Hite, B.H. (1899). The effect of pressure in the preservation of milk. *Bull. West Virg. Univ. Agric. Exp. Station, 58,* 15–35.

Hori, K., Manabe, Y., Kaneko, M., Sekimoto, T., Sugimoto, Y., & Yamane, T. (1992). The development of high pressure processor for food industries. In C. Balny, R. Hayashi, K. Heremans, & P. Masson (Eds.), *High Pressure and Biotechnology,* Colloque INSERM (Vol. 224, pp. 499–507). Montrouge: John Libbey Eurotext.

Ishii, T. (1991). Recent trends in HIP. *KOBELCO Techno. Rev.,* 1–5.

Kanda, T., Yamauchi, T., Naoi, T., & Inoue, Y. (1992). Present status and future prospects of high pressure food processing equipment. In C. Balny, R. Hayashi, K. Heremans, & P. Masson (Eds.), *High Pressure and Biotechnology,* Colloque INSERM (Vol. 224, pp. 521–524). Montrouge: John Libbey Eurotext.

Mermelstein, N.H. (1998). High-pressure processing begins. *Food Technol., 52,* 104–106.

Mertens, B., & Deplace, G. (1993). Engineering aspects of high-pressure technology in the food industry. *Food Technol., 47,* 164.

Mertens, B.A. (1994). High pressure equipment for the food industry. *High Pressure Res., 12,* 229–237.

Meyer, R. (2000). Ultra high pressure, high temperature food preservation process. U.S. Patent 6017572.

Ting E.Y., & Farkas, D.F. (1995). Cost of ultrahigh pressure food pasteurization. *American Society of Mechanical Engineers, Pressure Vessels and Piping Division (Publication), 297,* 75–83.

Tonello, C., Wurtz, P., Largeteau, A., Jolibert, F., Deschamps, A., Ducastaing, A., & Demazeau, G. (1994). Recent developments in the use of high hydrostatic pressures in food processing. *High Pressure Res., 12,* 205–213.

Träff, A., & Bergman, C. (1992). High pressure equipment for food processing. In C. Balny, R. Hayashi, K. Heremans, & P. Masson (Eds.), *High Pressure and Biotechnology,* Colloque INSERM (Vol. 224, pp. 509–514). Montrouge: John Libbey Eurotext.

van den Berg, R.W., Bartels, P.V., & Van Schepdael L.J.J.M. (1998). High pressure apparatus. Patent WO99/61146.

Waugh, P. (2000). Development of refrigerated, low acid products stabilized by high pressure. Workshop on nonthermal processing of food, Portland, Oregon, 25–27 January.

Westerlund, J. (1994). High pressure equipment for food processing. *High Pressure Res., 12,* 221–226.

Zimmerman, F., & Bergman, C. (1993). Isostatic high-pressure equipment for food preservation. *Food Technol., 47,* 162.

Index

About the Editors

MARC E.G. HENDRICKX

Marc Hendrickx was born in Ramsel, Belgium, in 1959. Hendrickx is currently Professor of Food Technology and Head of the Laboratory of Food Technology (within the Department of Food and Microbial Technology) at Katholieke Universiteit Leuven, in Leuven, Belgium, where he has served since 1997. He is Coordinator of the International Course Program "Postharvest and Food Preservation Engineering," and member of the Research Council at Katholieke Universiteit Leuven.

Professor Hendrickx earned an MS in chemical and agricultural industries (with an emphasis in technology and management of agricultural and food industries) in 1983 and a PhD in food technology in 1988 from Katholieke Universiteit Leuven. He has served as Research Assistant (1983–1989), Postdoctoral Fellow (1989–1992), Assistant Professor (1992–1994), Associate Professor (1994–1997), and Professor (since 1997) on the Katholieke Universiteit Leuven Faculty of Agricultural and Applied Biological Sciences.

He has served as Editor of *Innovative Food Science and Emerging Technologies* since 2000 and is a member of the Editoral Board of *Journal of Food Process Engineering.* He is a member of numerous professional societies including the European Federation of Food Science and Technology (where he serves on the Executive Committee), the Scientific Advisory Council of International Association of Refrigerated Warehouses, the Institute for Thermal Processing Specialists, and the Institute of Food Technologists.

The general research objective of his research group is "the development of scientifically based methods to quantitatively evaluate the impact of physical preservation and/or processing techniques of foods in terms of quality, safety and functionality and the use of these methods in design, evaluation and optimization of such processes." With regard to processing technologies, his research is directed toward conventional techniques, including thermal processes (sterilization, pasteurization, cooling, and freezing), and novel techniques such as high pressure/thermal treatments, high pressure–assisted freezing and thawing, and the use of high electric field pulses. His research focuses on kinetics of chemical and enzymatic changes of biological materials affecting technological and nutritional functionality.

In the past decade, Hendrickx has played an active role in European Commission (EC)–funded research including the coordination of two AIR projects, "The Development of New Time-Temperature-Integrators (product history indicators) for the Quantification of Thermal Processes in Terms of Food Safety and Quality" (1992–1996) and "A New Approach to Heat Distribution and Heat Penetration Studies for Thermal Processing of Foods in Modern Batch Type Overpressure Rotary Retorts" (1994–1997), and the FAIR project, "Combined High Pressure Thermal Treatments of Foods: A Kinetic Approach to Safety and Quality Evaluation" (1996–1999). He has published more than 120 papers in international peer-reviewed journals and books and has contributed more than 150 papers to scientific conferences and meetings. He is co-editor of two books. He received the1995 Institute of Food Technologists Samuel Cate Prescott Award for excellence in research. Under Hendrickx's leadership, the Laboratory of Food Technology has recently been recognized as an EC Marie Curie Training Site (2000–2003).

DIETRICH KNORR

Dietrich Knorr was born in Waidhofen, Austria, in 1944. Since 1989, he has been Professor of Food Technology and Head of the Department of Food Biotechnology and Food Process Engineering at the Berlin University of Technology, in Berlin, Germany. In addition, he is Research Professor of Food Processing and Biotechnology at the Department of Animal Science and Food Science of the University of Delaware, in Newark, Delaware, in the United States, since 1987.

He obtained an engineering degree in 1971, and in 1974 he received a PhD in food and fermentation technology from the University of Agriculture, in Vienna, Austria. He has served as Assistant Professor in the Department of Food Technology, University of Agriculture (1973–1978); Visiting Scientist at the Western Regional Research Center, U.S. Department of Agriculture, Berkeley, California, United States (1976–1977); Visiting Scientist at the Department of Food Science, Cornell University, Ithaca, New York, United States (1978); Visiting Associate Professor at the Department of Food Science and Human Nutrition, University of Delaware (1978–1980); Visiting Scientist at the Department of Food Science, University of Reading, Reading, United Kingdom (1979); Associate Professor at the Department of Food Science and Human Nutrition, University of Delaware (1980–1984); Visiting Professor at the Association for Biotechnological Research, GBF, Braunschweig, Germany (1983, 1985–1986); Full Professor at the Department of Food Science, University of Delaware (1984–1989); and Acting Chair in the same department (1987).

He has been Editor of *Innovative Food Science and Emerging Technologies* since 2000 and was Editor of *Food Biotechnology* from 1986 to 2000. He is a member of the Editoral Board of a number of other journals in the field of food science and biotechnology. He is a member of various professional societies including Institute of Food Technologists, American Association for the Advancement of Science, and Association of Austrian Food and Bio-Technologists. In addition, he is Secretary General of the European Federation of Food Science and Technology.

His current research activities range from plant biotechnology to food process engineering, and he is operating and contributing at the interface of food biotechnology, food engineering, food processing and preservation, and food technology. His current research interest is focused in three main areas: (1) food process engineering with an emphasis on nonthermal as well as thermal processing of foods; (2) food biotechnology with an emphasis on plant cell and tissue cultures model systems (biosynthesis and production and metabolites), biopolymers, and enzyme technology; and (3) food quality and functionality with an emphasis on minimal processing and waste-free technologies.

In the past decade, he has played an active role in EC-funded research including the coordination of the AIR project, "High Hydrostatic Pressure Treatment: Its Impact on Spoilage Organisms, Biopolymer Activity, Functionality and Nutrient Composition of Food Systems" (1992–1996), and the FAIR-projects, "High Electric Field Pulses: Food Safety, Quality, and Critical Process Parameters" (1997–2000) and "Nutritional Enhancement of Probiotics: Technology Aspects on Microbial Viability, Stability, Functionality, and Prebiotic Function" (2000–2004). He has edited or coedited four books, has published 75 book chapters and more than 200 articles in peer-reviewed journals, and has presented more than 250 papers at workshops and international conferences.

List of Sources

Chapter 1

Table 1–1 15 *Source:* Adapted with permission from G.W. Gould, ed., *Mechanisms of Action of Food Preservation Procedures,* © 1989.

Exhibit 1–1 17 *Source:* Adapted from G.W. Gould, ed., *New Methods of Food Preservation,* p. xvi, © 1995, Aspen Publishers, Inc.

Exhibit 1–2 18 *Source:* Adapted from G.W. Gould, ed., *New Methods of Food Preservation,* p. xvii, © 1995, Aspen Publkishers, Inc.

Table 1–2 22 *Source:* Adapted from N.J. Russell and G.W. Gould, eds., *Food Preservatives, 2nd Edition,* © 2001, Aspen Publishers, Inc.

Exhibit 1–3 24 *Source:* Reprinted from G.W. Gould, ed., *New Methods of Food Preservation,* pp. xvii–xviii, © 1995, Aspen Publishers, Inc.

Chapter 2

Figure 2–1 46 *Source:* Adapted from *Chemistry and Physics of Lipids,* Vol. 91, C. Czeslik, et al., Effect of High Pressure on the Structure of Dipalmitoylphosphatidylcholine Bilayer Membranes: A Synchrotron X-ray Diffraction and FT-IR Spectroscopy Study Using the Diamond Anvil Technique, pp. 135–144, Copyright 1998, with permission from Elsevier Science and with permission from D.D. Klug, et al., Transformation of Ice VIII to Amorphous Ice by "Melting" at Low Temperature, *Journal of Chemical Physics,* Vol. 90, pp. 2390–2392, Copyright 1989, American Institute of Physics.

Table 2–3 51 *Source:* Data from K. Suzuki, Studies on the Kinetics of Protein Denaturation Under High Pressure, *Review in Physical Chemistry Japan,* Vol. 29, pp. 91–98, © 1960.

Figure 2–2 54 *Source:* Reprinted from *Biochemical Biophysics Acta,* K. Heremans and L. Smeller, Protein Structure and Dynamics at High Pressure, pp. 353–370, Copyright 1998, with permission from Elsevier Science.

Figure 2–6 65 *Source:* Data from P. Rubens, Exploring the Pressure-Temperature Phase Diagrams of Biopolymers, Ph.D. Dissertation, 1999, Katholieke Universiteit, Lueven, Belgium.

Chapter 3

Table 3–1 92 *Source:* Reprinted from *Trends in Food Science Technology,* Vol. 9, J.P.P.M. Smelt, Recent Advances in the Microbiology of High Pressure Processing, pp. 152–158, Copyright 1998, with permission from Elsevier Science.

Chapter 4

Figure 4–2 123 *Source:* Data from V. Heinz, Wirkung Hoher Hydrostatischer Drucke Auf das Absterbe und Keimungsverhalten Sporenbildender Bakterien am Beispiel von Bacillus Subtilis ATCC 9372, Ph.D. Thesis, 1997, Berlin University of Technology, Berlin, Germany.

Figure 4–6 145 *Source:* Data from V. Heinz and D. Knorr, High Pressure Germination and Inactivation Kinetics of Bacterial Spores, in *High Pressure Food Science, Bioscience and Chemistry,* N.S. Isaacs, ed., pp. 435–441, © 1998.

Figure 4–7 146 *Source:* Reprinted with permission from S. Monch, et al., X-Ray Microscopy in Food Sciences, in AIP Conference Proceedings #507, *X-Ray Microscopy: Proceedings of the 6th International Conference,* Meyer-Ilse, Warwick, and Atwood, ed., Copyright 1999 American Institute of Physics.

Figure 4–8 150 *Source:* Data from T. Okazaki, et al., 1996, P. Rovere, et al., 1999, C.R. Stumbo, 1948; and S.K. Navani, et al., 1970.

Figure 4–9 151 *Source:* Data from P. Rovere, et al., Advances in Bacterial Spores Inactivation in Thermal Treatments Under Pressure in *Advances in High Pressure Bioscience and Biotechnology,* H. Ludwig, ed., pp. 114–120, © 1999, Springer-Verlag.

Figure 4–11 154 *Source:* Data from P. Rovere, et al., Advances in Bacterial Spores Inactivation in Thermal Treatments Under Pressure, in *Advances in High Pressure Bioscience and Biotechnology,* H. Ludwig, ed., pp. 114–120, © 1999, Springer-Verlag.

Figure 4–5 133 *Source:* Data from J.R. Esty and K.F. Meyer, The Heat Resistance of the Spores of B. Botulinus and Allied Anaerobes, *Journal of Infectious Disease,* Vol. 31, pp. 650–663, © 1922.

Figure 4–6 145 *Source:* Reprinted with permission from V. Heinz and D. Knorr, High Pressure Germination and Inactivation Kinetics of Bacterial Spores, in *High Pressure Food Science, Bioscience and Chemistry,* N.S. Isaacs, ed., pp. 435–441, © 1998.

Chapter 5

Table 5–1 169 *Source:* Data from J.R. Whitaker, ed., *Principles of Enzymology for the Food Sciences,* © 1972; A. Fersht, ed., *Enzyme Structure and Mechanism, 2nd Edition,* pp. 98–120, © 1985; and L. Stryer, ed., *Biochemistry,* © 1988.

Table 5–2 174 *Source:* Data from C.M. Robert, et al., Kinetic Study of the Irreversible Thermal Deactivation of Palmito (Acantophoenix rubra)

Polyphenol Oxidase and Effect of pH, *Journal of Agricultural and Food Chemistry,* Vol. 43, pp. 1143–1150, © 1995; A.F. Rizvi and C.H. Tong, Fractional Conversion for Determining Texture Degradation Kinetics of Vegetables, *Journal of Food Science,* Vol. 62, pp. 1–7, © 1997; and M. Hendrickx, et al., Effects of High Pressure on Enzymes Related to Food Quality, *Trends in Food Science Technology,* Vol. 9, pp. 197–203, © 1998.

Table 5–3 174 *Source:* Data from S. Arrhenius, Uber die Reactionsgeschwindigkeit bei der Inversion von Rohrzucker Durch Sauren, *Z. Phys. Chem.*, Vol. 4, pp. 226–248, 1889; W.D. Bigelow, The Logarithmic Nature of Thermal Death Time Curves, *Journal of Infectious Disease,* Vol. 29, No. 5, pp. 528–536, © 1921; and M. Hendrickx, et al., Effects of High Pressure on Enzymes Related to Food Quality, *Trends in Food Science Technology,* Vol. 9, pp. 197–203, © 1998.

Figure 5–2 179 *Source:* Data from C. Weemaes, et al., Effect of pH on Pressure and Thermal Inactivation of Avocado Polyphenoloxidase: A Kinetic Study, *Journal of Agricultural and Food Chemistry,* Vol. 46, pp. 2785–2792, © 1998.

Figure 5–4 180 *Source:* From Kinetics of Combined Pressure-Temperature Inactivation of Avocado Polyphenoloxidase, C. Weemaes, et al., *Biotechnology Bioengineering,* Vol. 60, pp. 292–300, Copyright 1998 John Wiley & Sons, Inc. Reprinted by permission of Wiley-Liss, Inc., a subsidiary of John Wiley & Sons, Inc.

Figure 5–5 184 *Source:* Reprinted with permission from I. Van den Broeck, L.R. Ludikhuyze, A.M. Van Loey, and M.E. Hendrickx, Inactivation of Orange Pectinesterase by Combined High Pressure and Temperature Treatments: A Kinetic Study, *Journal of Agricultural and Food Chemistry.* Copyright 2000 American Chemical Society.

Figure 5–6 184 *Source:* Reprinted with permission from I. Van den Broeck, L.R. Ludikhuyze, A.M. Van Loey, and M.E. Hendrickx, Inactivation of Orange Pectinesterase by Combined High Pressure and Temperature Treatments: A Kinetic Study, *Journal of Agricultural and Food Chemistry.* Copyright 2000 American Chemical Society.

Figure 5–7 185 *Source:* Reprinted with permission from I. Van den Broeck, L.R. Ludikhuyze, C.A. Weemaes, A.M. Van Loey, and M.E. Hendrickx, Effect of Temperature and Pressure on Tomato Pectinesterase Activity, *Journal of Agricultural and Food Chemistry,* Vol. 48, No. 2, pp. 551–558. Copyright 2000 American Chemical Society.

Figure 5–8 185 *Source:* Reprinted with permission from I. Van den Broeck, L.R. Ludikhuyze, C.A. Weemaes, A.M. Van Loey, and M.E. Hendrickx, Effect of Temperature and Pressure on Tomato Pectinesterase Activity, *Journal of Agricultural and Food Chemistry,* Vol. 48, No. 2, pp. 551–558. Copyright 2000 American Chemical Society.

Figure 5–9 187 *Source:* Reprinted with permission from L. Ludikhuyze, et al., Effect of Combined Pressure and Temperature on Soybean Lipoxygenase:

I. Influence of Extrinsic and Intrinsic Factors on Isobaric-isothermal Inactivation Kinetics, *Journal of Agricultural and Food Chemistry,* Vol. 46, No. 10, pp. 4074–4080. Copyright 1998 American Chemical Society.

Figure 5–10 190 *Source:* From L. Ludikhuyze, L. Rodrigo, and M. Hendrickx, The Activity of Myrosinase From Broccoli (Brassica oleracea L. cv. Italica): Influence of Intrinsic and Extrinsic Factors, *Journal of Food Protection,* © 2000. Reprinted with permission from *Journal of Food Protection.* Copyright held by the International Association for Food Protection, Des Moines, Iowa, U.S.A.

Figure 5–19 208 *Source:* Reprinted from *Innovative Food Science & Emerging Technologies,* Vol. 1, S. Denys, A. Van Loey, and M.E. Hendrickx, A Modeling Approach for Evaluating Process Uniformity During Batch High Hydrostatic Pressure Processing: Combination of a Numerical Heat Transfer Model and Enzyme Inactivation Kinetics, pp. 5–19, Copyright 2000, with permission from Elsevier Science.

Figure 5–20 209 *Source:* Reprinted from *Innovative Food Science & Emerging Technologies,* Vol. 1, S. Denys, A. Van Loey, and M.E. Hendrickx, A Modeling Approach for Evaluating Process Uniformity During Batch High Hydrostatic Pressure Processing: Combination of a Numerical Heat Transfer Model and Enzyme Inactivation Kinetics, pp. 5–19, Copyright 2000, with permission from Elsevier Science.

Chapter 6

Table 6–1 226 *Source:* Data from O.R. Fennema and S.R. Tannenbaum, Introduction to Food Chemistry, in *Food Chemistry, 2nd Edition,* O.R. Fennema, ed., pp. 1–21, © 1985.

Figure 6–1 229 *Source:* Reprinted with permission from A. Van Loey, et al., Thermal and Pressure-temperature Degradation of Chlorophyll in Broccoli (Brassica oleracea L. italica) Juice: A Kinetic Study, *Journal of Agricultural and Food Chemistry,* Vol. 46, No. 12, pp. 5289–5294. Copyright 1998 American Chemical Society.

Figure 6–2a & b 229 *Source:* Reprinted with permission from C. Weemaes, et al., Pressure-temperature Degradation of Green Color in Broccoli Juice, *Journal of Food Science,* Vol. 64, No. 3, pp. 504–508, © 1999, Institute of Food Technologists.

Figure 6–3A 237 *Source:* Reprinted with permission from S. Basak and H.S. Ramaswamy, Effect of High Pressure Processing on Texture of Selected Fruit and Vegetables, *Journal of Texture Studies,* Vol. 29, pp. 587–601, © 1998.

Figure 6–5 243 *Source:* Reprinted with permission from P. Butz, et al., Influence of High Hydrostatic Pressure on Aspartame: Instability at Neutral pH, *Journal of Agricultural and Food Chemistry,* Vol. 45, pp. 302–303. Copyright 1997 American Chemical Society.

Figure 6–4 241 *Source:* Reprinted with permission from I. Van den Broeck, et al., Kinetics and Isobaric-Isothermal Degradation of L-ascorbic Acid,

Journal of Agricultural and Food Chemistry, Vol. 46, No. 5, pp. 2001–2006. Copyright 1998 American Chemical Society.

Chapter 7

Figure 7–2 259 *Source:* Reprinted from *Innovative Food Science & Emerging Technologies,* Vol. 1, M. Stolt, S. Oinonen, and K. Autio, Effect of High Pressure on the Physical Properties of Barley Starch, pp. 167–175, Copyright 2000, with permission from Elsevier Science.

Figure 7–3 260 *Source:* K. Autio and M. Stolt, Effect of High Pressure on the Structure of Starch, in *Fresh Novel Foods by High Pressure,* K. Autio, ed., pp. 61–67, © 1998, VIT Information Service.

Table 7–3 263 Courtesy of Nestle, Ltd, Lausanne, Switzerland.

Table 7–1 256 *Source:* Reprinted with permission from S. Basak and H.S. Ramaswamy, Effect of High Pressure Processing on the Texture of Selected Fruits and Vegetables, *Journal of Texture Studies,* Vol. 29, pp. 578–601, © 1998.

Figure 7–1 258 *Source:* Reprinted from *Journal of Food Engineering,* M. Stolt, et al., Evaluation and Modelling of Rheological Properties of High Pressure Treated Waxy Maize Starch Dispersions, Copyright 1999, with permission from Elsevier Science.

Figure 7–11 273 Courtesy of Nestle, Ltd, Lausanne, Switzerland.

Figure 7–7 266 Courtesy of Nestle, Ltd, Lausanne, Switzerland.

Figure 7–8 266 Courtesy of Nestle, Ltd, Lausanne, Switzerland.

Chapter 8

Figure 8–1 287 *Source:* Data from P.W. Bridgman, Water, in the Liquid and Five Solid Forms, Under Pressure, *Proceedings of the American Academy of Arts Sciences XLVII,* Vol. 13, pp. 439–558, 1912 and W. Wagner, A. Saul, and A. Prub, International Equations for the Pressure Along the Melting and Along the Sublimation Curve of Ordinary Water Substance, *Journal of Chemical Reference Data,* Vol. 23, No. 3, pp. 515–527, © 1994.

Figure 8–6 306 *Source:* Reprinted with permission from S. Denys, L.R. Ludikhuyze, A.M. Van Loey, and M.E. Hendrickx, Modeling Conductive Heat Transfer and Process Uniformity During Batch High-Pressure Processing of Foods, *Biotechnology Progress,* Vol. 16, No. 1, pp. 92–101. Copyright 2000 American Chemical Society and American Institute of Chemical Engineers.

Figure 8–7 307 *Source:* Reprinted with permission from S. Denys, L.R. Ludikhuyze, A.M. Van Loey, and M.E. Hendrickx, Modeling Conductive Heat Transfer and Process Uniformity During Batch High-Pressure Processing of Foods, *Biotechnology Progress,* Vol. 16, No. 1, pp. 92–101. Copyright 2000 American Chemical Society and American Institute of Chemical Engineers.

Figure 8–8 308 *Source:* Reprinted with permission from S. Denys, L.R. Ludikhuyze, A.M. Van Loey, and M.E. Hendrickx, Modeling Conductive Heat Transfer and Process Uniformity During Batch High-Pressure Processing of Foods, *Biotechnology Progress,* Vol. 16, No. 1, pp. 92–101.

Copyright 2000 American Chemical Society and American Institute of Chemical Engineers.

Figure 8–11 312 *Source:* Reprinted from *Innovative Food Science & Emerging Technologies,* Vol. 1, S. Denys, A. Van Loey, and M.E. Hendrickx, A Modeling Approach for Evaluating Process Uniformity During Batch High Hydrostatic Pressure Processing: Combination of a Numerical Heat Transfer Model and Enzyme Inactivation Kinetics, pp. 5–19, Copyright 2000, with permission from Elsevier Science.

Figure 8–5 304 *Source:* Data from D.J. Cleland, Prediction of Freezing and Thawing Times for Food, Ph.D. Dissertation, 1985, Massey University, New Zealand.

Figure 8–2 288 *Source:* Reprinted with permission from D. Knorr, O. Schlueter, and V. Heinz, Impact of High Hydrostatic Pressure on Phase Transitions in Foods, *Food Technology,* Vol. 52, No. 9, pp. 42–45, © 1998, Institute of Food Technologists.

Table 8–2 315 *Source:* Reprinted from *Trends in Food Science Technology,* Vol. 6, M.T. Kalichevsky, D. Knorr, and P.J., Lillford, Potential Food Applications of High-Pressure Effects on Ice-Water Transitions, pp. 253–259, Copyright 1995, with permission from Elsevier Science.

Chapter 9

Figure 9–3 330 Courtesy of ABB Flow Pressure System, Vasteras, Sweden.

Figure 9–4 330 Courtesy of Esteban Espuna, S.A., Olot, Spain.

Figure 9–7 332 *Source:* Reprinted with permission from P. Rovere, et al., Modelling and Calculation of the Sterilizing Effect in High Pressure Heat-Treatments, *Industria Conserve,* Vol. 73, pp. 303–313, © 1998, Stazione per l'Industria delle Conserve Alimentari.

Figure 9–8 333 *Source:* Reprinted with permission from P. Rovere et al., Produzione e Stabilzzazione di Purea di Mela Mediante Trattamenti ad alta Pressione, in *Richerche e Innovazioni Nell l'industria Alimentare*—Vol. I, S. Porrettta, ed., pp. 455–468, © 1994, Chiriotti Editori.

Table 9–1 337 *Source:* Reprinted with permission from F. Bonomi, et al., Molecular and Functional Properties of HP-Treated Egg Components, in *High Pressure Food Science, Bioscience and Chemistry,* N. Isaacs, ed., pp. 160–165, © 1998.

Figure 9–10 338 *Source:* Reprinted with permission from S. Gola, et al., Innattivazione Microbica in Nettare di Albicocca e in Sistemi Medello Trattati con le alte Pressioni, *Industria Conserve,* Vol. 69, pp. 194–198, © 1994, Stazione per l'Industria delle Conserve Alimentari.

Figure 9–11 339 *Source:* Reprinted with permission from P. Rovere, Stabilization of Apricot Puree, *Prehrambeno-technol, Food Technology and Biotechnology Review,* Vol. 32, pp. 145–150, © 1994, University of Zagreb.

Figure 9–14 341 *Source:* Reprinted with permission from P. Rovere, et al., Chicken Mechanically Deboned Meat: High Pressure Processed, in *High*

Chapter 10

Figure 10–5 364 Courtesy of Institute of Food Research, Norwich, United Kingdom.

Table 10–4 364 *Source:* From E.C. Needs, et al., Comparison of Heat and Pressure Treatments of Skim Milk, Fortified with Whey Protein Concentrate for Set Yogurt Preparation: Effects on Milk Proteins and Gel Structure, *Journal of Diary Research,* Vol. 67, pp. 329–348, © 2000, Cambridge University Press. Reprinted with the permission of Cambridge University Press.

Figure 10–8 366 Courtesy of Institute of Food Research, Norwich, United Kingdom.

Table 10–5 369 *Source:* From E.C. Needs, et al., High-Pressure Treatment of Milk: Effects on Casein Micelle Structure and on Enzymatic Coagulation, *Journal of Dairy Research,* Vol. 67, pp. 31–42, © 2000, Cambridge University Press. Reprinted with the permission of Cambridge University Press.

Figure 10–10 370 Courtesy of Institute of Food Research, Norwich, United Kingdom.

Chapter 11

Figure 11–1 388 Courtesy of ACB Pressure Systems—Hyperbar, France.
Figure 11–2 391 Courtesy of ATO B.V., Wageningen, The Netherlands.
Figure 11–3 392 Courtesy of ATO B.V., Wageningen, The Netherlands.
Figure 11–4 393 Courtesy of ATO B.V., Wageningen, The Netherlands.
Figure 11–5 394 Courtesy of ATO B.V., Wageningen, The Netherlands.

Kolibri 360 GmbH
Ferdinand-Jühlke-Straße 7
99095 Erfurt, Deutschland
produktsicherheit@kolibri360.de